The Last Refuge of the Mt. Graham Red Squirrel

The Last Refuge of the Mt. Graham Red Squirrel

Ecology of Endangerment

EDITED BY
H. REED SANDERSON AND JOHN L. KOPROWSKI

FOREWORD BY JACK WARD THOMAS

The University of Arizona Press   Tucson

The University of Arizona Press

www.uapress.arizona.edu

Library of Congress Cataloging-in-Publication Data

The last refuge of the Mt. Graham red squirrel : ecology of endangerment / edited by H. Reed Sanderson and John L. Koprowski ; foreword by Jack Ward Thomas.

p. cm.

Includes bibliographical references and index.

ISBN 978-0-8165-2768-7

1. Tamiasciurus hudsonicus—Arizona—Graham, Mount. 2. Endangered species—Arizona—Graham, Mount. 3. Wildlife conservation—Arizona—Graham, Mount. 4. Ecology—Arizona—Graham, Mount. I. Sanderson, H. Reed, 1932–. II. Koprowski, John L.

QL737.R68L368 2009
599.36'3—dc22 2008051124

Publication of this book is made possible in part by grants from the University of Arizona Office of the Vice President for Research, Graduate Studies, and Economic Development, and the Coronado National Forest.

Manufactured in the United States of America on acid-free, archival-quality paper containing a minimum of 30 percent post-consumer waste and processed chlorine free.

14 13 12 11 10 09 6 5 4 3 2 1

Ignorance more frequently begets confidence than does knowledge.
—Charles Darwin

Contents

Foreword

During my tenure as chief of the U.S. Forest Service (1993–1996), the politics surrounding the management of Mt. Graham and the endangered Mt. Graham red squirrel came up on the administration's political radar screen. The University of Arizona and its partners in the state-of-the-art observatory being constructed atop Mt. Graham were becoming increasingly agitated about delays, and the significant economic and political costs, in approvals of construction plans due to concerns over the welfare of the Mt. Graham red squirrel—an endangered subspecies. Those sincerely interested in the welfare of the squirrel and those simply opposed to the construction of the telescopes and all that would go with it grasped the potential weapon of the Endangered Species Act to effect change. It would take an act of law to break the impasse.

As a result of that interest on the part of the administration and concerns of involved U.S. Forest Service employees, I spent a week reviewing the situation—both at the University of Arizona and on Mt. Graham itself. By the end of that week, it was clear—to me at least—that the project would proceed with some minor modifications. Further, as a quid pro quo, that money would be made available, from a variety of sources, to undertake monitoring and associated research relative to the ecology, numbers, and welfare of the Mt. Graham red squirrel. And, so it was to be, as reflected in this volume.

This volume includes summaries of most of the studies carried out in the formal and informal agreements to further understand the ecology of the Mt. Graham red squirrel, the value of the observatory, the changing ecological conditions on Mt. Graham itself, and the potential consequences for the squirrel and the ecosystem it serves as a surrogate. In reading over the results of the various studies that make up this volume, it is well to remember that one of the clearly stated purposes of the Endangered Species Act is "to provide a means whereby the ecosystems upon which endangered species and threatened species may be conserved."

A careful reading of the many studies reported here indicates ongoing interactions involving changing climate, changes in plant communities, introduction of exotic species, increasing fuel loadings, increased likelihood of stand-replacement fire, increased insect and disease responses to changed conditions, etc., may very well be producing overall ecological changes that will lead to the demise of the "indicator species" in question—the Mt. Graham red squirrel. If so, that could, and likely should, lead to serious review of the efficacy of the Endangered Species Act itself for the purpose so clearly stated above.

At the very least, what is reported here should lead land managers and scientists alike to understand that all management decisions and subsequent actions should be subject to two, repeat two, risk assessments—the short-term risk of actions taken and the long-term risk of actions delayed or not taken at all. There is much to learn from the pages that follow.

Jack Ward Thomas
Chief Emeritus, U.S. Forest Service
Emeritus Boone and Crockett Professor
College of Forestry, University of Montana

Acknowledgments

> Doing what little one can to increase the general stock of knowledge is as respectable an object of life, as one can in any likelihood pursue.
> —Charles Darwin

We wish to acknowledge the following people who spent their time and energy reviewing the chapters presented in this volume. Some reviewed more than a single chapter. We also wish to recognize the two anonymous people who reviewed the entire manuscript. We sincerely thank them and appreciate their efforts, which led to many improvements and corrections. Further, we express our appreciation of Allyson Carter's diligence at the University of Arizona Press to accomplish the publication of the Mt. Graham red squirrel symposium proceedings as well as the assistance of Keith LaBaw, Al Schroder, and Harrison Shaffer. The keen eye of Sharon Hunt made the final copy a clean and clear manuscript, and Seth Miller of Matrix Productions, Inc. ably directed us through the final editing. Gloria Tierney accepted the challenge of the index. We appreciate the time, efforts, and patience of many.

William Allred
Karen Arabas
Ken Armitage
Keith Aubry
Paul Barrett
Ellen Campbell
Norm Cimon
Russ Davis
Norris Dodd
Sam Drake
John Edwards
Laurie Fernwood
Bobbe Fitzgibbon
Genice Froehlich
Thetis Gamberg
John Gurnell
Bill Healy
Susan Kephart
Roy Kirkpatrick
Karl Larsen

Mike Leonard
Peter Lurz
John Madsen
Chris Maser
Bill Matter
John Morgart
Jeremy Moss
Chuck Nixon
Dave Patton
Steve Rushton
Leslie Sage
Carmen Salsbury
Gary San Julian
Peter Smallwood
Winston Smith
Lia Speigel
Mike Steele
Bob Steidl
Donna Stockrahm
Tad Theimer
Karl Vernes
Richard Wadleigh
Luc Wauters
Peter Weigl
Gary White
J. T. Williams

Most important, we thank our wives, Georgie Sanderson and Nancy Koprowski, for enduring our absence while we pursued our research and our children, Chris and Allen Sanderson, and Emma and Zachary Koprowski, for sharing us with many wild things in wild places.

The Last Refuge of the Mt. Graham Red Squirrel

Introduction

> No matter how sophisticated you may be, a huge granite mountain cannot be denied—it speaks in silence to the very core of your being.
> —Ansel Adams

MT. GRAHAM WAS THRUST into the international consciousness for the first time in the late 1980s with the proposed construction of an astrophysical site on the high peaks of the Pinaleño Mountain Range nearly simultaneously with the listing of the endemic Mt. Graham red squirrel (*Tamiasciurus hudsonicus grahamensis*) as federally endangered. What followed has been a textbook case on the challenges faced by modern conservation efforts. We believe that the lessons learned through the subsequent two decades provide important insights for conservation biologists, wildlife managers, state and federal land management agencies, social scientists, and the general public. Humans sometimes fail to appreciate natural ecosystems as changing and dynamic; however, changes often occur rapidly when a species is imperiled. In the pages that follow, we document the changing landscape of the Pinaleños and provide a compendium of perspectives, critical data and analysis, and projections for the endangered endemic squirrel and its dwindling habitat.

Like most geologic origins, the Pinaleños were a violent new addition to the landscape whether literally or figuratively. The Pinaleño Mountains are known to different groups as Mt. Graham or the Graham Mountains, the Pinaleños, and Dzil nchaa si'an. Many who traversed the deserts and arid grasslands that surround this sky island in southeastern Arizona value these cool mesic mountains that rise dramatically from desert valley floor. The summit of Mt. Graham (also known as High Peak) is 3,267 m high, a rise of almost 2,438 m above the Gila River basin

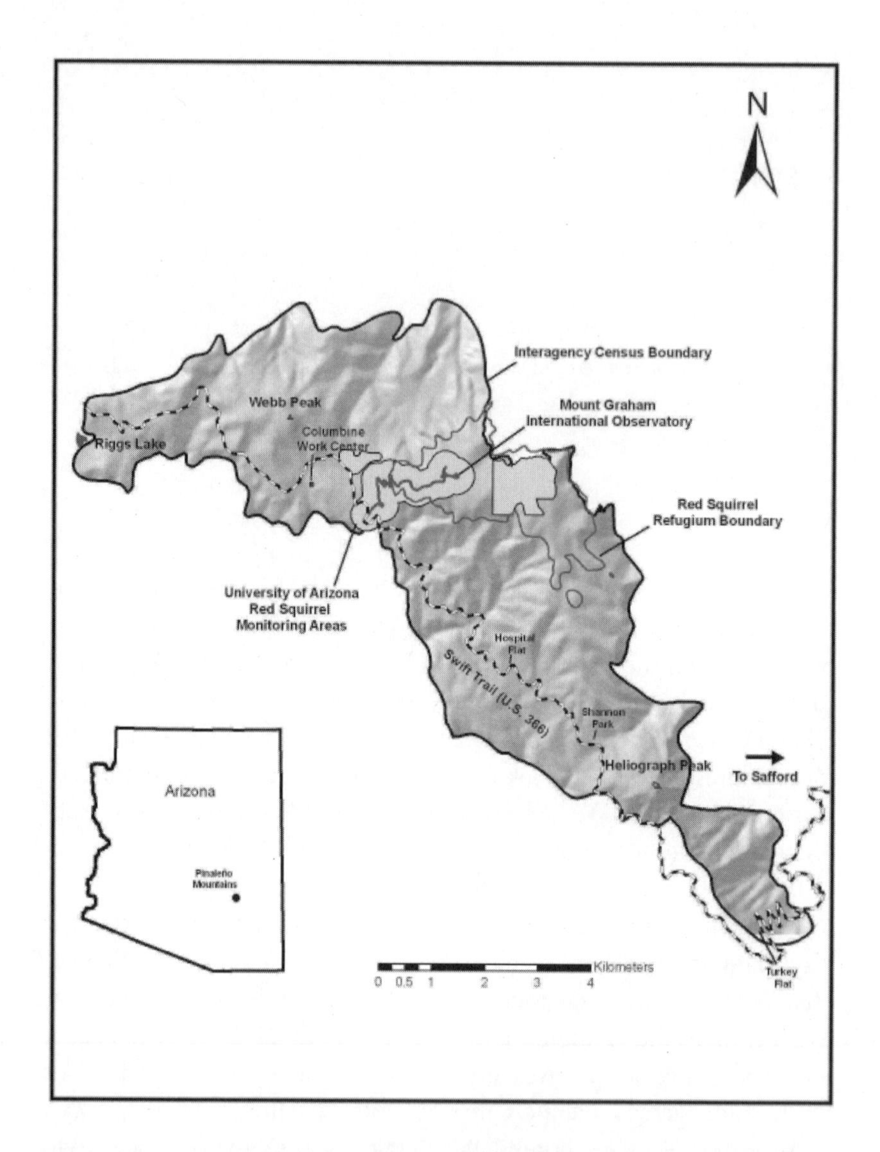

Figure 0.1. Location of the Pinaleño Mountains, illustrating the Interagency Census Area, the University of Arizona Red Squirrel Monitoring Areas, the Red Squirrel Refugium, and the location of the Mount Graham International Observatory. Illustrated by M. Merrick, UA.

north and east of the mountain range (fig. 0.1). Mt. Graham is on the Safford Ranger District, Coronado National Forest and is managed by the U.S. Forest Service.

Mt. Graham was named in 1846 during the war with Mexico when the Army of the West, led by General Stephen Watts Kearney, came down the Gila River on its way to California. The accompanying topographer and scientist, Lieutenant William Emory, located Mt. Graham on his map and named the mountain for his friend Lieutenant Colonel James Duncan Graham, a senior officer in the U.S. Army Corps of Topographical Engineers (Coronado National Forest 1991; Anonymous 1996).

Mt. Graham is a marked contrast to the other peaks in the area. Most of the others are lava deposits; on Mt. Graham, a metamorphic core complex was revealed as faulting uplifted the ranges and erosion stripped off the overlying rocks, and it rose to an elevation of 3,267 m above sea level. Unlike the lava-based mountains that are barren or foliated by low shrubs, Mt. Graham has a profusion of trees that date back to antiquity.

Researchers from the University of Arizona's Laboratory of Tree-Ring Research have discovered living trees that date back to AD 1257 and 1270. Botanists say the Douglas-fir (*Pseudotsuga menziesii*) has survived because the rocky cliffs of the mountains have served as a fire barrier. The scientists also found dead firs that dated as far back as AD 1102 (Lowe 1991).

While the Mt. Graham area has been inhabited more or less continually, the Gila River Valley can trace its written history back to the Spanish conquistadores who passed through in the 1500s while searching for the Seven Cities of Cibola, fabled settlements where streets were paved with gold. Many of them stayed in what was once known as Pueblo Viejo, near what is now called San Jose, on the northeast slopes of Mt. Graham. Treasure hunters still turn up armor, weapons, and equipment from that era.

In the early 1880s, settlers came to the area in large numbers. These settlers found that the nearby mountains were a great benefit. Lumbering provided materials for building as well as employment for growing communities. The settlers also found Mt. Graham to be a cool retreat from the hot desert. Summer homes were built around the sawmill areas, so families could be with the men who were working there (Anonymous 1996).

The first roads up the mountain were only wide enough for people to ride up on horseback. The need to transport lumber down from the mills required improved roads for wagons and later for large trucks. The government road-building project started in the 1920s and reached Arcadia by about 1927 (Anonymous 1996). This road was named Swift Trail after Thomas T. Swift, a forest supervisor at the Crook National Forest at that time. The forest was named for General George Crook, who was involved with the Indian Wars in the late 1800s. In 1953, the Crook National Forest came under the jurisdiction of the Coronado National Forest, and the Safford Ranger District was established.

In the 1930s and early 1940s, the Civilian Conservation Corps (CCC) was in operation to provide employment for young men during the Great Depression, and several camps were located on Mt. Graham (Anonymous 1996). Treasure Park and Columbine were used during the summer months; Arcadia, Noon Creek, and other sites were used during the winter months when the higher portions of the mountain were covered with snow. Many campgrounds, hiking trails, roads, and other facilities built by the CCC personnel are now enjoyed by visitors to Mt. Graham (Anonymous 1996). Today, Mt. Graham is still a popular retreat from the hot summer desert floor for a variety of recreation activities from leisurely sightseeing to hiking and camping, rock climbing, cross-country skiing, hunting, and fishing (Anonymous 1996). As Ansel Adams suggests, all visitors are influenced by the sheer enormity of Mt. Graham, but the impact and value vary considerably.

The Pinaleños are the southernmost locality of the red squirrel (*Tamiasciurus hudsonicus*), which is found northward through the Rockies into Alaska, across the northern territories and prairie provinces of Canada, and southward through the Great Lakes and eastern seaboard of the United States as far as the Appalachian Mountains of Georgia (Hall 1981). The endemic Mt. Graham red squirrel subspecies represents a population isolated for some 10,000 years (Sullivan and Yates 1995). The Mt. Graham red squirrel was first described by Allen (1894), based primarily on subtle morphological differences and geographic isolation. Subsequently, an initial survey of protein electromorphs suggested a fixed unique isozyme in the population on the Pinaleños (Sullivan and Yates 1995). More recently, the calls of the Mt. Graham red squirrel have been shown to have a unique composition (Yamamoto et al. 2001).

Mt. Graham red squirrels and their habitats occur in stands above about 2,450 m and are found primarily in three biotic communities: the subalpine spruce-fir, mixed-conifer, and a transitional community, or ecotone, that falls in between the higher- and lower-elevation forests.

The spruce-fir forest occurs above 2,900 m and is dominated by Engelmann spruce (*Picea engelmannii*) and corkbark fir (*Abies lasiocarpa* var. *arizonica*). Dominant canopy species of the mixed-conifer zone (ca. 2,450–2,900 m) include Douglas-fir, white fir (*Abies concolor*), and southwestern white pine (*Pinus strobiformis*), with ponderosa pine (*Pinus ponderosa*) subdominants at the lower elevational range, and quaking aspen (*Populus tremuloides*), Engelmann spruce, and corkbark fir being subdominants at the upper range. The transitional forest falls in between and contains a mixture of all conifer species, but contains a high proportion of Engelmann spruce and corkbark fir. These forests provide the primary foods of the red squirrel: conifer seeds and fungi (U.S. Fish and Wildlife Service 1993). Squirrels nest in tree cavities, leaf nests (dreys), and ground nests within these forest types (Steele 1998). The most conspicuous sign of the larderhoarding Mt. Graham red squirrels are the large cone scale piles known as "middens" in which cones are stored—often at traditional sites lasting for decades (Finley 1969). Conifer seeds are stored for winter use by burying intact cones in groups of up to 24 each at a depth of 4 cm inside their middens (Smith 1968; Hurly and Robertson 1990). The average red squirrel midden has an area of 35 m^2, a depth of 47 cm, and may contain up to 5,000 stored cones (Hurly and Lourie 1997).

The Mt. Graham red squirrel population was believed to be in decline in the 1950s (Hoffmeister 1956). During the 1960s, the squirrel was believed to be extirpated from the mountain (Minckley 1968), but was reported again in the 1970s (Brown 1984). In 1987, the Mt. Graham red squirrel was added to the U.S. Fish and Wildlife Service Endangered and Threatened Wildlife and Plants listing (*Federal Register* 52(106), 3 June 1987)—a level of protection that is retained up to the present day.

Several of the research projects reported herein were funded by the University of Arizona because of the final mitigation to build an astrophysical site on Mt. Graham. Funds were provided at the annual rate of $50,000 to $60,000 from 1989 through 1998 (totaling $533,000). The

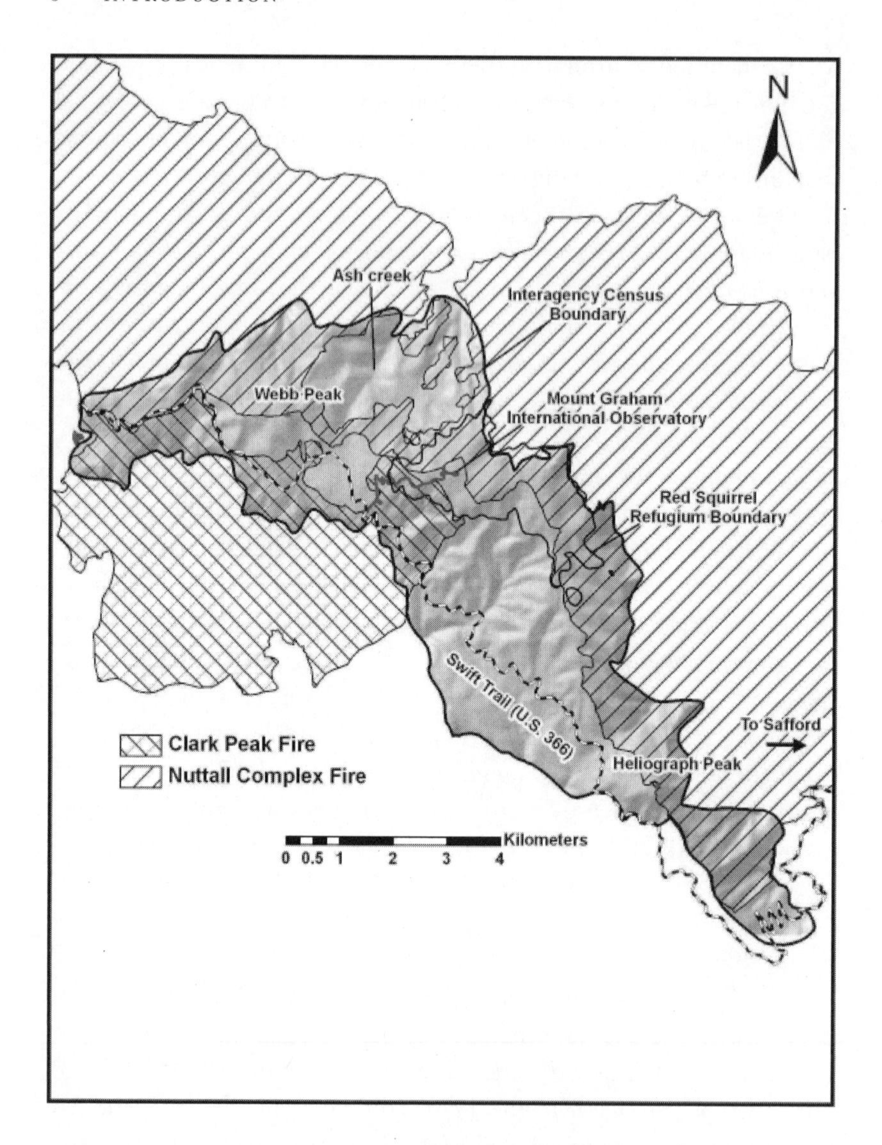

Figure 0.2. Range of the Mt. Graham red squirrel overlaid with the Clark Peak and Nuttall Complex fires. Illustrated by M. Merrick, UA.

funds were transferred to the U.S. Forest Service and were administered by the Mt. Graham Red Squirrel Study Committee, which consisted of representatives from the Arizona Game and Fish Department, the University of Arizona, the U.S. Forest Service, and the U.S. Fish and Wildlife Service. The University of Arizona, Vice President for Research sponsored a symposium on the Mt. Graham red squirrel that was held in Safford, Arizona, May 20–22, 2003 and was cosponsored by the Arizona Game and Fish Department, the Coronado National Forest, the U.S. Fish and Wildlife Service, and the Arizona Chapter and Southwest Section of the Wildlife Society. This book represents the proceedings of that symposium. The symposium was designed to represent a wide spectrum of activities on Mt. Graham ranging from astronomy, management activities and plans, interagency census results, the University of Arizona monitoring results, and current research on the Mt. Graham red squirrel and its habitat.

Since the symposium, the Nuttall Complex fire in 2004 moderately or severely burned about 556 ha of red squirrel habitat (U.S. Forest Service 2004) and threatened the telescope complex (fig. 0.2). Also, the U.S. Forest Service has proposed a thinning project designed to reduce the volume of fuel and potential for catastrophic forest fire that includes an unknown amount of red squirrel habitat. Both of these events are discussed in the respective chapters.

The challenges are many and the solutions are elusive, but worthy of the chase. The efforts of the many contributors here detail the chase that has turned into a race to avoid the extinction of the Mt. Graham red squirrel from the snow-capped peaks of the Pinaleños. We trust that the works that follow not only will inform those interested in this lone mountaintop but also will prove useful on the many other "islands," be they literal or figurative, where conservation efforts are focused.

Literature Cited

Allen, J. A. 1894. Descriptions of five North American mammals. *Bulletin American Museum of Natural History* 6:347–350.

Anonymous. 1996. *Eastern Arizona almanac. Eastern Arizona Courier,* Safford, Arizona.

Brown, D. E. 1984. *Arizona's tree squirrels.* Arizona Game and Fish Department, Phoenix.

Coronado National Forest. 1991. *Who named the mountains?* Tucson, Arizona.

Finley, R. B., Jr. 1969. Cone caches and middens of *Tamiasciurus* in the Rocky Mountain region. Pages 233–273 in J. K. Jones Jr., ed. *Contributions in mammalogy.* University of Kansas Museum of Natural History Miscellaneous Publication 51.

Hall, E. R. 1981. *Mammals of North America.* John Wiley and Sons, New York.

Hoffmeister, D. F. 1956. Mammals of the Graham (Pinaleño) Mountains, Arizona. *American Midland Naturalist* 55:257–288.

Hurly, T. A., and S. A. Lourie. 1997. Scatterhoarding and larderhoarding by red squirrels: Size, dispersion, and allocation of hoards. *Journal of Mammalogy* 78:529–537.

Hurly, T. A., and R. J. Robertson. 1990. Variation in the food hoarding behaviour of red squirrels. *Behavioral Ecology and Sociobiology* 26:91–97.

Lowe, S. 1991. *History, legends and tales—Old West Highway, where history still lives.* The Old West Highway Committee, Globe, Arizona.

Minckley, W. L. 1968. Possible extirpation of the spruce squirrel from the Pinaleño (Graham) Mountains of south-central Arizona. *Journal of the Arizona Academy of Science* 5:110.

Smith, C. C. 1968. The adaptive nature of social organization in the genus of three [*sic*] squirrels *Tamiasciurus. Ecological Monographs* 38:31–63.

Steele, M. A. 1998. *Tamiasciurus hudsonicus. Mammalian Species* 586:1–9.

Sullivan, R. M., and T. L. Yates. 1995. Population genetics and conservation biology of relict populations of red squirrels. Pages 193–209 in C. A. Istock and R. S. Hoffmann, eds. *Storm over a mountain island: Conservation biology and the Mt. Graham affair.* University of Arizona Press, Tucson.

U.S. Fish and Wildlife Service. 1993. *Mount Graham Red Squirrel* (Tamiasciurus hudsonicus grahamensis) *Recovery Plan*. Albuquerque, New Mexico.

U.S. Forest Service. 2004. *Nuttall Complex. Burned area emergency response (BAER) team executive summary.* Coronado National Forest, Safford, Arizona.

Yamamoto, O., B. Moore, and L. Brand. 2001. Variation in the bark call of the red squirrel (*Tamiasciurus hudsonicus*). *Western North American Naturalist* 61:395–402.

Part I

The Mountain

> The mountain and the squirrel
> Had a quarrel,
> And the former called the latter "Little Prig"
>
> —Ralph Waldo Emerson

CULTURAL, SPIRITUAL, ECONOMIC, AESTHETIC, ECOLOGICAL—examples of the values ascribed to Mt. Graham are many. Various groups have used the Pinaleños in a myriad of ways throughout history and into the present day and will continue to do so into the future. The challenges created by this diversity of groups and interests in the Pinaleños have been the source of important confrontations in the past and perhaps will be the source of important collaborations in the future. The efficacy of conservation and management efforts in the Pinaleños will depend on the quality of the dialogue and respect that is fostered. As Ralph Waldo Emerson suggests in our opening quote, perspectives on the value of the mountain and its inhabitants vary and can result in conflicting opinions.

Part I provides a diversity of perspectives on the Pinaleño Mountains. Pat Spoerl explores the use of the Pinaleños in historical times by humans, with a particular emphasis on Native American use and the significance of the mountain known to them as Dzil nchaa si'an. For many involved with the mountain, this historical perspective and insight into the significance of the peak to Western Apache peoples will be enlightening. The eye opening continues as John Ratje elucidates the astronomical value of Mt. Graham, and the value that many physical scientists place upon the high peaks. Many readers will be familiar with the conflict that raged over development of an astrophysical site on Mt. Graham and that brought the peaks into the national and international eye. Finally, Tom Swetnam and colleagues view the mountain through the lens of an ecologist. Using historical information gleaned from an analysis of tree rings, we are given a sense of the mountain as a dynamic living ecosystem that has changed markedly over recent centuries.

CHAPTER 1

The Cultural Significance of Mt. Graham (Dzil nchaa si'an) in Western Apache Tradition

PATRICIA M. SPOERL

AT FIRST GLANCE, it may appear that there are few connections among the Mt. Graham Red Squirrel Refugium, the Mt. Graham International Observatory (MGIO), the Mt. Graham (Dzil nchaa si'an) Traditional Cultural Property, and the Mt. Graham Wilderness Study Area other than they are all specially designated geographical areas in the Pinaleño Mountains of the Coronado National Forest. This chapter focuses on one of the specially designated areas—the Mt. Graham (Dzil nchaa si'an) Traditional Cultural Property—and the cultural significance to Western Apaches that makes it eligible for listing in the National Register of Historic Places. I provide a brief history of the Pinaleño Mountains to aid in understanding the relationship of Mt. Graham to Western Apache history and culture, and then describe the significant cultural themes associated with Dzil nchaa si'an. Finally, National Register eligibility status as a traditional cultural property and implications of special designations for this geographic area are considered.

I cannot speak as an Apache person would about the traditional cultural significance of Mt. Graham. I can merely document that which is written and conveyed orally by Apache people and then structure that information in terms of the criteria established by the National Park Service for eligibility and listing in the National Register of Historic Places (*National Register Bulletin* 15) (Spoerl 2001). Apache people traditionally do not often speak or write about sacred places and practices; thus, relatively little information exists in documentary records. Describing the importance of Mt. Graham to Western Apache communities also is clouded by the presence of the MGIO. Congress authorized construction of the observatory in 1988 under the Arizona-Idaho Conservation

Act (Public Law 100–696). Since that time, statements regarding Mt. Graham's role in Apache traditional history and culture have centered on opposition to the observatory. In describing the significance of Mt. Graham, I make a concerted effort to separate statements made about Apache values associated with Mt. Graham that were, and are, tied to opposition to the University of Arizona astrophysical site from other documented traditional Apache values, although it should be recognized clearly that much of the information provided to the U.S. Forest Service (USFS) and the University of Arizona regarding Mt. Graham has been brought forward only because of the MGIO.

Historical Background

Archaeological evidence for human use and occupation of the Pinaleño Mountains extends back thousands of years, and some Apache stories include ties to Mt. Graham since time immemorial. Prehistoric farming villages dating from the AD 900s to the AD 1400s are located throughout the Gila River Valley north of the Pinaleño Mountains (Sheridan 1995; Gilman 1997), a period during which the high peaks of Mt. Graham were also visited (Spoerl 1992). After about AD 1400, there is archaeological evidence of Apachean movement into southern Arizona (Wilcox 1981). During the 1600s and 1700s, when Spain ruled southern Arizona, Spanish documents describe Apaches in what was then called the "Sierra de Florida" or "Sierra de Santa Rosa de la Florida" (Wilson 1995). The first Spanish military campaigns to Mt. Graham to round up Apaches and bring them to peace settlements near Tucson took place in the early 1700s (Gillespie 2000).

In 1821, Mexico gained independence from Spain, and thereafter, only sporadic forays into the Gila Valley by outsiders occurred. The Gadsden Purchase followed in 1853, giving the United States control of Mexican territory south of the Gila River. Subsequent settlement and exploration of the Gila River Valley by European Americans began in earnest and was hindered only by the presence of Apaches who already occupied lands on either side of the river.

The decades of the 1860s, 1870s, and 1880s are commonly known as the "Apache Wars," during which time the United States military established military posts to force Apaches to give up their nomadic and raiding lifestyle as well as their claims to large mountainous areas in central and southern Arizona. As part of congregating Apaches in specific locales, the U.S. government established the White Mountain Apache Reservation in 1870 and added the San Carlos Division the following year. Its southern boundary was described as running in a line 15 miles south of and parallel to the Gila River (Royce 1899), which would have included the northern portion of the Pinaleño Mountains currently included in the Coronado National Forest. This division was reduced in size repeatedly during the following five years.

In 1874, the U.S. Army began a program to round up all Western, Chiricahua, and Yavapai Apaches not already confined to reservations and place them on the San Carlos Apache Reservation (Basso 1971:22). The reservation became a forced location for many Apaches who did not claim the area as their homeland. Apaches from diverse and unrelated groups were forced together with little regard for cultural, linguistic, or political differences. Band and territory distinctions were divided and lost. In 1897, the San Carlos and Fort Apache reservations were separated for administrative purposes, with the division at the Black River. This boundary formed a north-south distinction, while the traditional band territories fell more closely along east-west lines (Basso 1983) (fig. 1.1). Goodwin noted (1942:9) that the location of various groups on the San Carlos Apache Reservation made it a Western Apache melting pot. By the 1920s, group and band distinctions there had broken down more than on the Fort Apache Reservation, where people still were able to retain some of their old territorial affiliations.

The conversion of Apaches to Christianity had not been a primary mission of the Spanish, Mexican, or U.S. governments, although Christian faiths became incorporated into community life after establishment of the reservations. Various denominations were established beginning in the 1890s, and even though many Apaches now attend numerous churches on the reservations, it is apparent that they also have maintained Native beliefs and practices throughout more than a century of reservation life.

While Western Apaches were being moved to, and held on, reservations, the Gila River Valley and surrounding mountains were being discovered by Anglo-American settlers moving west. The first documented expedition to the top of Mt. Graham occurred in 1871 when a government topographical survey party reached the summit. The town of

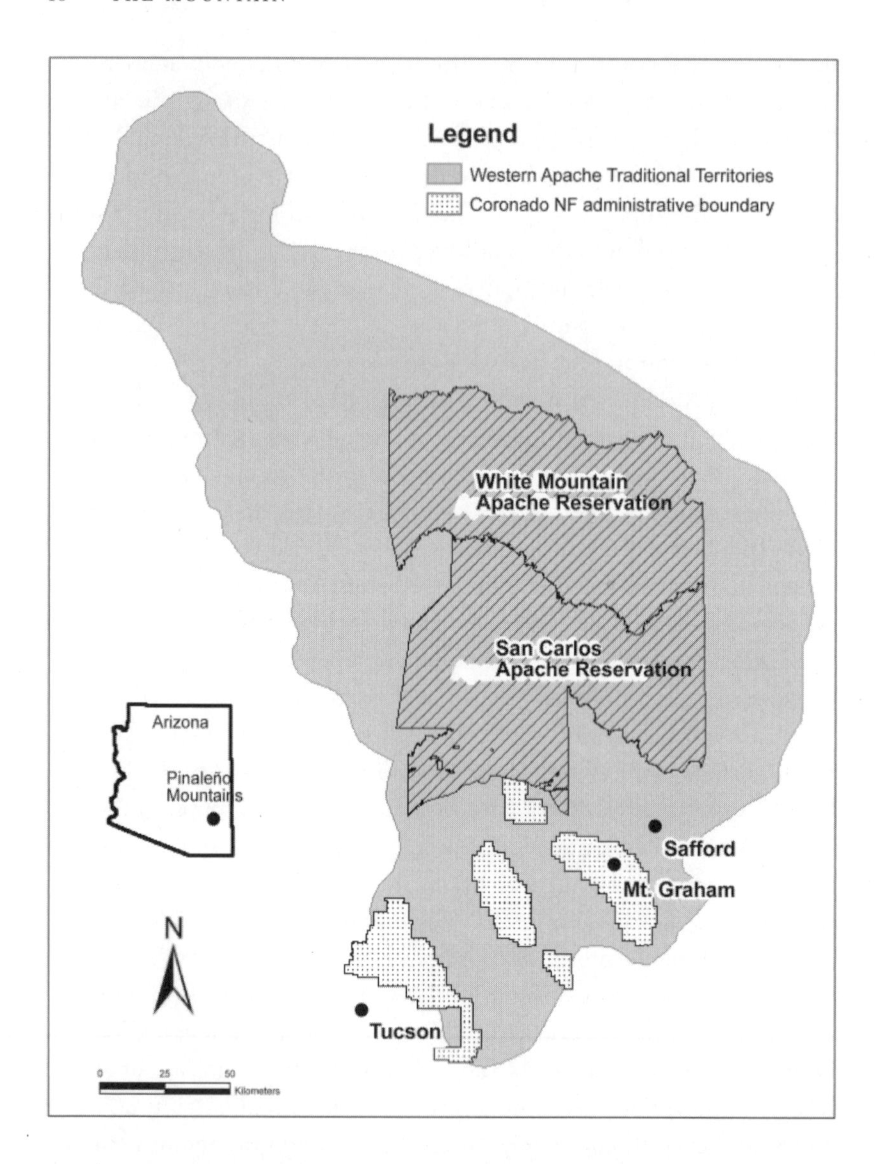

Figure 1.1. Western Apache Tribal Territories. Illustration by Chris Le Blanc, U.S. Forest Service.

Safford was founded in 1876, and by the 1890s, Mt. Graham had become a popular summer retreat from the desert heat for Gila River Valley residents. Substantial amounts of timber were also removed from the high elevations in the 1880s for military use at Fort Grant and for commercial purposes in the developing communities along the Gila River. In 1902, the Mt. Graham Forest Reserve was established; protection of the mountain's water supply and the ability to regulate timber harvest were the main reasons for establishment.

The United States granted Apaches the right to self-governance in 1934, after half a century of restriction to the reservations; thus, the ability of individuals to leave the reservations increased after that time. We do not know the extent to which Mt. Graham has been visited since the 1930s. Today, some tribal elders recall trips to Mt. Graham made during their youth where they learned of special places and plants. Other tribal members note that even though few trips may have been made to the mountain in recent years, Apache people do not need to physically visit a sacred mountain for it to remain important in their traditions.

We also know very little about Apache use of Mt. Graham in pre-reservation times. Records are few and scattered. Information obtained for describing Apache traditional uses of Mt. Graham and for evaluating National Register eligibility came from a variety of sources. These sources include USFS files, published documents, unpublished manuscripts and notes, litigation records, and verbal communications with elders and other individuals from the San Carlos, White Mountain, and Yavapai Apache tribes. Much of the published literature about Western Apaches presents an outsider's view of their culture. There are numerous publications about the "Apache Wars" of the late 1800s, and much of Apache history from this period has been romanticized or portrayed from the Euro-American perspective of the time. Little has been written from Apache viewpoints or with the recognition that there are a number of distinct Apachean groups. We are fortunate that several anthropologists took an interest in Western Apache history and culture. Pliny Goddard, one of the founding fathers of the American Anthropological Association, was the first to record traditional stories of the San Carlos and White Mountain Apaches (Goddard 1918, 1919).

Ethnographic research and documentation of traditional Western Apache culture is perhaps best known from the work of anthropologist Grenville Goodwin, and much of what we know about Apache people who lived in the vicinity of Mt. Graham comes from his work. Goodwin lived in Bylas and San Carlos for extended periods of time between 1930 and 1936, and his purpose in being there was to record Western Apache culture, particularly the stories of the "old days" as told by tribal elders, that is, from before the establishment of Indian reservations. Goodwin's work is especially important in helping identify traditional uses and areas because he interviewed individuals who, in the 1930s, were in their seventies, eighties, and even nineties, thus extending our direct knowledge of Western Apache culture back to the 1850s. Unfortunately, Goodwin died in 1940 at the age of 33, just as he was completing the first of a series of planned monographs resulting from his fieldwork. In 1968, the Goodwin family donated his field journals, notebooks, photographs, and watercolor illustrations to the Arizona State Museum at the University of Arizona; the notes about Apache religion that include references to Mt. Graham were not filed with the museum until 1989. These notes (Goodwin 1929–1939), along with Goodwin's published accounts (Goodwin 1942, 1994), are important sources of information.

Other documentary sources include federal court records from two lawsuits (*Apache Survival Coalition v. United States*, decided in 1994, and *Apache Survival Coalition v. United States*, decided in 1997). In both cases, individuals from the San Carlos Apache Tribe provided written declarations for the court regarding their traditional beliefs about Mt. Graham.

Another important source of information is a series of interviews conducted with Western Apaches in 1998 and 1999 as part of a Coronado National Forest American Indian traditional uses assessment. Individuals from the San Carlos and White Mountain Apache tribes were interviewed by a member of the White Mountain Apache Tribe under contract to the USFS (transcripts on file, Coronado National Forest). Field visits to Mt. Graham also took place.

Themes of Significance

From the available data, I identified four closely related themes to describe Mt. Graham in terms of the values ascribed to it by Western Apaches, so the themes could then be used to evaluate its significance in determining eligibility for the National Register of Historic Places (Spoerl 2001). These themes are (1) home of the mountain spirits (*gaan*), (2) source of natural resources and traditional medicine for ceremonial uses, (3) place of prayer, and (4) source of supernatural power.

Western Apache culture includes supernatural, or spiritual, beings that have a role in influencing an individual's life. One important class of beings is the *gaan*, who are believed to live within certain mountains. They have been described (Goodwin 1929–1939:xxiii) as "a class of supernaturals living inside the mountains and certain caves and who may be equated with the Pueblo kachinas. They were a people living on this earth long ago, but went away never to return." The gaan are embodied on earth by masked dancers and appear in certain ceremonies. Goodwin's informants said the gaan at one time lived in caves in the mountains and that men would go to the mountains to sing to the gaan and dance with them. The gaan continue to serve an important role in Apache spiritual healing and appear in religious ceremonies as well as at sunrise ceremonies.

Mt. Graham provides plants and other natural resources for spiritual, ceremonial, and medicinal uses. In the 1930s, Goodwin's informants described plants found at a variety of elevations. For example, one said that while learning ceremonies a man is taught what roots to use for medicine, and he is taken to the mountains to learn how to dig up roots and about the ceremony that goes along with it (Goodwin 1929-1939, MS 17:84). Other documents and interviews conducted by Coronado National Forest personnel also note the importance of medicinal plants. Specific information regarding these resources and their uses is restricted among Western Apaches and not readily provided outside the tribes.

Sacred powers are attributed to Mt. Graham, and it is often considered the southern sacred mountain of the Western Apache. It has been, and continues to be, visited for prayer. Traditionally, the mountain was associated in prayer with the powers of thunder and lightning to provide rain for crops (Goodwin 1929–1939, MS 17:89), and to promote good health and long life (Goodwin 1929–1939, MS 17:85). Goodwin's (1929–1939, MS 17:89) informants in the 1930s spoke about the sacred nature of Mt. Graham: "Graham Mount, Turnbull, Chiricahua Peak, the White Mountains, together with the Blue Range and . . . are all holy mountains, and can be prayed to because the clouds hang on them sometimes and Lightning People are on them, pray for *crops*, *life*, and *hunting*." Another component of its sacred nature involves the wind and air, both within the mountain and above it. Prayers pass through the mountain to help keep it strong so that it may protect Apache people.

The White Mountain Apache Tribal Resolution of July 1999 (Resolution No. 07–99–153) states: "*Dzil nchaa si'an*, the vast and unique mountain range known to non-Apaches as Mt. Graham, is a sacred site of long-standing and ongoing historical, cultural, and religious importance to many Apaches. The sacred character of the range encompasses all landforms, minerals, plants, and waters associated with or flowing from *Dzil nchaa si'an*."

The fourth theme is that of power. The concept of supernatural, or spiritual, power among the Western Apache comes from the beliefs that everything that exists has life and that all life has certain powers (*diyih*). Power must be respected in Apache beliefs so that it can be put to good use. Major sources of power include fire, lightning, thunder, water, wind, deer, bears, horses, mountain lions, bats, eagles, snakes, and lizards. It is not these natural features, occurrences, or animals that are sacred, but rather the power they contain (Goodwin 1929–1939, MS 17:85). Certain places, such as mountains, may be associated with a number of forms of power such as the location of ceremonies or where plants are gathered. Lightning in the mountains is generally considered an important source of power. Power is controlled by individuals through prayers and chants and can be dangerous if not properly controlled (Basso 1983:477). Its major use is to prevent and cure illnesses. A Goodwin informant states: "Mt. Graham was mentioned in hunting for deer, and the songs and power associated with deer" (1929–1939, MS 17:46). Another says: "Four mountains are mentioned in horse songs. One of these is Mt. Graham. They said they possessed horses, so were sung to. A giant blue horse and other ordinary horses lived on Graham Mountain, but no one ever saw them, except horse sign" (1929–1939, MS 17:89).

There is no available recorded information regarding the possible role of squirrels in Western Apache traditions and practices, although White Mountain Apache tales include a story that the squirrel originally controlled fire until it was stolen from him by the coyote (Goodwin 1994:147). The red squirrel has certainly existed in the mixed-conifer and spruce-fir forests of Mt. Graham as long as Apache peoples have visited it and was apparently much more abundant in the mid- to late 1800s than today.

I identified three additional themes from the available information that are relevant to Mt. Graham's importance in traditional Western Apache culture. These themes, however, are not considered essential to demonstrate that Mt. Graham meets the eligibility criteria for the National

Register of Historic Places as a traditional cultural property (*National Register Bulletin* 15). They relate more to pragmatic daily living practices than to the maintenance of cultural relationships and traditions.

First, the mountain was used as a refuge at various times in the past. A number of stories tell of individuals hiding in the mountains, particularly during the decades preceding and just after the establishment of reservations. Second, the mountain is said to be the location of Apache burials. Specific burial areas have not been identified to or by the USFS; however, the belief that there are burials on Mt. Graham is repeatedly noted by Apache individuals. And third, the lower slopes of Mt. Graham were used for food-gathering purposes. Acorn and agave gathering are mentioned frequently in historical accounts and in interviews. Mesquite, manzanita, and yucca are also mentioned as important food plants grown on the slopes of Mt. Graham. To date, specific hunting or gathering locales have not been identified through archaeological surveys, although general geographic areas have been described by Apache informants (Spoerl 2001).

The significance of Mt. Graham to Western Apaches is perhaps best summarized by anthropologist Keith Basso (1992:7), who states in a court declaration: "Despite the many changes that have affected Apache society in recent decades, Mt. Graham continues to be a vital part of the people's natural and moral universe. As a cherished feature of their original homeland, as dwelling place of the supernatural *gaahn'*, as homes to forms of life required for traditional ceremonial activities, as object of personal prayers, and finally as ancient burial ground, Mt. Graham stands as a critical component in the Apache way of life."

A Traditional Cultural Property and the National Register

Given the traditional cultural and religious values associated with Mt. Graham by Western Apache people, as described above, how do we translate these values into eligibility criteria for the National Register of Historic Places? A variety of property types is and can be listed on the National Register; the most familiar ones are historic structures and archaeological sites. A traditional cultural property is a type of property that is far less common (*National Register Bulletin* 38), and there is still much to learn about how best to describe and manage such properties (King 1998).

A traditional cultural property is defined as a location considered eligible for listing in the National Register of Historic Places "because of its association with cultural practices or beliefs of a living community that (a) are rooted in that community's history, and (b) are important in maintaining the continuing cultural identity of the community" (*National Register Bulletin* 38:1).

The traditional cultural significance of a property is defined as "significance derived from the role the property plays in a community's historically rooted beliefs, customs, and practices" (*National Register Bulletin* 38). "Traditional" in this context refers to "those beliefs, customs, and practices of a living community of people that have been passed down through the generations, usually orally or through practice." Examples of properties possessing traditional cultural significance include a rural community whose organization, buildings and structures, or patterns of land use reflect the cultural traditions valued by its long-term residents; a location where Native American religious practitioners have historically gone and are known, or thought, to go today to perform ceremonial activities in accordance with traditional cultural rules of practice; a location where a community has traditionally carried out economic, artistic, or other cultural practices important in maintaining its identity; and a location associated with the traditional beliefs of a Native American group about its origins, its cultural history, or the nature of the world.

A property may be eligible for the National Register under one or more of four criteria. Mt. Graham is considered eligible under criterion "A," a property associated with events that have made a significant contribution to the broad patterns of our history in the area of Native American ethnic heritage.

Another criterion for evaluation of National Register eligibility is the integrity of a property; that is, its ability to convey its significance through both the integrity of "relationship" with traditional cultural practices and beliefs, and its integrity of "condition" such that it retains integrity of setting and place in contemporary times that are sufficient for traditional relationships to continue. The integrity of Mt. Graham's relationship to Western Apaches is well documented in interviews and historical records

(Basso 1992; Welch 1997; Spoerl 2001); its integrity of condition is not as clear.

Mt. Graham has been affected by a wide variety of activities since the 1870s when Indian reservations were established, including timber harvest, highway construction, development of numerous campgrounds and picnic areas, creation of Riggs Flat Lake, an electronic site at Heliograph Peak, and fire lookout towers. The Pinaleños contain almost 80 miles of road open to public use, about 150 miles of trails, 88 summer homes (Turkey Flat and Columbine), 17 grazing allotments, and a variety of other USFS administrative and permitted facilities. The Mt. Graham Wilderness Study Area, established in 1984, includes 62,000 acres (24,900 ha) of the steep slopes of the mountain range where roads and other evidence of human use are minimal. The Mt. Graham Red Squirrel Refugium was created in 1989 for the protection of an endangered species and includes 1,750 acres (700 ha) closed to entry without authorization. Most recently, the MGIO authorized by Congress (Arizona-Idaho Conservation Act) was constructed on 8.6 acres (3.4 ha) within the red squirrel refugium.

Much of Apache opposition to the observatory revolves around the belief that the astrophysical site is disrupting or destroying the mountain's strength in Western Apache traditional beliefs and practices (Cassadore Davis 1992). However, the White Mountain Apache Tribe's resolution stating that Mt. Graham is a traditional cultural property eligible for the National Register was passed in 1999, after construction of telescopes and related support facilities; therefore, it is evident that the tribal council feels the mountain retains its integrity. Integrity is most important in the views of the traditional practitioners (King 1998), and according to many Apaches, Mt. Graham still retains the ability to convey its significance in traditional practices in spite of the presence of certain modern facilities and features.

In sum, Mt. Graham is significant in Western Apache spiritual beliefs and practices (*National Register Bulletin* 38). The mountain is associated with their oral history and plays a role in stories, songs, and myths that reflect ties to it, both in historical and contemporary traditional cultural activities. Sources that document its significance include ethnographic reconstructions of pre-reservation lifeways and spiritual practices that involve visitation to Mt. Graham, myths and songs about the mountain, and contemporary tribal interviews that describe its use and importance today. The mountain is associated with a pattern of events both spiritual and historical as evidenced by information provided during the 1930s and 1990s. Cultural themes for the mountain include a home of the mountain spirits (gaan); a source of natural resources and traditional medicine gathering for ceremonial uses; a place of prayer (spiritual significance, sacred place); and a source of spiritual power (certain places, presence on the horizon, specific animals).

Mt. Graham's historical association with traditional cultural practices is greater than 50 years old, thus meeting National Register eligibility criteria (*National Register Bulletin* 15). Such practices can be documented from at least the 1850s to the present, and probably extending back in time to the early 1600s and possibly 1500s (Gillespie 2000). The mountain may not have been visited frequently during most of this century after establishment of the reservations and before the attention focused on it by construction of the MGIO. The lack of visitation during certain time periods, however, does not affect its National Register eligibility status (*National Register Bulletin* 38:16). In addition, Mt. Graham may be important to certain tribal individuals and may not be to others; however, the available information indicates that it remains important to the Western Apache community, particularly in terms of its historical relationship to traditional Apache lifeways.

Although much of the significance of Mt. Graham to Western Apaches may be thought of in terms of "religion," this component does not make the property ineligible for the National Register. As stated in *National Register Bulletin* 38:13: "the fact that a property is used for religious purposes by a traditional group, such as seeking supernatural visions, collecting or preparing native medicines, or carrying out ceremonies, or is described by the group in terms that are classified by the outside observer as 'religious' should not by itself be taken to make the property ineligible, since these activities may be expressions of traditional cultural beliefs and may be intrinsic to the continuation of traditional cultural practices."

The boundaries of a traditional cultural property can range from an hectare or less to thousands of hectares. Identifying boundaries for the Mt. Graham (Dzil nchaa si'an) property was particularly difficult because defining where a mountain begins and ends in Apache traditions cannot be accomplished readily by drawing lines on a map. The White

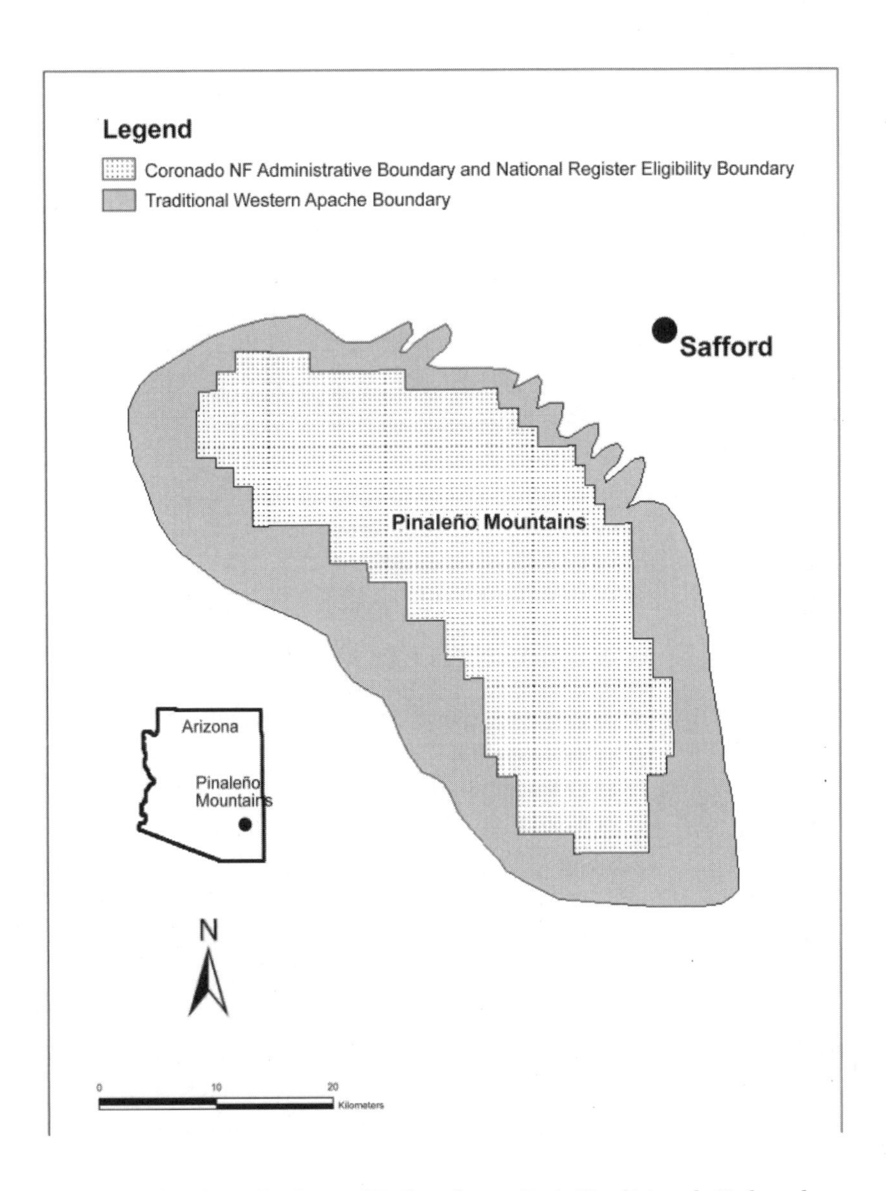

Figure 1.2. Mt. Graham (Dzil nchaa si'an) Traditional Cultural Property Boundaries. Illustration by Chris Le Blanc, U.S. Forest Service.

Mountain Apache Tribal Council Resolution (No. 07–99–153) made its position very clear in stating: "Be it further resolved by the Tribal Council of the White Mountain Apache Tribe that any determination of eligibility or National Register nomination that fails to identify the entirety of the mountain range known as Dzil nchaa si'an as having substantial and indivisible historical, cultural, and religious importance to Apaches is unlikely to be acceptable to or endorsed by the White Mountain Apache Tribe." Ramon Riley, Cultural Resources Director for the White Mountain Apache Tribe, outlined a general boundary that encompasses the entire mountain mass and demonstrates the importance of drainages (fig. 1.2).

The Western Apache tribes were unwilling to identify specific places of importance because of the belief that these places would lose power or be destroyed once identified. Although only a few sites are described, it is clear that specific locations on the mountain are visited for healing, prayer, and collection of natural resources for ceremonial and medicinal uses.

After numerous discussions and review of maps, the USFS recommended to the Keeper of the National Register, with the agreement of the Western Apache tribes, that the traditional cultural property boundary should be the USFS administrative boundary of the Pinaleño Mountains unit (fig. 1.2). The Keeper's determination of April 29, 2002 states: "The Forest Service Pinaleño Mountains unit is part of a larger regional landscape that is sacred to the Western Apache tribes. Potentially, this larger regional landscape, or certain parts of it, may be eligible for listing with further identification, evaluation, and documentation" (National Park Service Determination of Eligibility Notification of April 29, 2002. Unpublished document transmitted to John McGee, Forest Supervisor, Coronado National Forest). The Mt. Graham (Dzil nchaa si'an) traditional cultural property includes slightly over 198,800 acres (79,520 ha).

Recognition of the role Mt. Graham has played, and continues to play, in Western Apache culture does not diminish its importance to other cultures or ethnic groups. There are documented Chiricahua Apache associations with the mountain, and the Hopi Tribe claims associations with specific places and areas. The Pueblo of Zuni was actively involved in evaluating alternatives for the location of the MGIO during the environmental analysis process and provided information about traditional associations with specific high-elevation areas (Spoerl 1992). Archaeological sites and districts have been determined eligible for, or listed in, the National Register of Historic Places, including historic fire lookouts, prehistoric habitation districts (e.g., Marijilda Canyon, Oak Draw), and CCC buildings and features (e.g., Columbine Administrative Site).

Special Designations and Management

In addition to the variety of properties listed in, or eligible for, the National Register of Historic Places, representing well over 1,000 years of human use and occupation, the Pinaleños also contain specific and sometimes overlapping natural resource geographical designations. These specially designated areas serve purposes quite different from recognition and protection of traditional and historical values and involve different management protocols and legal mandates.

The Goudy Research Natural Area, encompassing 300 ha at the high elevations between Columbine and Riggs Flat, was designated in 1980 to preserve a representative sample of an ecological community and provide for scientific research and education. Use is permitted for only these purposes.

The 24,800 ha Mt. Graham Wilderness Study Area, containing steep and rugged slopes surrounding the Swift Trail (State Highway 366), was established in 1984 and recommended for wilderness designation in the 1986 Coronado National Forest Land and Resource Management Plan. Although Congress has not established it as a wilderness, motorized and mechanized equipment is not permitted, and the area is managed for its wilderness values "where the earth and its community of life are untrammeled by man, where man himself is a visitor who does not remain" (Wilderness Act of 1964).

The Mt. Graham Red Squirrel Refugium, located in the spruce-fir forest at the summit of the Pinaleños, is closed to all human entry to protect the endangered Mt. Graham red squirrel and its habitat. Recreation opportunities and facilities, including camping areas, hiking trails, and roadways, were closed, and in many cases obliterated, when the refugium was established as part of the Arizona-Idaho Conservation Act in 1989. The closure applies to 700 ha, and entry is only for red squirrel monitoring and research and USFS administrative purposes. The MGIO is the sole exception within the refugium, comprising 3.4 ha authorized for astrophysical use.

The Mt. Graham (Dzil nchaa si'an) traditional cultural property encompasses all of the specially designated areas within its 7,955 ha. Delineation of the traditional cultural property is part of Western Apache efforts to reassert control over their history and culture (Welch 2002) and thus ensure that their traditional knowledge, values, and practices are carefully considered and incorporated in management activities. There are no specific restrictions on access or use associated with its status as a National Register–eligible property. Actual listing of the property in the National Register has been proposed by the White Mountain Apache Tribe as part of a multiple-property nomination that would recognize the four major sacred mountains of the Western Apache (Resolution No. 12–2003–296). National Register–listed status does not modify management protocols above those of National Register–eligible status.

The specially designated areas in the Pinaleños include varying restrictions on use and access. The Goudy Research Natural Area and the Mt. Graham Wilderness Study Area have specific restrictions on the kinds of uses (e.g., scientific and educational) and activities (e.g., non-motorized, non-mechanized) considered appropriate, and both are open for public entry. The most restrictive area is the Mt. Graham Red Squirrel Refugium where conditions of entry and use are specifically identified in the congressional authorization. Opinions of Western Apache peoples differ on the importance of the refugium area in traditional beliefs and practices. It has been stated (Cassadore Davis 2000; Nosie 2000) that the refugium is part of Western Apache sacred area and includes mountain peaks, cultural sites, vegetation, waters, springs, and geological formations important in Western Apache traditions. There are specific places within the refugium that are special, particularly because the refugium encompasses the highest-elevation areas of the mountain. Access is guaranteed for tribal individuals who want to use the area and its resources of traditional purposes; however, certain Apache people feel that merely having the designation restricts their ability to carry out activities as they wish and violates their religious freedom.

The Mt. Graham traditional cultural property designation is not a restrictive one in the same respect as other specially designated areas: it is designation of a place that holds significant qualities to certain communities and cultures and merits special consideration because of these values. For the traditional cultural property, an essential component of management is the need to coordinate, communicate, and consult with Apache communities, and then carefully consider and incorporate their cultural and spiritual values in the decision-making process. Authorized uses and activities must not adversely affect the qualities that make the property eligible for the National Register of Historic Places (National Historic Preservation Act) and the values associated with a sacred site (Executive Order 13007).

Summary

These proceedings focus on the biological components of Mt. Graham and the recovery of an endangered species. It should be kept in mind, however, that it is impossible to separate the natural and cultural aspects of our environment. People have been part of the ecosystems being studied and managed for many thousands of years. Recognition of the human component and the knowledge associated with it will only enrich the biological and ecological work being accomplished.

The values and themes of cultural significance associated with Mt. Graham (home of the mountain spirits [gaan], source of natural resources used for ceremonial and medicinal purposes, a place of prayer, and a source of supernatural power) are integrally part of the natural environment. For the Mt. Graham traditional cultural property (Dzil nchaa si'an), USFS management and decision making will go forward with consideration and respect for Western Apache traditions and values. For the Western Apache tribes, the status of Mt. Graham as a traditional cultural property is viewed as one aspect of gaining recognition and respect, and protection of a sacred mountain.

Literature Cited

Basso, K. H., ed. 1971. *Western Apache raiding and warfare: From the notes of Grenville Goodwin*. University of Arizona Press, Tucson.

——. 1983. Western Apache. Pages 462–488 in A. Ortiz, ed. *Southwest: Handbook of North American Indians*, vol. 10. Smithsonian Institution, Washington, D.C.

——. 1992. Declaration of Keith Basso in support of preliminary injunction. *Apache Survival Coalition v. United States*. CIV. No. 91–1350 PHX WPC.

Cassadore Davis, D. O. 1992. Declaration of Ola Cassadore Davis in support of preliminary injunction. *Apache Survival Coalition et al. v. USA*. CIV. No. 91–1350 PHX WPC.

——. 2000. Letter from Ola Cassadore Davis, Apache Survival Coalition to Safford Ranger District, July 15, 2000. On file, Coronado National Forest, Tucson, Arizona.

Gillespie, W. B. 2000. Apaches and Mount Graham: A review of the historical record. Unpublished manuscript. Coronado National Forest, Tucson, Arizona.

Gilman, P. A. 1997. Wandering villagers: Pit structures, mobility, and agriculture in southeastern Arizona. *Anthropological Research Papers* no. 49. Department of Anthropology, Arizona State University, Tempe.

Goddard, P. E. 1918. Myths and tales from the San Carlos Apache. *Anthropological Papers of the American Museum of Natural History*, 24:1–86.

——. 1919. Myths and tales of the White Mountain Apache. *Anthropological Papers of the American Museum of Natural History*, 24:93–139.

Goodwin, G. 1929–1939. Archival Material MS 17. Unpublished field notes on file in the Arizona State Museum, The University of Arizona, Tucson.

Folder 26. Notebook Series D, Vol. 1: Place names in the Coyotero area.

Folder 37. Biographies. Sherman Curley, John Rope.

Folder 46. Hunting, food gathering.

Folder 84. "Notes on various topics: clothing, ornaments, hair, Divining Power, Learning Power, Deer Power Dream, Hawk Power and Curing Ceremonies, Lightning Power and Curing." 1–59.

Folder 85. "Concepts of earth, nature, universe." 60–210.

Folder 86. "Various sicknesses and cures; Learning Curing Power." 213–284.

Folder 88. "*gaan* dance." 308–397.

Folder 89. "Various topics: songs, prayers, cardinal points, caps, causes of various sicknesses, Silas John, Love Power, Pollen-Giving Dance 1932, cosmology, concepts of life." 398–456.

——. 1942. *Social organization of the Western Apache*. University of Chicago Press, Chicago. Reprinted as 2nd ed., University of Arizona Press, Tucson, 1969.

——. 1994. *Myths and tales of the White Mountain Apache*. University of Arizona Press, Tucson.

King, T. F. 1998. *Cultural resource laws and practice, an introductory guide*. AltaMira Press, Division of Sage Publications, Inc., Walnut Creek, California.

National Register Bulletin 38. 1990. *Guidelines for evaluating and documenting traditional cultural properties*. P. L. Parker and T. F. King. Interagency Resources Division, National Park Service, Washington, D.C.

National Register Bulletin 15. 1991. *How to apply the National Register criteria for evaluation*. Interagency Resources Division, National Park Service, Washington, D.C.

Nosie, W. 2000. Letter from Wendlser Nosie, Apaches for Cultural Preservation to Safford Ranger District, July 15, 2000. On file, Coronado National Forest, Tucson, Arizona.

Royce, C. C., compiler. 1899. Indian land cessions in the United States. In *Eighteenth annual report of the Bureau of American Ethnology 1896–97*, pt. 2. U.S. Government Printing Office, Washington, D.C.

Sheridan, T. E. 1995. *Arizona: A history*. University of Arizona Press, Tucson.

Spoerl, P. M. 1992. Shrine sites on Mount Graham: Their significance and to whom. Paper presented at the 7th Mogollon Conference, September 25–26, Las Cruces, New Mexico.

——. 2001. *Mount Graham (Dzil nchaa si'an), A Western Apache traditional cultural property, or determination of eligibility for the National Register of Historic Places, Mount Graham (Dzil nchaa si'an), Safford Ranger District, Coronado National Forest, Arizona*. Report on file, Coronado National Forest, Tucson, Arizona.

Welch, J. R. 1997. White eyes' lies and the battle for Dzil Nchaa Si'an. *American Indian Quarterly* 21:75–109.

——. 2002. The White Mountain Apache Tribe Heritage Program: Origins, operations, and challenges. Pages 67–83 in K. E. Dongoske, M. Aldenderfer, and K. Doehner, eds. *Working together: Native Americans and archaeologists*. Society for American Archaeology, Washington, D.C.

Wilcox, D. R. 1981. The entry of Athapaskans into the American Southwest: The problem today. Pages 213–256 in D. R. Wilcox and W. B. Masse, eds. *The protohistoric period in the North American Southwest, A.D. 1450–1700*. Arizona State University Anthropological Research Paper No. 24. Tempe.

Wilson, J. P. 1995. *Islands in the desert: A history of the uplands of southeastern Arizona*. University of New Mexico Press, Albuquerque.

## CHAPTER 2

Exploring the Larger Environment
Astrophysical Explorations from Mt. Graham

JOHN R. RATJE

STEWARD OBSERVATORY, the research arm for the Department of Astronomy at the University of Arizona, has been exploring the larger environment, the cosmos, since 1916. It is one of the world's leading astronomical research centers, which has grown since its inception and now operates more than a half-dozen research instruments throughout southern Arizona. One of the newer mountain research stations, the Mt. Graham International Observatory (MGIO), is located in the Pinaleño Mountains of southeastern Arizona.

In addition to the ground-based research efforts, Steward Observatory plays key roles in major space astronomy missions, including the Hubble Telescope, launched into orbit in 1990, and the Spitzer Space Telescope (formerly called the Space Infrared Telescope Facility), launched during the summer of 2003.

Astronomical "Seeing" and Mt. Graham

Mt. Graham was an obvious choice for an observatory, even dating back to the mid-1950s when a search was underway for a location for the new national observatory. That particular search resulted in the selection of Kitt Peak. Then, in the late 1970s, the search began for the next generation of telescopes. The major search criteria were high elevation (low water vapor), clear skies (cloud-free transparent atmosphere), dark skies (low light pollution), proximity to the Tucson campus (extremely important for student access), and relatively low environmental impact. Mt. Graham met those requirements. It even had a road to the summit that was mostly paved.

The only astronomical-site characteristic not considered in the initial survey was image sharpness (astronomical seeing) because there is no generally accepted method of doing this other than by making empirical in situ measurements. Astronomical seeing can be generally thought of as consisting of three components: local seeing or dome seeing related to convection currents created by the telescope and its enclosure, ground or boundary layer seeing occurring directly above the telescope for approximately 25 m, and atmospheric turbulence. Site testing, using small telescopes and micro-thermal towers, concentrates on measuring the boundary layer and atmospheric turbulence contributions to astronomical seeing.

Astronomical seeing is a term used to quantify the steadiness or turbulence of the atmosphere. A telescope image that jiggles, twinkles, swarms, or otherwise moves in a manner that destroys visible detail is a result of astronomical seeing. The atmosphere contains a great number of micro-thermal air cells or bubbles that effectively act as lenses. A pinpoint of starlight (or more correctly a wave front) coming through the atmosphere is distorted by these lenses such that the image produced is fuzzy or blurred. Because the micro-thermal cells move around in the atmosphere, the refracted image dances about the focal plane of the telescope (or your eye), producing an image that appears to boil or twinkle. Seeing is determined by measuring the diameter of the fuzzy image in arcseconds—a larger-diameter image correlates with an increase in atmospheric turbulence (i.e., poor seeing). The quest is to find the perfect observatory site where the seeing is limited by the telescope optics and not the atmosphere.

The ideal situation is to place the telescope in orbit above Earth's atmosphere. However, that is an extremely expensive proposition—from both a capital and a maintenance point of view. The capital cost ratio between orbital and ground-based telescopes is on the general order of 25 to 50 times, and maintenance cost for orbital telescopes is out of this world.

One way to achieve excellent ground-based observations is to clear away the atmosphere. Steward Observatory's Center for Adaptive Optics is researching and developing adaptive optics techniques to reduce the effects of atmospheric turbulence. This feat is accomplished by placing a smaller mirror or lens in the optical path of the telescope and bending that optical element in real time to remove the atmospheric distortion. The Large Binocular Telescope (LBT) will greatly benefit from adaptive optics technology.

Site Testing on Mt. Graham

Once a site has passed the initial requirements, it must be tested empirically to evaluate the seeing characteristics. To determine the image-sharpness characteristics of Mt. Graham, Dr. Richard Cromwell, Dr. Nick Woolf, and others carried out a ten-year study. These studies first identified the Emerald Peak region as the best general area on Mt. Graham for an observatory because it provided good image sharpness and space for telescopes. In the second study phase, measurements were made of seeing differences between various locations within the 67 ha Mt. Graham Astronomical Research Area as identified by Congress in the 1988 Arizona-Idaho Conservation Act (fig. 2.1).

All of the telescope sites that Steward Observatory has tested extensively within the Mt. Graham Astrophysical Research Area, around Emerald Peak, provided excellent astronomical seeing. Several years of data show that the typical image size for a large telescope located 25 m above the ground is about 0.6 arcseconds: this compares favorably with the image quality at the Multiple Mirror Telescope (MMT) on Mt. Hopkins in the Santa Rita Mountains in southern Arizona. The Emerald Peak sites are also competitive on a worldwide basis (Woolf and Cromwell 1988) (table 2.1; data provided by the observatories concerned).

Studies have demonstrated that two atmospheric layers contribute most of the image blurring or seeing. A high-level, 10 km layer contributes about 0.5–0.53 arcseconds, and a thin 20 m boundary layer at roughly tree height contributes most of the remainder and accounts for local seeing differences. By suitable design of the telescope, the effects of this boundary layer can be minimized. This has, of course, been accommodated in the LBT design.

Among the Emerald Peak sites, the site of the 1.8 m Vatican Advanced Technology Telescope (VATT), the smallest of the telescopes of the MGIO, yields the poorest image sharpness (median 0.63 arcseconds at a height of 25 m, which even so is very competitive internationally). The VATT, operational since 1993, has a more than 10-year record measuring

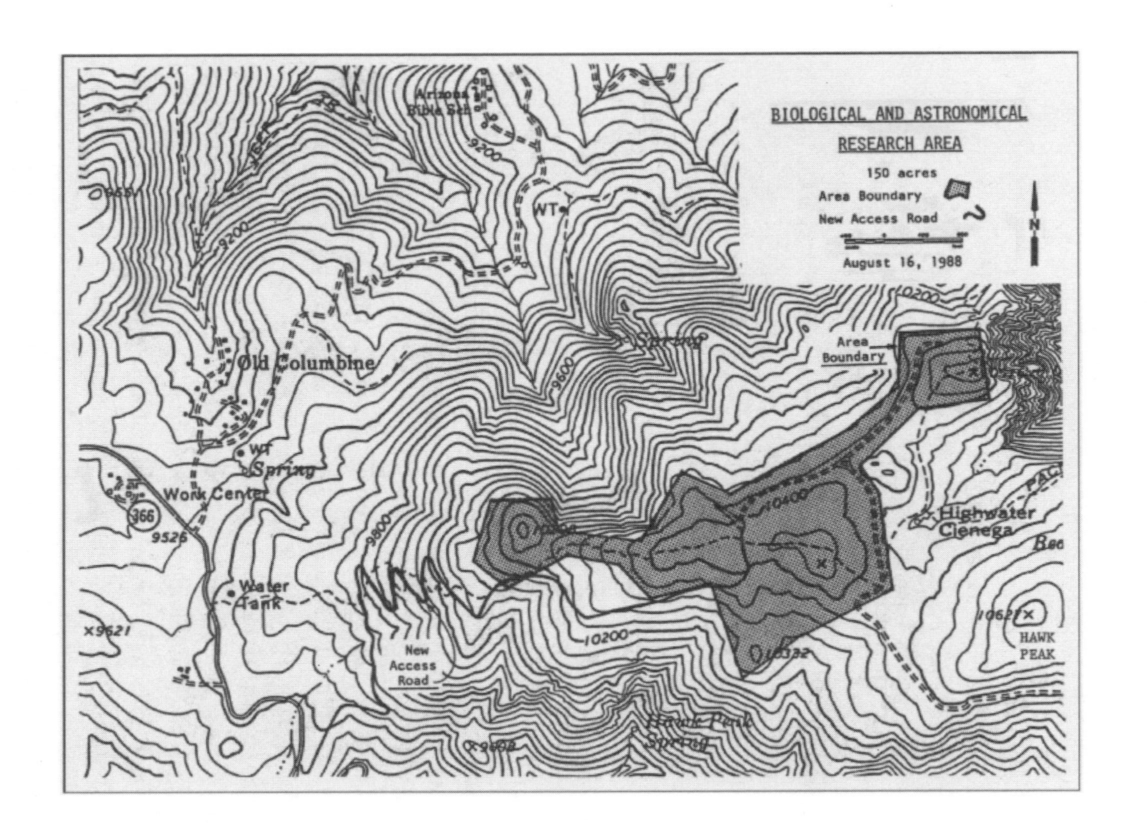

Figure 2.1. Biological and Astronomical Research Area.

Table 2.1. Comparative Image Sharpness

Site	Median Seeing
Mauna Kea Observatory, Hawaii	0.45
Emerald Peak sites (range), Mt. Graham	0.55–0.63
Multiple Mirror Telescope (Mount Hopkins/summit, Santa Rita Mountains, Arizona)	0.60
Very Large Telescope site (European Southern Observatory, Chile)	0.66
Magellan telescope (Las Campanas Observatory, Chile) and La Silla site, Chile (European Organisation for Astronomical Research in the Southern Hemisphere)	0.76

actual image sharpness at that telescope and providing empirical calibration of the test data. The VATT, which is placed 10 m above ground, is achieving regularly image sharpness around 0.75 arcseconds. (This measurement was taken without thermal control of the mirror.) After adjustment to 25 m, the predicted image sharpness for LBT may be slightly better than that shown in table 2.1.

Public Law 100–696 authorized the first three telescopes on Mt. Graham under the auspices of the MGIO. It further specified the locations of each of the telescopes according to a sketch accompanying the U.S. Fish and Wildlife Service's Biological Opinion (1989).

The LBT would be located about 180 m west of the entry road to the observatory site (fig. 2.2). A request to place the LBT, the largest telescope in the first phase of the MGIO, at the best potential site (peak 10298, median seeing 0.55 arcseconds at 25 m) was turned down by the U.S. Fish and Wildlife Service (USFWS) because they felt the Phase I telescopes should remain clustered, as shown in the Biological Opinion. At a later date, but prior to construction of the LBT, biologists approached observatory administration about relocating the LBT from the location specified in the Biological Opinion to peak 10471 (also known as Emerald Peak), a site that would have much less impact on the endangered Mt. Graham red squirrel. Observatory administration agreed, and Emerald Peak, with a seeing of 0.59 arcseconds median at

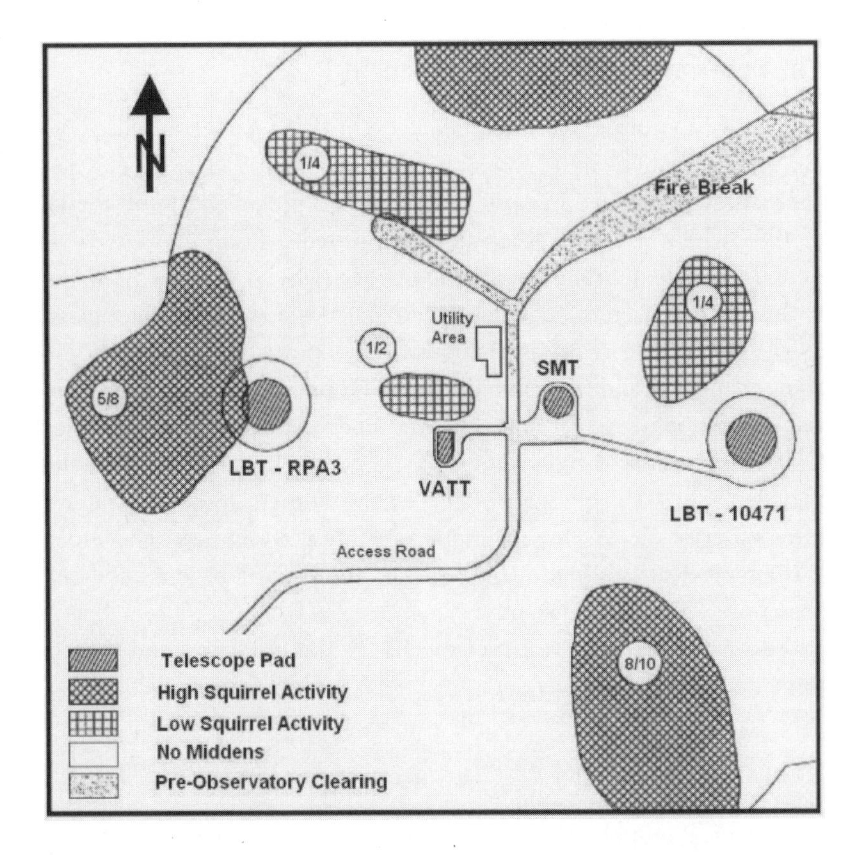

Figure 2.2. MGIO Schematic Layout. Circled numbers indicate the number of active red squirrel middens out of the total number of middens found in each respective area in 1994 during the dispute over the LBT site location.

25 m, was approved in December 1993. While peak 10298 was preferred on the basis of image quality alone, the LBT will be fully competitive at Emerald Peak, which better satisfied other important criteria besides image quality.

The Mt. Graham site thus combines the familiar advantages of existing southwestern U.S. sites with very good image sharpness, low light pollution, an altitude high enough for infrared work, and an existing paved access highway. It is essentially unique in its astronomical qualities, at least in the continental United States (Woolf and Cromwell 1988).

The Environmental Impact Statement

The final Environmental Impact Statement (EIS) for the proposed Mt. Graham Astrophysical Area was issued in November 1988, and the Record of Decision was issued in April 1989. The EIS process began in the fall of 1984 when Steward Observatory submitted a Preliminary Analysis Actions document to the U.S. Forest Service (USFS) for evaluation.

In all, there were six alternatives evaluated—three of the alternatives allowed various levels of astrophysical development, and three did not allow astrophysical development. The USFS preferred alternative G provided for astrophysical development as outlined in the USFWS's Biological Opinion's reasonable and prudent alternative #3 (U.S. Fish and Wildlife Service 1989). This alternative allowed the immediate development of three observatories, necessary support facilities, and a new access road. Astrophysical site testing for the possible future siting of four additional telescopes would be allowed.

The EIS covered nine issues specific to the Pinaleños and the first three telescopes. They were

1. plant and animal diversity;
2. watershed management;
3. recreation uses and opportunities;
4. wilderness and special area designations;
5. visual quality;
6. cultural resources and Native American use;
7. astrophysical values and benefits;
8. socioeconomic benefits;
9. safety and protection.

Legislation

During the EIS work, three separate legislative bills clarified or impacted the studies (Anonymous 2005):

- The Arizona Wilderness Act of 1984
- The Arizona-Idaho Conservation Act of 1988
- The Omnibus Appropriations Bill of 1996

The Arizona Wilderness Act of 1984

The Arizona Wilderness Act of 1984 (AWA) established the wilderness study area (~62,000 acres) around the summit areas of Mt. Graham. At the time the AWA was being written, Steward Observatory was preparing to present their Mt. Graham observatory proposal to the USFS. The concern was that the wilderness study area might preclude access for utility lines for any future project on the mountain. Steward Observatory personnel working with various environmental groups successfully negotiated a 500-foot-wide corridor up through the proposed wilderness study area along the southwest slope of the Pinaleño Mountains, thus allowing for possible future utility extensions. Following the construction of the first two telescopes in 1993 and with the LBT construction requiring additional electrical power, the underground power line was completed in 2001, and commercial electricity first powered the site in November of that year.

The Arizona-Idaho Conservation Act of 1988

The Arizona-Idaho Conservation Act of 1988 (AICA) established three observatories on 8.6 acres. It has provisions for an additional four observatories pending further environmental studies. President Ronald Reagan signed it into law November 18, 1988 (Public Law 100–696).

AICA incorporated the terms of the USFWS Biological Opinion, agreed to by the USFS and the University of Arizona (UA), into federal law. It authorized the UA and its partners to build three telescopes, necessary support facilities, and an access road on Emerald Peak as the first phase of the observatory. Following the initial three telescopes, AICA further authorized the construction of four additional telescopes, subject to the Endangered Species Act and other applicable law. The entire seven-instrument observatory, including support buildings and roads, is limited to no more than 24 acres. The Biological Opinion required the UA to construct a new, narrow, two-mile access road that required 6.2 acres, leaving only 2.4 acres for the initial MGIO site consisting of the first three telescopes. Further, Steward Observatory restored to its natural state the original five-mile road to the site. The road restoration project comprised about 60 acres. The Biological Opinion required the USFS and/or the UA to implement a 10-year study to define the population dynamics, habitat, and microclimatic factors concerning the red squirrel. The USFWS further specified that the red squirrel monitoring program adjacent to the observatory and access road would continue for the life of the astrophysical complex.

Omnibus Appropriations Bill of 1996

In December 1993, after a year of studies and evaluation of four different sites for the LBT, the USFS approved construction on site 10471 (also known as Emerald Peak). Biologists from the USFWS, the USFS, and the UA agreed that the site would have the least impact on the endangered red squirrel population.

Site 10471, to the east of the Submillimeter Telescope (SMT) (fig. 2.2), is about 1,300 ft to the east of the site referred to as RPA3 (reasonable and prudent alternative #3) on the map. Hatched areas on the map indicate the varying levels of squirrel activity. Site RPA3 was shown as the site for the LBT telescope in a map appended to the Biological Opinion that was cited by Congress in its authorization of construction of three telescopes on Mt. Graham in 1988. The USFS and the UA have always interpreted the bill since its passage to mean that the construction location of RPA3 was approximate, and that the USFS had discretion to approve changes based on the available biological information.

In May 1994, the Mt. Graham Coalition filed suit alleging that the location change of the LBT (the last of the three telescopes initially authorized by AICA) was improper and requesting a temporary restraining order. The district court in Tucson ruled against the federal agencies, and the UA in July 1994 entered a permanent injunction against further LBT work on Mt. Graham pending further environmental studies. The UA appealed to the 9th U.S. Circuit Court of Appeals, and oral arguments were heard in November 1994. In April 1995, the 9th Circuit Court of Appeals upheld the district court by agreeing that, under AICA, the USFS did not have the prerogative of changing the LBT site. While the dissenting opinion argues that the USFS is entrusted by the Arizona-Idaho Protection Act to administer the terms of the act and, in this case, insure the least disruption to the squirrels on Mt. Graham, the 9th Circuit Court ruled in favor of the district court's interpretation.

On July 20, 1995, both a petition by the UA for rehearing and a suggestion by the UA for rehearing en banc were denied. From this, strong community support developed to resolve the issues concerning the site for the LBT. Several townships in the Mt. Graham area and the boards of supervisors in Graham and Cochise counties passed resolutions calling for immediate action by Congress to allow for telescope construction. Many felt that Congress should clarify its intention on the Arizona-Idaho Conservation Act that was passed in 1988.

The Omnibus Appropriations Bill, containing the Mt. Graham rider offered by Congressman James Kolbe, was passed on April 25, 1996, by a vote of 399–24 in the House of Representatives and a vote of 88–11 in the Senate. President Clinton signed it into law the following day. The Omnibus Appropriations Bill of 1996 made it possible to continue construction of the LBT on the site preferred by biologists and astronomers. The signing of the bill provided the needed clarification to allow the LBT construction to proceed at the 10471 site.

The rider stated, "The United States Forest Service approval of Alternative site 2 (ALT 2), issued on December 6, 1993, is hereby authorized and approved and shall be deemed to be consistent with, and permissible under, the terms of Public Law 100–696 (Arizona-Idaho Conservation Act of 1988)."

The Observatories

The Mt. Graham International Observatory (MGIO) is an "umbrella" organization, operated by Steward Observatory at the University of Arizona. It supports the scientific research efforts of the various user organizations. Currently, MGIO consists of three user groups. The Vatican Advanced Technology Telescope is operated by the Vatican Observatory Foundation and Steward Observatory, and the Submillimeter Telescope is operated by the Arizona Radio Observatory, the University of Arizona. The Large Binocular Telescope Observatory is operated by a consortium comprised of the University of Arizona; LBT Beteiligungsgesellschaft, Germany; the Italian National Institute for Astrophysics, Italy; Ohio State University; and the Research Corporation in Tucson, Arizona.

Figure 2.3. The Vatican Advanced Technology Telescope.

The Vatican Advanced Technology Telescope

The Vatican Observatory, one of the oldest astronomical research institutions in the world, has its headquarters located at the papal summer residence in Castel Gandolfo, Italy, outside Rome. It operates the 1.8 m Alice P. Lennon Telescope (fig. 2.3), with its Thomas J. Bannan Astrophysics Facility, known together as the Vatican Advanced Technology Telescope (VATT). The heart of the telescope is an f/1.0–honeycombed construction, borosilicate primary mirror. This was manufactured at Steward Observatory Mirror Laboratory (SOML), the University of Arizona, which pioneered both the spin-casting and the stressed lap polishing techniques that are being used for telescope mirrors up to 8.4 m in diameter.

The Submillimeter Telescope

The Submillimeter Telescope, operated by the Arizona Radio Observatory, a division of Steward Observatory, is the world's most accurate radio telescope (fig. 2.4). It probes the heavens at the shortest wavelengths of light that can pass relatively unimpeded through the earth's atmosphere. Until very recently, this region of the electromagnetic spectrum had not been explored. The telescope is used for around-the-clock observing. Special prototype radio receivers that employ superconducting electronic devices are used to probe areas of star formation in our galaxy and beyond. From studies of planets, asteroids, and comets, to studies of gas and dust that pervade interstellar space, this telescope is virtually expanding our knowledge in every facet of astronomical research.

The Large Binocular Telescope—A Giant Step Forward

The LBT is like no other telescope in the world. Engineers designed each feature to achieve performance never before possible. The LBT's unique design overcomes technological barriers that limited ground-based astronomy, and it does so at a relatively low cost. This breakthrough was achieved by using the binocular design and innovative mirror technology.

The LBT is equipped with two giant mirrors, each 8.4 m across and

Figure 2.4. Heinrich Hertz Submillimeter Telescope in operation.

positioned side by side on a single mount spanning 23 m. Each of the LBT's mirrors is larger than any ever before achieved.

The mechanical design of the telescope is revolutionary, combining enormous stiffness with relatively lightweight compactness. The telescope has an elevation over an azimuth mounting. The optical support structure moves on two large "C"–shaped rings, and the compact azimuth platform transmits the load directly to the pier. This combination yields excellent performance and the necessary stability to combine the two light beams interferometrically. The short focal length of the primary mirrors (f/1.14) and the binocular layout also permit a very compact enclosure.

LBT project organization. By its very nature, astronomical exploration is highly interdisciplinary. It involves physicists, chemists, mathematicians, opticians, software specialists, and mechanical and electrical engineers, as well as astronomers.

An international consortium was formed in 1992 to meet the challenges of designing, constructing, and operating the LBT and to exploit its unique capabilities for exploring the cosmos. Members share responsibility for the scientific planning and for contributions to the cost of construction and operations. An outstanding international team assembled from member institutions brings together the best minds in academia and industry from both the United States and Europe. The consortium, through the LBT Corporation, oversees the LBT Observatory, which is a department of the University of Arizona.

The LBT consortium members include

- The University of Arizona (Tucson, Arizona);
- Istituto Nazionale di Astrofisica (Rome, Italy);
- LBT Beteiligungsgesellschaft (Heidelberg, Germany);
- Research Corporation (Tucson, Arizona);
- Ohio State University (Columbus, Ohio).

A *unique telescope.* The LBT (fig. 2.5), as a telescope, was fully operational as of 2008 and has the light-gathering power of a single 11.8 m mirror—greater than any other single telescope. The LBT instrument

Figure 2.5. The Large Binocular Telescope under construction – one mirror blank in place.

complement is progressing and will be fully instrumented in 2009–2010. However, instruments will be designed and fabricated over the lifetime of the telescope—over the next 100+ years. At its combined focus, the LBT will produce images that are 10 times sharper than the Hubble Space Telescope. The combined mirrors can also be used as a nulling interferometer so that light from a bright star is eliminated, while light from a nearby planet or interplanetary disk is enhanced. Because of its design, the LBT can switch rapidly between observing modes using swing-arm-mounted secondary and tertiary optics, permitting optimum use of different observing conditions. This telescope's unsurpassed light-gathering power will enable astronomers to observe extremely distant and faint objects—such as ancient galaxies and quasars—that will help unravel the evolutionary history of the universe.

The 16-story enclosure (fig. 2.6) for the LBT is a unique structure. The top 10 stories rotate along with the telescope as it scans the heavens.

Figure 2.6. The Large Binocular Telescope enclosure.

The rotating structure weighs about 1,450 tons and has doors on all sides that open to the night breezes. This feature allows the telescope to operate at ambient temperature, thereby reducing image degradation due to convection currents caused by temperature differences. The building provides protection for the telescope and associated instrumentation during the day and at all times during poor weather conditions. Additionally, it provides for the needs of the technical and scientific research staff.

The primary optics. Tennessee Ernie Ford sang: "You load 16 tons, and what do you get? Another day older and deeper in debt." In the case of the LBT's primary mirror (fig. 2.7), Steward Observatory Mirror Laboratory (SOML) loads 21 tons of borosilicate glass into a rotating furnace, then melts the glass and spins the furnace to create a parabolic shape on the top surface of the glass mirror blank.

Following casting, to produce a surface useful for astronomical re-

Figure 2.7. LBT's primary mirror blank fresh from the spin-casting furnace.

search, the blank must be first ground and then polished to a surface accuracy of one-millionth of an inch of a true parabola. This feat required the development and fabrication of a "stressed lap" polishing machine. The completed mirror blank has a final weight of about 18 tons. The spin-casting technique and the unique stressed lap polishing technique, used to make the world's largest cast telescope mirrors, were perfected by the SOML, directed by Dr. J. Roger P. Angel.

Once the polishing has been completed, to create the mirror, the polished front surface of the blank is coated with a very thin layer of aluminum (~900 angstroms thick or 3.5 millionths of an inch—a thousand times smaller than the diameter of a human hair). The LBT's primary mirrors are coated with aluminum in situ—while they rest in their mirror cells attached to the telescope structure.

A unique feature in the design of the honeycombed structure allows active temperature control of the mirror—air-conditioned mirrors! Classical telescopes typically have large thermal masses for their primary mirrors. These thermal masses are always changing with ambient temperature changes but always with some thermal lag. The temperature difference results in local convection currents causing blurred images. By actively maintaining the honeycombed mirror at outside ambient air temperatures, the quality of the images are significantly better than that achieved with classical telescopes.

An Analogy to the Human Eye

A scientist categorizes the capabilities of a telescope by looking at its light-collection ability and its ability to resolve fine details. Light collection is a function of the area of the primary mirror or lens. A telescope that has a primary mirror twice as large as another has four times the light grasp. The resolution of a telescope is based on the outer diameter of the primary optics. A telescope mirror with twice the diameter of another has twice the resolution.

One way to look at the immense size of the LBT is to compare it to a binocular we are familiar with—the eyes. How many eyes would it take to match the light-collecting performance of the LBT? Assuming the pupil of a dark-adapted eye opens to a diameter of 7 mm, calculations reveal that 1,440,000 eyes are equivalent to an 8.4 m diameter telescope mirror. For both mirrors, we need 2,880,000 eyes or 1,440,000 pairs of eyes. And to compare the resolution, a circle drawn around the outside edge of a pair of human eyes is ~65 mm diameter, while a similar circle drawn around the LBT's primary optics is 23,000 mm diameter. Thus, the LBT has a resolution that is about 350 times that of a pair of human eyes (and 10 times better than the Hubble Space Telescope)!

Astronomical Research at MGIO

Areas of current and future research conducted at MGIO include

- exploration of star-forming regions;
- study of the evolution of galaxies and the universe;
- study of the life cycles of stars;
- study of our solar system and planetary systems around nearby stars;
- image planets around nearby stars;
- study of enigmatic objects such as quasars and black holes;
- search for life outside our solar system;
- study of asteroids, Kuiper Belt objects, and the Oort Cloud.

The year 2003 marked the tenth anniversary of the dedication of the VATT and SMT. Significant research has been accomplished over the last decade. Some examples of that research are presented below. In addition, we elaborate on the future research opportunities as the LBT comes to the forefront of scientific operations.

Vatican Advanced Technology Telescope

Researchers Rich Boyle and Chris Corbally collaborated with Arlin Crotts, the principal investigator, on the search for MACHOs, an acronym for Massively Condensed Halo Objects. Among the first observations made with the VATT were time-sequence images of the nearby galaxy Andromeda (M31). Multiple images of fields were compared, seeking variations in the brightness of individual stars caused by Earth- to Sun-sized bodies in the halo of M31 passing between the VATT and the star. A MACHO will momentarily brighten the image of the star as its gravity bends and focuses the light of a star toward the observer. These "microlensing" events reveal otherwise invisible MACHOs, thought to be part of the 90 percent of a galaxy's mass that does not shine, but exerts a gravitational force on the stars of the galaxy, making the galaxy rotate faster than can be accounted for by the glowing matter. In more than seven years of the project, microlensing events indicating MACHOs around the size of the sun have been detected in our neighbor galaxy but, interestingly, in not nearly enough quantity to be the non-shining mass (Anonymous 2003).

In another example of VATT research, a number of observations confirming discoveries of distant supernovae have contributed to the work of the High Redshift (High-Z) Supernova Search. These bright stellar explosions are astronomers' best "standard candle" for measuring the distances to the farthest galaxies. Much to the surprise of the search team, the distances to the supernovae have been greater than those predicted by the big bang theory. It seems that the universe is expanding faster now than it did in the past; hence, it is accelerating in its expansion. This has lead to a theory of a "dark energy" much like the "dark matter" of which MACHOs form a part. The dark energy is a repulsive force that acted during the initial inflation of the universe when it expanded from the size of a proton to the size of a basketball in the first 1×10^{-35} seconds after the big bang. (If it had started as the size of a pea, it would have expanded to about the size of Saturn's orbit, 1.41 billion km, in that time!)

The density of matter with its mutual gravitational attraction has kept this repulsive force in check for over 10 billion years, but it has begun asserting itself again as matter spreads more and more thinly through space because of the continuing expansion of the cosmos.

The VATT has shown its potentiality much nearer home in the discovery of a binary asteroid. The object, 3782 Celle, was known to be one of the fragments blown off the asteroid Vesta by some collision in the past. William Ryan and his team have recently shown, by carefully monitoring the object's variations in brightness over 14 nights of observing, that this asteroid has a "moon" circling around it, about two and a half times smaller in diameter than the main body. This is the first "Vesta chip" known to have a companion, and it helps us understand what kind of debris comes out of collisions between asteroids and therefore the makeup of the asteroids themselves. This all adds to our knowledge of how planets formed around our sun and presumably other "suns" also.

Submillimeter Telescope—A Snapshot of Past Scientific Results

The SMT, with its highly accurate surface and excellent pointing, is truly a sub-mm dish with high-frequency (terahertz) capabilities. This characteristic was clearly demonstrated by the first observations of the molecule carbon monoxide (CO) at 1.037 THz (terahertz) (Kawamura et al. 2002). Spectra were recorded at several positions along the ridge of the Orion molecular cloud.

Remarkably, this transition of CO was detected over a 4' region, indicating very hot extended gas in the vicinity of Orion-KL. These observations showed the presence of two distinct clouds along the line of sight.

The high-frequency capability of this telescope is also evidenced by numerous observations of the $J=7\rightarrow6$ transition of CO near 810 GHz, the $J=6\rightarrow5$ line of 13CO at 661 GHz, and the neutral atomic carbon (CI) lines at 492 GHz and 809 GHz. Various galactic molecular clouds with star formation have been mapped at 810 GHz, including Orion, Cepheus B, Sharpless 106, and ON-1 (Wilson et al. 2001). The Orion outflow has been imaged in the 13CO: $J=6\rightarrow5$ line. These data show exactly where the hot gas is located and indicate the relative ages of the various star-forming regions. A similar differentiation in star formation timescales has been evident in CI observations (492 and 809 GHz) taken towards the

W3 complex (Tieftrunk et al. 1998). Most interesting has been mapping data obtained towards the outflow source IRAS 20126+4104, where the blue- and red-shifted CO wings ($J=7 \rightarrow 6$) meet at the precise position of the infrared source—an accomplishment only possible because of the pointing accuracy of the SMT telescope (Kawamura et al. 1999).

Lower frequency observing at the CO: $J=3 \rightarrow 2$ transition frequency near 345 GHz is routine at the SMT. Studies of this emission towards nearby face-on galaxies such as M51, NGC 6946, and M83 show significant structure in the spiral arms and the presence of a warm ($T>30K$) galactic disk (Mauersberger et al. 1999; Dumke et al. 2001).

One of the most recent successes, albeit primarily a technical one, has been the Very Long Baseline Interferometer (VLBI) experiment carried out April 2002 at 2 mm, using the SMT–Kitt Peak 12 m baseline. Fringes were cleanly detected at 147 GHz towards the bright AGN 3C279 and at 129 GHz towards the SiO maser source VY CMa—the highest-frequency spectral line VLBI to date. The IRAM 30 m telescope in Spain joined the SMT–Kitt Peak 12 m array, and 3C279 was detected from the 30 m to both Arizona sites. A fringe spacing of ≤50 μarcsec was achieved in this case—the highest angular resolution ever obtained by ANY astronomical technique. The record achievement of April 2002 was superseded by new observations in April 2003 as a similar interferometric array working at 230 GHz achieved a fringe spacing of about 30 μarcseconds!

The Large Binocular Telescope and the Search for Origins (Spitz and Neefe 2004)

The timing of the LBT could not be more favorable, for it coincides with major advances in our thinking about the origin of the universe and of life itself. These achievements, together with the immense potential of instruments such as the LBT, are causing a major shift in the direction of astronomical research. The goal of tracing the entire evolution of the universe—from its big bang origins, through the formation of galaxies, stars, and planets, and the chemical elements that are essential to life—is now reachable.

Looking out over the vast distances of intergalactic space, astronomers are now detecting galaxies at their epoch of formation—some 13 to 14 billion years ago. With LBT, they will be able to study the dynamical processes that create individual galaxies and clusters. Combined with detailed studies of the cosmic microwave background radiation—a relic of the earliest phases of the big bang universe—these observations should yield a major advance in our understanding of the physical origins of the universe and of the mysterious dark matter and dark energy that seem to play a crucial role, but still escape direct detection.

The power of the LBT may provide answers to questions that have long eluded astronomers. What is the origin of the incredibly luminous quasars and radio sources—galaxies with giant particle accelerators and almost certainly massive black holes in their centers? How and when are the chemical elements other than hydrogen and helium formed? How are they distributed in space? How are they incorporated in successive generations of stars and planets?

In our own and nearby galaxies, the LBT will also help unravel the mysteries of star and solar system formation. With the LBT's unique imaging capabilities at infrared wavelengths, astronomers will analyze the process of star formation in cool, dusty, interstellar clouds. The LBT will play a critical role in understanding proto-planetary disks around individual stars, direct evidence for which has only recently emerged.

After years of painstaking effort, astronomers have established, by indirect means, the likely presence of planets around nearby stars. The next step is to detect such extra-solar planets by direct imaging, an enormously difficult task given the relative brightness difference between star and planet and their close proximity. The LBT is uniquely designed to meet the challenge: its "nulling" mode suppresses light from the star while enhancing that from the nearby planet and any interplanetary material in a circumstellar disk. The mystery of planet formation at last seems solvable.

These prospects raise hopes of addressing perhaps the most tantalizing question of all: Can we find the presence of non-terrestrial life forms? The National Aeronautics and Space Administration and the European Space Agency have made the search the focus of their planetary exploration programs. The search for life elsewhere in the solar system has received strong impetus from the recent discovery of water on other solar system bodies. The LBT will pioneer the nulling technology required for future searches for life outside the solar system. The ultimate goal of understanding the origins of life itself, not only on Earth but also perhaps elsewhere in the cosmos, is within reach.

Literature Cited

Anonymous. 2003. *Vatican Observatory Newsletter*, Spring 2003 <http://clavius.as.arizona.edu/vo/R1024/News/vofnews03Spring.pdf> (accessed September 11, 2008).

Anonymous. 2005. A long time ago . . . Legal issues, Mt. Graham International Observatory <http://mgpc3.as.arizona.edu> (accessed September 11, 2008).

Dumke, M., C. Nieten, G. Thuma, R. Wielebinski, and W. Walsh. 2001. Warm gas in the central regions of nearby galaxies: Extended mapping of CO(3–2) emission. *Astronomy and Astrophysics* 373:853–880.

Kawamura, J. H., T. R. Hunter, C.-Y. E. Tong, R. Blundell, C. A. Katz, D. C. Papa, and T. K. Sridharan. 1999. First image with the CfA Superconductive HEB Receiver: The protostellar outflow from IRAS 20126+4104 in CO (J = 7–6). *Publications of the Astronomical Society of the Pacific* 111:1088–1094.

Kawamura, J. H., T. R. Hunter, C.-Y. E. Tong, R. Blundell, D. C. Papa, F. Patt, W. Peters, T. L. Wilson, C. Henkel, G. Gol'tsman, and E. Gershenzon. 2002. Ground-based terahertz CO spectroscopy towards Orion. *Astronomy and Astrophysics* 394:271–274.

Mauersberger, R., C. Henkel, W. Walsh, and A. Schultz. 1999. Dense gas in nearby galaxies XII: A survey for CO J = 3–2 emission. *Astronomy and Astrophysics* 341:256–263.

Spitz, A., and L. Neefe, eds. 2004. *The Large Binocular Telescope brochure*, dedication edition. The University of Arizona, Tucson.

Tieftrunk, A. R., S. T. Megeath, T. L. Wilson, and J. T. Rayner. 1998. A survey for dense cores and young stellar clusters in the W3 giant molecular cloud. *Astronomy and Astrophysics* 336:991–1006.

U.S. Fish and Wildlife Service. 1989. *Biological Opinion on development of the astrophysical area*. Phoenix Ecological Services Office, Arizona.

Wilson, T. L., D. Muders, C. Kramer, and C. Henkel. 2001. Sub-mm CO line emission from Orion. *The Astrophysical Journal* 557:240–249.

Woolf, N., and R. Cromwell. 1988. A long time ago . . . site testing eras <http://mgpc3.as.arizona.edu/theearly.htm> (accessed September 11, 2008).

CHAPTER 3

Tree-Ring Perspectives on Fire Regimes and Forest Dynamics in Mixed-Conifer and Spruce-Fir Forests on Mt. Graham

THOMAS W. SWETNAM, CHRISTOPHER H. BAISAN, AND HENRI D. GRISSINO-MAYER

RECENT WILDFIRES IN SOUTHERN ARIZONA and other regions of North America have sharply focused our attention on the condition of forests and landscapes. Large, high-severity wildfires are occurring in the context of both drought and cumulative changes in forests due to past human land uses. Although the role of humans in changing forest structure and fire regimes in the past century seems relatively clear in mid-elevation ponderosa pine forests (e.g., Allen et al. 2002), it is arguable that fewer changes have occurred in high-elevation spruce-fir forests (e.g., Romme et al. 2003; Schoennagel et al. 2004). Insect and disease outbreaks are also occurring in many western U.S. forests, and likewise, the relative roles of humans and climate in changing these disturbance regimes are subjects of much debate and concern (Logan et al. 2003).

In many ways, Mt. Graham is a microcosm of these issues. The forests of Mt. Graham have been altered by more than a century of livestock grazing, road building, logging, and active fire suppression (Wilson 1985; Bahre 1991, 1998). In recent years, the addition of astronomical facilities to the mountain has greatly increased the concern about the state of these forest ecosystems (Hoffman and Istock 1995). Concerns were further increased following a large, destructive fire in 1996 and during the current drought and massive outbreak of phytophagous insects on the mountain. The purposes of this chapter are to review the fire history of mixed-conifer forests and the age structure and fire history of spruce-fir forests and to use this knowledge to interpret past and current forest and fire-regime changes in the Pinaleño Mountains.

Forests and Fuels on Mt. Graham

Typical fire regimes and vegetation in the Pinaleño Mountains are associated with the elevational gradients found here (figs. 3.1 and 3.2). The topography is generally quite steep, and on the south side of the mountain, where our studies were conducted, there are few major barriers to fire spread (e.g., there are no broad rivers and few unbroken cliff faces or large talus slopes that might impede spreading fires). The consequence of these features is that fires starting in chaparral and woodlands on the steep slopes below about 2,800 m tend to burn upward into the mixed-conifer zone at about 2,800 to 3,000 m. The mixed-conifer forest covers the relatively level benches and ridges at the mid-elevations, and spruce-fir forests occupy the steep slopes and peaks above the benches and ridges (from about 3,000 to 3,300 m).

Summary of Mixed-Conifer Fire History

We began our fire history and age structure investigations on Mt. Graham in the early 1990s with support from the University of Arizona and the U.S. Forest Service (USFS). As in many other investigations of fire history in southern Arizona and elsewhere in the western United States (see summaries in Swetnam et al. 2001; Swetnam and Baisan 1996, 2003), we utilized tree-ring-dated fire-scar samples from living and dead trees to reconstruct a multi-century record of fire history (Grissino-Mayer and Fritts 1995; Grissino-Mayer et al. 1996). A total of 90 fire-scarred trees from three different sites on the mountain (fig. 3.1) were obtained, crossdated, and assembled into fire chronology charts (fig. 3.3). These collections were from relatively small areas (about 50 to 100 ha each) to the south and west of Webb Peak, and on Webb Peak (fig. 3.1). The fire-scar collection sites were in mixed-conifer forests composed of ponderosa pine, southwestern white pine, Douglas-fir, and white fir.

The fire-scar chronologies from Mt. Graham (fig. 3.3) illustrate a historical pattern commonly found elsewhere in the Southwest—a remarkable abundance of surface fires during the period before the 1890s, and a striking decrease in fire frequency during the twentieth century. This drop-off in fire occurrence is most likely related to intensive livestock

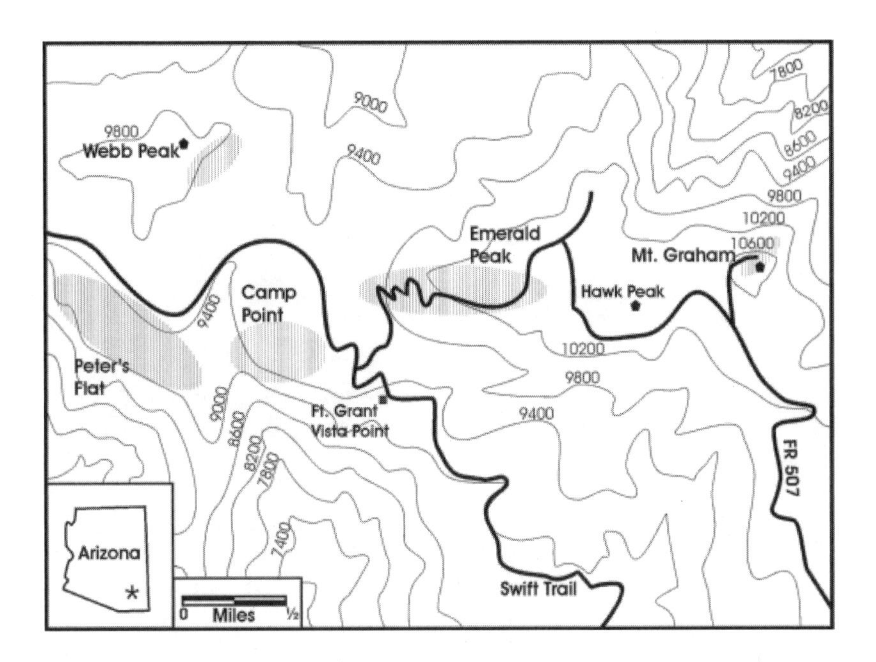

Figure 3.1. Map of tree-ring sample areas in the Pinaleño Mountains, Arizona. Fire scars were sampled in the shaded areas (vertical bars) at Peter's Flat, Camp Point, and Webb Peak. Cross section samples for age structure analysis were obtained from the cleared roadway and telescope construction sites near Emerald Peak and Mt. Graham.

grazing, which began in the 1880s or 1890s, followed by extensive road building and organized fire suppression by government agencies in the early 1900s. The evidence for livestock grazing as the initial cause of decline in the frequency of surface fires in the Southwest was originally noted by Aldo Leopold (1924) and was also commented on by early USFS rangers and others who witnessed these changes (Swetnam et al. 2001). The chief effect of intensive livestock grazing appears to have been the removal of grass fuels that were a primary carrier of the frequent surface fires. Another factor was the disruption of continuous ground fuels by livestock driveways and trails.

An interesting feature of the Peter's Flat and Camp Point fire-scar collection areas was the relatively high fire frequencies in these mixed-conifer sites during the pre-1893 period. The frequency of fires occurring

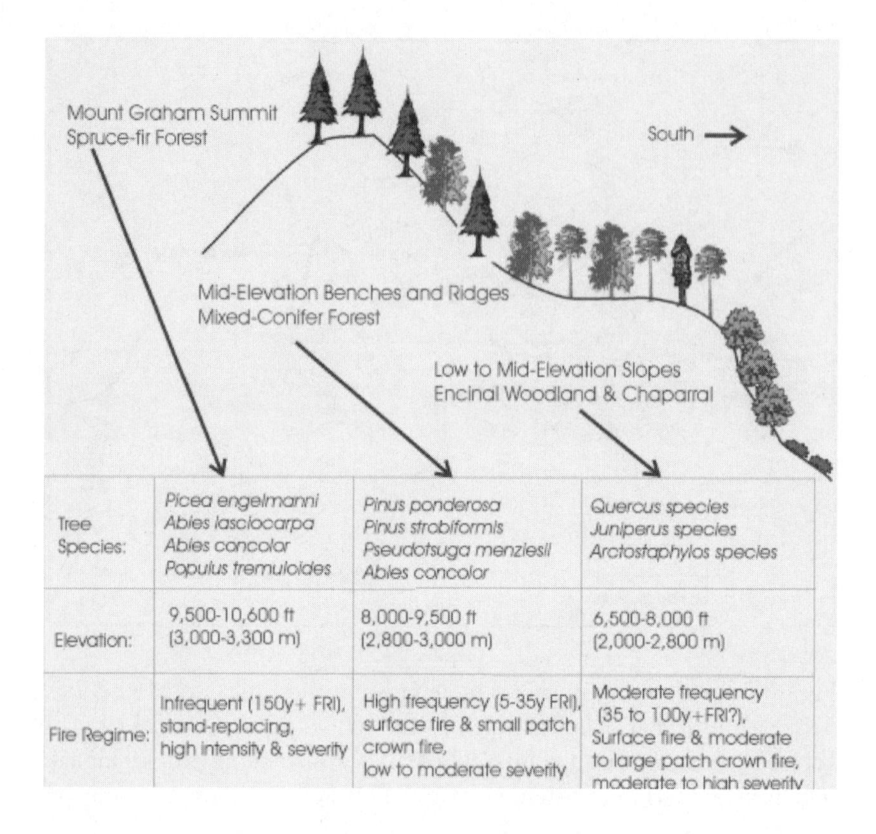

Tree Species:	*Picea engelmanni* *Abies lasciocarpa* *Abies concolor* *Populus tremuloides*	*Pinus ponderosa* *Pinus strobiformis* *Pseudotsuga menziesii* *Abies concolor*	*Quercus species* *Juniperus species* *Arctostaphylos species*
Elevation:	9,500-10,600 ft (3,000-3,300 m)	8,000-9,500 ft (2,800-3,000 m)	6,500-8,000 ft (2,000-2,800 m)
Fire Regime:	Infrequent (150y+ FRI), stand-replacing, high intensity & severity	High frequency (5-35y FRI), surface fire & small patch crown fire, low to moderate severity	Moderate frequency (35 to 100y+FRI?), Surface fire & moderate to large patch crown fire, moderate to high severity

Figure 3.2. Schematic of elevational gradient on Mt. Graham, showing dominant tree species in forest and woodland over stories, and inferred pre-1900 fire regimes (FRI= fire return interval).

anywhere within each of these sites was at least one fire per decade, and sometimes two or three fires per decade. There were a few longer periods without fire, lasting 13 years or longer. These high fire frequencies were more typical of lower-elevation ponderosa pine–dominated forests studied at numerous locations in the Southwest (Swetnam and Baisan 1996, 2003). Typical fire frequencies in southwestern mixed-conifer forests above 2,700 m were about one fire occurring every 15 to 30 years within similar size areas (i.e., about 50 to 100 ha sites; Swetnam and Baisan 1996). Pre-1900 fire intervals in the Webb Peak site (at around 3,000 m elevation) were much longer, ranging from about 30 to 60 years. When sampled, this site was essentially mixed-conifer, surrounded by spruce-fir forest.

The unusually high fire frequency for the Mt. Graham mixed-conifer stands at 2,700 m elevation may be the result of the topographic setting. Steep slopes drop away to the south and west of these sites (fig. 3.1), and it was likely that fire occurring at lower elevations within several very large watersheds would spread into these forests. Likewise, fires igniting on the benches and ridges at mid-elevations to the north or east would have spread unhindered because there were no major barriers to fire spread from these directions. The high fire frequencies in mixed-conifer forests during the pre-1893 period led to some important interpretations regarding the history and dynamics of the spruce-fir forests that existed upslope on the higher peaks (i.e., Webb, Emerald, Hawk, and Mt. Graham). We will return to these interpretations and implications in the summary and conclusions section.

There were a few other interesting changes in fire occurrence in the combined Peter's Flat and Camp Point fire chronologies (fig. 3.3). Specifically, in addition to the sharply decreased fire frequency after the 1890s, other decadal-scale shifts in fire frequency can be discerned between different periods during the 1700s and 1800s (see notations at the top of the fire chronology chart in fig. 3.3). A relatively higher frequency of fire-scar dates occurred during the 1700s, and a relatively lower frequency during the early to mid-1800s. Moreover, the frequent fires during the 1700s were somewhat less synchronous (i.e., less widespread) between trees than the less-frequent fire dates during the early to mid-1800s. A slight increase in fire frequency occurred in the late 1800s.

Similar decadal-scale changes have been detected in several other locations in the Southwest (e.g., the Santa Catalina Mountains in southern Arizona, the Mogollon Mountains of southern New Mexico, El Malpais National Monument in west-central New Mexico, and the Jemez Mountains of northern New Mexico) (Grissino-Mayer and Swetnam 2000; Swetnam and Baisan 2003). The shifts in fire frequency are less pronounced in the Mt. Graham chronologies than in these other mountains, where the reduced fire frequency during the early 1800s was sometimes marked by an exceptionally long fire-free period lasting two or more decades.

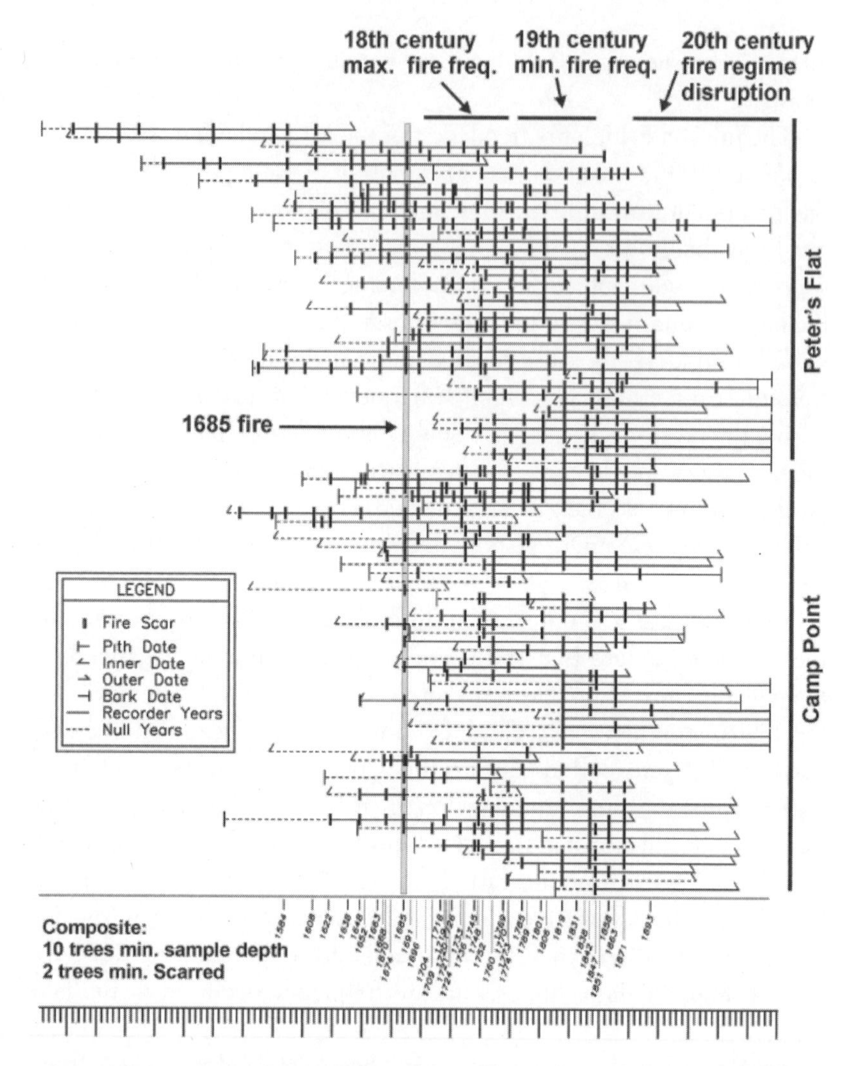

Figure 3.3. Combined fire scar chronology charts from Peter's Flat and Camp Point. Horizontal lines are the time spans covered by the tree-ring record from individual sampled trees, and the vertical tick marks are the fire dates recorded on those trees. The dates listed in the composite at the bottom are the fire events that were probably relatively widespread. (Hint: An effective way to better visualize fire frequency shifts and degree of synchrony of fire dates on the chart is to tilt the page to a low angle relative to your eye – looking toward the top of the page from the bottom.)

Summary of Spruce-Fir Age Structure and Fire History

Spruce-fir forests generally burn only as crown fires and at intervals of centuries (Schoennagel et al. 2004). Surviving fire-scarred trees are exceedingly rare in these forests. Therefore, to determine the fire history and dynamics of spruce-fir forests, it is necessary to study tree ages. Numerous trees were felled during the clearing of roadways and telescope construction sites on Mt. Graham, and so we took this opportunity to obtain cross-section samples from the stumps near ground level. We crossdated the tree rings and determined the innermost ring dates at the pith on 156 spruce trees and 117 fir trees. These dates were probably within 5 to 10 years of the actual germination dates of the trees. The resulting temporal distribution of innermost ring dates indicates that recruitment of spruce and fir trees was nearly continuous since the late 1600s, but with several surges in recruitment by spruce or fir (fig. 3.4).

Several aspects of the spruce-fir age distribution were notable. First, it appears that at least part of the spruce-fir forest on Mt. Graham was largely the result of a long-term process of succession following an extensive crown fire in 1685. Both spruce and fir were present early in the succession process. This interpretation derives from the facts that (1) we identified a widespread 1685 fire event from fire-scar dates within mixed-conifer forests downslope (see fig. 3.3), (2) none of the innermost tree-ring dates from the spruce or fir trees extended before the 1685 fire date, (3) growth rates of surviving Douglas-fir trees at the spruce-fir ecotonal border were severely suppressed, suggesting they may have been damaged (e.g., crown scorch or root killing) by a severe fire, and (4) the average growth rates of spruce and fir trees during the 1690s and early 1700s were rapid, indicating a relatively open stand condition (fig. 3.4). Stromberg and Patton (1991), who estimated ages for numerous spruce-fir stands on the mountain, also inferred a widespread disturbance just prior to 1700.

In addition to the evidence suggesting a post-1685 fire origin for the sampled spruce-fir stands, the tree-age structure and fire-scar evidence from Webb Peak provides other clues about the dynamics of these forests. Periods of increased and decreased tree recruitment of the two species may reflect responses to the combined effects of climatic variations and

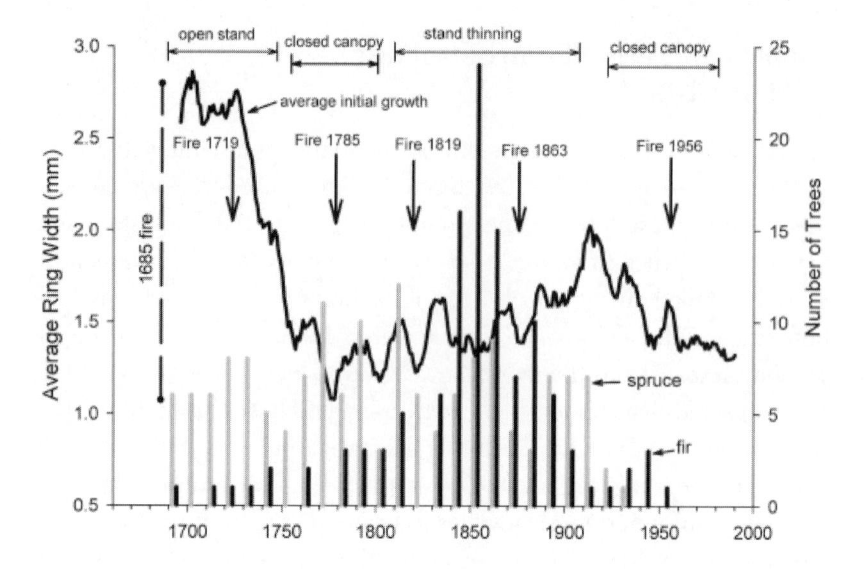

Figure 3.4. Spruce and fir age structure compared with average initial growth rates. The earliest tree-ring dates from the spruce and fir cross sections are plotted in 10-year groups. The average initial growth is the running average of the first 20 years of ring widths among all of the spruce and fir trees. The fire events indicated were recorded at Webb Peak as well as Peter's Flat and Camp Point fire history sites.

fire events (fig. 3.4). Surface fires were very rare or non-existent in spruce-fir forests on Mt. Graham, as indicated from the lack of fire-scarred trees in these stands (personal observations by the authors from extensive searching in these areas). The few fire-scar samples we have from Webb Peak were from a warm, southern exposure supporting a mixed-conifer stand (mainly Douglas-fir and southwestern white pine), rather than spruce-fir. However, because this stand was nestled within the spruce-fir zone, the fire history from this location reflects the cooler, more-mesic environment at this elevation. It was also likely that localized, intense crown-fire events of small to moderate patch size also occurred in the spruce-fir forests during recent centuries. Such localized crown fires were suggested by the presence of small patches of aspen trees within the spruce-fir stands on Mt. Graham in various locations. We currently do not have tree-ring dates or size estimates from these aspen patches, but research is underway on these patches in the Pinaleños and elsewhere in the Southwest (Ellis Margolis, Laboratory of Tree-Ring Research, the University of Arizona, personal communication).

The combined evidence of age structure, growth rates, and the few fire-scar dates from the Webb Peak site suggests the following scenario. The 1685 fire burned through mixed-conifer stands at the 9,000-foot level as a very large surface fire, and possibly created openings by torching individual trees and small patches of forest. This fire likely occurred as a widespread crown fire within spruce-fir stands on the higher-elevation peaks. After 1685, spruce and fir trees re-established simultaneously in the severely burned areas at the high elevations. Spruce was probably a larger proportion of the initial regeneration than fir. The open nature of the stands during the initial conditions was reflected by the typically large ring widths of trees that established at this time (fig. 3.4). Occasional fires (e.g., 1719, 1785) may have ignited and spread into parts of the young spruce-fir stands, but by the mid- and late 1700s, the spruce-fir forest had largely re-established. An increasing proportion of fir trees established during the mid-1800s, with a notable pulse occurring between 1830 and 1860.

We think it is likely that a combination of factors led to the increased fir recruitment during the nineteenth century, but the exact mechanisms are unclear. Climatic conditions in the Southwest were generally cool during the early to mid-1800s (Briffa et al. 1992; Salzer 2000), and drought conditions prevailed during the 1820s to 1830s (Cook et. al. 1999). Fire regimes shifted from relatively frequent to less frequent in the mixed-conifer forests downslope. The spruce-fir forests on the high-elevation peaks were approaching middle age (i.e., 120–150 years), and some stand thinning probably occurred via individual and group tree mortality (e.g., via lightning, beetle attack, etc.) and wind throw. Perhaps the effects of a sharp and sustained drought between 1817 and 1826, combined with beetle-mediated mortality, led to stand openings in the upper-elevation forests. By the late nineteenth century and early twentieth century, the canopies of most spruce-fir stands at the high elevations were relatively closed, with a broad mixture of old growth, middle-aged, and young trees (fig. 3.4).

Summary and Conclusions

A generalized picture of the variations in fire regimes along elevational gradients on Mt. Graham (fig. 3.2) was one of infrequent surface and crown fires in the lower elevations, frequent surface fires (5- to 35-year intervals) at middle to high elevations, and very infrequent crown fires at the highest elevations (i.e., 150- to 300-plus-year intervals). It is important to note that these are very general characterizations of fire frequencies and severities, and there was considerable spatial and temporal variability. For example, although fires were frequent in the mixed-conifer zone and generally of low severity, it was also likely that high-severity fire occurred in some variable-sized patches within this type. Also, very infrequent, large, high-severity crown fires were the norm within the spruce-fir zone, but it was possible that occasional surface or ground fires crept into portions of this forest from adjacent mixed-conifer, and high-severity, small patch size (individual trees or groups) events probably also occurred in this zone.

One of the more interesting implications of the pre-1900, high-frequency surface-fire regime of the mixed-conifer forests on Mt. Graham is the possibility that this fire regime promoted some degree of long-term stability to the higher-elevation spruce-fir forests. Frequent surface fires in the mixed-conifer zone probably maintained relatively open stands with low woody-fuel accumulations, grassy understories, and elevated tree-canopy layers. Fires igniting in the mixed-conifer, or at lower elevations, would have spread through the mixed-conifer zone at relatively low intensities/severities, so that when fires reached the high-elevation spruce-fir zone they were unlikely to spread into the canopy and develop into crown fires. A typical observation of surface-fire spread from mixed-conifer to spruce-fir (under low to moderate wind conditions) is that fire spread slows dramatically in the tightly packed needles of the closed-canopy spruce-fir. In the shady, cool, and moist conditions of spruce-fir, there is generally little herbaceous cover, and surface fires typically become smoldering ground fires that do not spread great distances (personal observations of authors Swetnam and Baisan).

In contrast, a century of greatly reduced fire frequencies since 1893 (fig. 3.3) has undoubtedly led to increased accumulations of woody fuels and forest densities in the mid-elevation mixed-conifer forests. As a result, fires igniting in this zone, or downslope of this zone, have a high probability of becoming crown fires before they reach the spruce-fir zone. This kind of transmission of crown fire from the low and mid-elevations to the high elevations was the circumstance for the 2,400+ ha Clark Peak fire of 1996. The fire started near Riggs Lake in pine and mixed-conifer and subsequently burned up into the spruce-fir zone (Erickson 1997). This late-April fire might have been a much larger event if it had not occurred so early in the season, when fuel moistures were still relatively high in the spruce-fir zone.

Ultimately, the preservation of extensive spruce-fir ecosystems (including red squirrels) on Mt. Graham will depend upon restoring forest structures and surface-fire regimes in the mid-elevation mixed-conifer forests. If these structures and fire regimes are not restored at a minimal level, we think that extensive crown fires will continue to occur on Mt. Graham until most of the spruce-fir forest is reduced to isolated, small patches and much-younger successional stands with widespread aspen. The mixed-conifer forests are also likely to be increasingly converted to aspen and shrub fields.

The current insect outbreaks on Mt. Graham certainly complicate the trajectory of change (Lynch in this vol.). It is likely that beetle outbreaks have occurred before on the mountain, but their frequency and extent is unknown. Conceivably, droughts of the late 1600s contributed to tree death and mortality, including beetle-killed trees, and this may have been a predisposing factor for the 1685 fire.

Acknowledgments

Funding for this research was provided by the U.S. Forest Service, Rocky Mountain Research Station and by the University of Arizona through the Mt. Graham Red Squirrel Study Committee, which is comprised of representatives from the Arizona Game and Fish Department, the University of Arizona, the U.S. Forest Service, and the U.S. Fish and Wildlife Service. We thank the Steward Observatory and the U.S. Forest Service for assistance and cooperation during the collection of tree-ring specimens.

Literature Cited

Allen, C. D., M. Savage, D. A. Falk, K. F. Suckling, T. W. Swetnam, T. Schulke, P. B. Stacey, P. Morgan, M. Hoffman, and J. Klingel. 2002. Ecological restoration of southwestern ponderosa pine ecosystems: A broad perspective. *Ecological Applications* 12(5):1418–1433.

Bahre, C. J. 1991. *A legacy of change: Historic human impact on vegetation of the Arizona borderlands.* University of Arizona Press, Tucson.

——. 1998. Late 19th century human impacts on the woodlands and forests of southeastern Arizona's sky islands. *Desert Plants* 14(1):8–21.

Baisan, C. H., and T. W. Swetnam. 1995. Historical fire occurrence patterns in remote mountains of southwestern New Mexico and northern Mexico. Pages 153–156 in J. K. Brown, R. W. Mutch, C. W. Spoon, and R. H. Wakimoto, technical coordinators. *Proceedings: Symposium on Fire in Wilderness and Park Management: Missoula, MT, March 30–April 1, 1993.* General Technical Report INT-320. U.S. Dept. of Agriculture, Forest Service, Intermountain Research Station, Ogden, Utah.

Briffa, K. R., P. D. Jones, and F. H. Schweingruber. 1992. Tree-ring density reconstructions of summer temperature patterns across western North America since 1600. *Journal of Climate* 5(7):735–754.

Cook, E. R., D. M. Meko, D. W. Stahle, and M. K. Cleaveland. 1999. Drought reconstructions for the continental United States. *Journal of Climate* 12(4):1145–1162.

Erickson, J. 1997. Mt. Graham firestorm: Anniversary brings plans, reflection on safeguards. *The Arizona Daily Star,* April 27, page 1B.

Grissino-Mayer, H. D., C. H. Baisan, and T. W. Swetnam. 1996. Fire history in the Pinaleño Mountains of southeastern Arizona: Effects of human-related disturbances. Pages 399–407 in L. Debano, P. F. Ffolliott, A. Ortega-Rubio, G. J. Gottfried, R. H. Hamre, and C. B. Edminster, technical coordinators. *Biodiversity and management of the Madrean Archipelago: The sky islands of southwestern United States and northwestern New Mexico: September 19–23, 1994, Tucson, Arizona.* General Technical Report RM-GTR-264. U.S. Dept. of Agriculture, Forest Service, Rocky Mountain Forest and Range Experiment Station, Fort Collins, Colorado.

Grissino-Mayer, H. D., and H. C. Fritts. 1995. Dendroclimatology and dendroecology in the Pinaleño Mountains. Pages 100–120 in R. Hoffman and C. Istock, eds. *Storm over a mountain island: Conservation biology and the Mount Graham affair.* University of Arizona Press, Tucson.

Grissino-Mayer, H. D., and T. W. Swetnam. 2000. Century-scale climate forcing of fire regimes in the American Southwest. *Holocene* 10(2):213–220.

Hoffman, R., and C. Istock, eds. 1995. *Storm over a mountain island: Conservation biology and the Mount Graham affair.* University of Arizona Press, Tucson.

Leopold, A. 1924. Grass, brush, timber, and fire in southern Arizona. *Journal of Forestry* 22:1–10.

Logan, J. A., J. Régnière, and J. A. Powell. 2003. Assessing the impacts of global warming on forest pest dynamics. *Frontiers in Ecology and the Environment* 1(3):130–137.

Romme, W. H., T. T. Veblen, M. R. Kaufmann, R. Sherriff, and C. M. Regan. 2003. Part 1: Historical (pre-1860) and current (1860–2002) fire regimes. Pages 181–195 in R. T. Graham, tech. ed. *Hayman Fire case study*. General Technical Report RMRS-GTR-114 (revision). U.S. Dept. of Agriculture, Forest Service, Rocky Mountain Research Station, Ogden, Utah.

Salzer, M. W. 2000. Dendroclimatology in the San Francisco Peaks region of northern Arizona, USA. PhD diss., Department of Geosciences, The University of Arizona, Tucson.

Schoennagel, T., T. T. Veblen, and W. Romme. 2004. The interaction of fire, fuels, and climate across Rocky Mountain landscapes. *Bioscience* 54:661–676.

Stromberg, J. C., and Patten, D. T. 1991. Dynamics of the spruce-fir forests on the Pinaleño Mountains, Graham Co., Arizona. *The Southwestern Naturalist* 36:37–48.

Swetnam, T. W., and C. H. Baisan. 1996. Historical fire regime patterns in the southwestern United States since AD 1700. Pages 11–32 in C. Allen, ed. *Fire effects in southwestern forests: Proceedings of the Second La Mesa Fire Symposium, Los Alamos, New Mexico, March 29–31, 1994*. General Technical Report RM-GTR-286. U.S. Dept. of Agriculture, Forest Service, Rocky Mountain Forest and Range Experiment Station, Fort Collins, Colorado.

———. 2003. Tree-ring reconstructions of fire and climate history in the Sierra Nevada and southwestern United States. Pages 158–195 in T. T. Veblen, W. Baker, G. Montenegro, and T. W. Swetnam, eds. *Fire and climatic change in temperate ecosystems of the western Americas*. Ecological Studies Volume 160. Springer, New York.

Swetnam, T. W., C. H. Baisan, and J. M. Kaib. 2001. Forest fire histories in the sky islands of La Frontera. Chapter 7. Pages 95–119 in G. L. Webster and C. J. Bahre, eds. *Changing plant life of la frontera: Observations on vegetation in the United States/Mexico borderlands*. University of New Mexico Press, Albuquerque.

Wilson, J. P. 1995. *Islands in the desert: A history of the uplands of southeastern Arizona*. University of New Mexico Press, Albuquerque.

Part II

Management

> Conservation is the foresighted utilization, preservation, and/or renewal of forests, waters, lands, and minerals, for the greatest good of the greatest number for the longest time.
> —Gifford Pinchot

GIFFORD PINCHOT served as the first chief of the U.S. Forest Service (USFS) under Theodore Roosevelt from 1905–1910. While a number of government agencies, tribes, and private entities have legal ties to the mountains, the USFS serves as the principal land management agency responsible for the Pinaleños and the haunt of the Mt. Graham red squirrel. From our opening quote, Pinchot clearly expressed the multiple uses and myriad challenges associated with natural resources management and conservation. These are perhaps no more evident than on Mt. Graham.

In Part II, we provide the perspective of two primary agencies responsible for management of natural resources in the Pinaleños. Larry Jones provides a review of the natural history of the mountain and outlines USFS management strategies past, present, and future. Paul Barrett then reviews the process and challenges associated with U.S. Fish and Wildlife Service's management of the federally endangered Mt. Graham red squirrel, the mountain's most well-known, though by no means only, protected species.

CHAPTER 4

Natural History and Management of the Pinaleño Mountains with Emphasis on Mt. Graham Red Squirrel Habitats

LAWRENCE L. C. JONES

THE MT. GRAHAM RED SQUIRREL has an extremely limited range, being found only at the upper elevations of one small mountain range of one ranger district of one national forest. This chapter provides information on the natural history of the Pinaleño Mountains, the historic and contemporary forest health conditions, and the management activities of the Coronado National Forest, emphasizing areas occupied by the Mt. Graham red squirrel. It also incorporates important changes that have occurred since the symposium for which it was written, including a 121 km^2 fire, a large-scale proposed fuel reduction, and a proposed forest restoration project.

Five federally listed threatened and endangered species or subspecies occur on the Pinaleño Mountains: the Mexican spotted owl (*Strix occidentalis lucida*); Apache trout (*Oncorhynchus apache*), which was introduced to the Pinaleños; lesser long-nosed bat (*Leptonycteris curasoae yerbabuenae*); bald eagle (*Haliaeetus leucocephalus*); and the Mt. Graham red squirrel. The Mexican spotted owl is the only other taxon with significant management concerns within the range of the Mt. Graham red squirrel. There are approximately 18 endemics known from the Pinaleños (D. Hodges, Sky Island Alliance, personal communication), but certain taxa, especially invertebrates, have not been adequately surveyed. Currently, there are 11 taxa on the Management Indicator Species list (U.S. Forest Service 1986, as amended) and 57 taxa on the Regional Forester's Sensitive Species list (U.S. Forest Service 1999) that are known to occur in the Pinaleños.

Historic Changes in Structure and Composition of the Montane Forests

Because of the mesic nature of this forest type, its natural fire regime is one of small, patchy fires, punctuated every 300–400 years by large stand replacement from insect outbreaks, fire, or both (Stromberg and Patten 1991; Grissino-Mayer et al. 1994).

The Pinaleño Mountains, as with other mountain ranges in the southwestern United States, have undergone significant changes during historic times. These changes have affected the natural processes and ecosystem equilibrium and are collectively referred to as forest health issues. The primary forest health issues that affect the Pinaleños and other sky islands include excessive fuel loads (the amount of combustible dead and down material that contributes to the spread and intensity of fire), insect pathogens, plant pathogens, non-native species, and climatological events (Fulé and Covington 1995; Sackett and Haase 1996; U.S. Forest Service 2002; Lynch in this vol.; Swetnam et al. in this vol.).

The path to the current forest health situation originated when European settlers brought cattle to the area. A lack of regulation led to overgrazing in the valleys and lower mountain slopes. This not only changed the structure and composition of the biotic communities but also began a cycle of passive fire suppression. Fires ignited during spring and summer lightning storms could not spread in overgrazed areas because the reduced grass cover would not allow fires to carry. Naturally occurring, low-intensity wildfires act to limit new growth of certain species (like invasive shrubs), but encourage the proliferation of fire-adapted species (like native grasses and forbs). The situation was exacerbated by the active fire suppression policies of land management agencies. This had a profound effect by altering the natural fire regime. While the subalpine forest type is not fire-adapted per se, it can be influenced by downslope effects, especially during drought years. The mixed-conifer-zone downslope acts as a buffer, protecting the subalpine forest from fires originating in the valleys or lower slopes.

Large-diameter trees were selectively harvested in the mixed-conifer zone of the Pinaleños, which encouraged the growth of saplings in the understory. The combination of fire suppression and logging led to a forest structure typified by excessive numbers of small-diameter trees and

Figure 4.1. Excessive fuel loads of forest litter, shrubs, tree limbs, and densely packed small-diameter trees and snags set the stage for catastrophic insect outbreaks, pathogens, parasites, and ultimately, crown fires.

high fuel loads (fig. 4.1) (Grissino-Mayer et al. 1994; Allen 1996; Sackett and Haase 1996).

Mixed-conifer and subalpine forests, especially the latter, have been significantly affected by insect pathogens (Koprowski et al. 2005). Looper moths (*Nepytia janetae*) are defoliators that caused an initial wave of

Figure 4.2. The subalpine forest above about 2,900 m in the Pinaleños experienced a stand-replacing insect epidemic that primarily affected Engelmann spruce. Later, the Nuttall Complex fire severely burned much of the area.

damage to Engelmann spruce in 1996–1998. After populations of that insect crashed, the exotic spruce aphid (*Elatobium abietinum* [Walker] [Homoptera: Aphididae]) further stressed spruce trees by feeding on leaves and causing leaves to drop early. Stands of Engelmann spruce, already stressed by these two insects and other factors, were then attacked by the spruce bark beetle (*Dendroctonus rufipennis* [Kirby]). Corkbark fir has been parasitized by western balsam bark beetle (*Dryocoetes confusus* [Swaine]), but the loss is not as severe as with Engelmann spruce. The result is that the highest elevations of the Pinaleños have had a stand-level insect outbreak (fig. 4.2). Lynch (in this vol.) discusses pathogenic insect ecology of the Pinaleños.

These threats set the stage for catastrophic stand-replacing events from pathogen outbreaks and crown fires, as well as chronic forest health issues. The subalpine forest has largely been converted to primary succession following insect epidemics and wildfire. Currently, the mixed-conifer zone has not suffered as greatly from catastrophic events.

Management Responsibilities

The entire range of the Mt. Graham red squirrel falls within public land administered by the Coronado National Forest, Safford Ranger District, and is managed as the Pinaleño Ecosystem Management Area (EMA). From a wildlife perspective, the primary charge of the U.S. Forest Service (USFS) is to manage their habitats and ecosystems, balanced with other needs and values. The subalpine forest biotic community in the Pinaleños has been designated as critical habitat for the Mt. Graham red squirrel, and much of the Pinaleños is designated critical habitat for the Mexican spotted owl.

Wildlife species are primarily managed by the Arizona Game and Fish Department (AGFD) by (1) setting state regulations for hunting, fishing, and scientific collecting, (2) maintaining a list of state sensitive species (Wildlife Species of Concern in Arizona), (3) maintaining a database for these species, and (4) conducting surveys and inventories for select species. The U.S. Fish and Wildlife Service (USFWS) concentrates on federally listed taxa and the authority of the Endangered Species Act. Because wildlife and habitats are intrinsically linked, the USFS, AGFD, USFWS, other government agencies, and non-governmental organizations work closely together on wildlife conservation issues, including those of the Mt. Graham red squirrel (Barrett in this vol.; Granillo and Froehlich in this vol.; Snow in this vol.).

Overview of Management Activities in the Pinaleño EMA

The USFS has undertaken a variety of management activities in recent years, mostly related to fire management, livestock grazing, recreation, wildlife, and special uses. Currently, there is no commercial timber harvesting. Values identified by the USFS, local community, Native Americans, ranchers, environmentalists, summer home owners, recreationers, and other stakeholders have strongly influenced project development. Public safety has always been of paramount consideration.

Recreation

Quality recreation has always been an important value of the Coronado National Forest, although new developments are not a priority, and none are currently planned. According to the USFS (Safford Ranger District, Safford, AZ, unpublished fact sheet [ca.] 1981), there are 52,000 acres of wilderness and 340 miles of trails on the four sky islands in the Safford Ranger District. There are about a dozen major trails in the Pinaleños. Trails and roads in one of the Mt. Graham red squirrel critical habitat patches are closed to the public. There are 10 developed and 8 undeveloped campgrounds, plus three picnic areas in the Pinaleños. Most of these are within the range of the Mt. Graham red squirrel. The Columbine Visitors Center and Columbine Administrative Site are also in the range of the Mt. Graham red squirrel. The latter typically serves as a staging area for interagency squirrel censuses.

There are hunting opportunities in the Pinaleños, which are in AGFD's Management Unit 31. AGFD establishes seasons and limits on a variety of species, including mule deer (*Odocoileus hemionus*), white-tailed deer (*Odocoileus virginianus*), black bear (*Ursus americanus*), lagomorphs, squirrels (except the Mt. Graham red squirrel), doves, quail, waterfowl, and reptiles. Fishing opportunities are limited in the Pinaleños, due to the paucity of perennial streams, but these streams have been stocked with non-native trouts, which often hybridize with the introduced Apache trout (Porath and Nielsen 2003). There is one developed fishing lake (Riggs Flat Lake) and one undeveloped fishing lake (Frye Mesa Reservoir). Although the Coronado National Forest does not have authority on hunting and fishing regulations, the USFS and AGFD meet annually to discuss fish and wildlife management issues.

Transportation

The Safford Ranger District is not heavily roaded, having 14,100 acres of roadless area (USFS Safford Ranger District, Safford, AZ, unpublished fact sheet [ca.] 1981). There are 57 roads totaling 270 miles. There are only two paved roads on the district, both in the Pinaleños: Swift Trail (Hwy. 366) and Stockton Pass Road (Hwy. 266), which are managed by the Arizona Department of Transportation (ADOT). Swift Trail is the primary road that bisects the Pinaleños along the upper axis (but not the entire length). Mt. Graham red squirrel habitat can be accessed by the Swift Trail, but most occupied habitat is northeast of the road. The pavement ends in the vicinity of the Shannon campground, with most of the squirrel habitat beyond that point, and is closed to the public there from 15 November to 15 April for protection of the Mt. Graham red squirrel and other resources. However, ADOT is considering paving the road from Shannon campground to the Columbine Visitors Center (M. Maiefski, ADOT, personal communication). Road maintenance includes resurfacing, rockfall containment, and hazard-tree removal. Any work outside the road prism, including brushing and hazard-tree removal, requires the appropriate environmental documentation to address threatened and endangered species. In recent years, numerous roads have been decommissioned, especially in Mt. Graham red squirrel critical habitat.

Range Management

Grazing has been reduced by about 12,000 animal-months on the Safford Ranger District in the past 25 years (C. Duncan, Safford Ranger District, personal communication). Currently, there are 17 grazing allotments in the Pinaleños, but there are none within Mt. Graham red squirrel habitat.

Special Uses and Administrative Sites

Special uses are activities that occur on national forest lands by non-government entities; these require a special use permit from the USFS. Examples include electronic sites, summer homes, mining claims, astrophysical sites, waterlines, fuelwood harvest, specialized forest-product collection, research activities, plant and animal collecting (except under valid state hunting and fishing licenses), and outfitter-guide services.

There are two summer home areas in the Pinaleños: 68 summer homes in the Turkey Flat area and 16 in the Old Columbine area. The Mt. Graham red squirrel ranges across the Old Columbine summer home area and above the Turkey Flat area. Two of the major threats that face summer home owners in the Pinaleños are falling trees and loss by fire, so hazard-tree removal and fuel-reduction projects are conducted in response to wildland-urban interface management direction. In the Old Columbine area, a 15 m radius no-touch zone protects each red squirrel midden.

The Mt. Graham International Observatory is operated by the University of Arizona (UA) and its affiliates (Ratje in this vol.). Their special use permit allowed them to build, maintain, and operate three telescopes, as well as support research on, and monitor the effects of, the observatory construction and operation on the Mt. Graham red squirrel. Currently, the UA is conducting radiotelemetry studies on natural history, habitat use, and demographics of Mt. Graham red squirrels; many studies in this volume reflect that research.

There is a special use permit for an electronics site at Heliograph Peak. It occupies a small footprint on the peak with an access road that is closed to the public. This area is largely on the periphery of occupied Mt. Graham red squirrel habitat.

There have been numerous studies on forest and woodland ecology of the Pinaleños (e.g., Anderson and Shafer 1991; Stromberg and Patten 1991; McLaughlin 1993; Grissino-Mayer et al. 1994, 1995; Rehfeldt 1999; Anderson and Smith in this vol.; Swetnam et al. in this vol.). Although the Mt. Graham red squirrel has been the primary focus of most wildlife research activities in recent years, there have also been investigations on other organisms, including mountain spiny lizards (*Sceloporus jarrovii*) (Brown 1976; Ruby and Dunham 1984), hummingbirds, band-tailed pigeons (*Patagioenas fasciata*), wandering garter snakes (*Thamnophis elegans*), small mammals, insects, spiders, and plants.

Watershed, Fish, Wildlife, Air, and Rare Plants

The district biologists on the Coronado National Forest, with support from the supervisor's office, work with the Watershed, Fish, Wildlife, Air, and Rare Plants (WFW) program. The WFW program primarily supports project-level management activities of other programs (especially range, fuels, vegetation management, and fire), but there are also WRW projects. In the Safford Ranger District, most of the WRW field activities involve surveys for threatened and endangered species, sensitive species, and management indicator species (MIS) at the project level.

There are currently five primary target species: Mexican spotted owl, northern goshawk (*Accipiter gentilis*), peregrine falcon (*Falco peregrinus*), Chiricahua leopard frog (*Rana chiricahuensis*), and Mt. Graham red squirrel. The owl is federally listed as threatened, and the squirrel is listed as endangered. Mexican spotted owls are primarily monitored along Swift Trail, as this is where most proposed actions occur and where most of the owls and their habitat exist. Known nesting and post-fledging areas of northern goshawks are surveyed annually. Peregrine falcons are monitored as part of an interagency post-delisting program. Chiricahua and other leopard frogs have not been detected in the Pinaleños in many years, but survey efforts have not been extensive. The Mt. Graham red squirrel has been intensively surveyed since the 1980s (Snow in this vol.; Young in this vol.). The University of Arizona, USFS, AGFD, and USFWS are involved with these annual surveys. The Coronado National Forest is contributing to the Mount Graham Red Squirrel Recovery Plan revision.

Wildfire Management

The Basin and Range Province of southeastern Arizona and southwestern New Mexico is among the most fire-prone ecosystems in the United States. The basins in this region are extremely arid, being primarily desert and desert grassland communities. Consequently, the woodlands and forests are also arid, as the surface area of mesic habitats in a sea of arid valleys is small. The recent fire history in the Southwest is largely one of suppression, and until summer 2005, this was the only option for wildfire management in the Pinaleños. Some of the wildfire concerns in the Pinaleños are protection of life and property, as well as threat of catastrophic crown fires. A recent Coronado National Forest Land and Resource Management Plan amendment (U.S. Forest Service 2005) gives fire managers the option to allow low-intensity wildland fire use in most of the Coronado National Forest, including the Pinaleños, but protection of life and property, and cultural and natural resource issues must be addressed.

Because of the high fuel loads and drought conditions, two large, high-severity crown fires have occurred in the Pinaleños in recent years (fig. 4.3). In 1996, the Clark Peak fire burned about 2,630 ha. This human-caused fire burned on the southwest-facing slope in the mixed-conifer and

Figure 4.3. Catastrophic crown fires destroyed much of the subalpine and mixed-conifer forests. In this photograph, the subalpine forest in this area had a high-severity outcome from the crown fire.

subalpine zones, roughly between Merrill Peak and Moonshine Creek. In 2004, lightning strikes caused the Nuttall Complex fire. Smaller fires merged into one, burning large areas of the mixed-conifer and subalpine zones. The fires burned nearly 121 km^2 on the northeast-facing slope of the Pinaleños, from about Twilight Creek to Carter Canyon, although much of the Old Columbine area was spared. Both of these fires substantially reduced the amount of available habitat for the Mt. Graham red squirrel (fig. 0.2), although not all squirrel habitat within the burn perimeters was adversely affected.

Forest Health Projects

Currently, forest and woodland health issues are of paramount concern in the Pinaleños. Management activities to improve forest health conditions have been identified as the highest priority in the National Forest System. Prescribed (Rx) burns and thinning projects are probably the most effective tools available to help offset over a century of fire suppression and manage for wildlife habitat (e.g., Sackett et al. 1996). Wildland fire use may also be an option.

There is currently a large project (about 405 ha) in the mixed-conifer zone of the Pinaleños: the Pinaleño Ecosystem Management (PEM) demonstration project (fig. 4.4). The PEM project was primarily undertaken to (1) decrease the threat of catastrophic wildfire by reducing fuel loads of large woody debris, (2) create a green fuel break along the Swift Trail (in places), and (3) protect and improve wildlife habitat. The PEM design consists of a series of blocks with differing densities of Mt. Graham red squirrel middens. High-density blocks are thinned (downed logs <41 cm, and standing trees and snags <23 cm diameter at breast height can be removed), and low-density blocks are similarly thinned and then broadcast burned. In all cases, a 15 m radius no-modification zone protects squirrel middens. The PEM project is a milestone for the Safford Ranger District because it involves thinning, piling, burning, and sometimes broadcast burning in an area occupied by the Mt. Graham red squirrel and Mexican spotted owl, as well as numerous USFS Sensitive Species and MIS.

In summer 2002, a small area (46 to 61 m radius) around the Mt. Graham International Observatory was treated for fire hazard/risk reduction. Small-diameter (<15 cm) logs were removed from the forest floor, and living and dead trees were limbed up to 3 m in height. No living or standing dead trees were cut. In 2003, hazard trees were removed from the vicinity of the observatory, Swift Trail, Old Columbine summer home area, and elsewhere. In 2004, standing dead trees were removed during Nuttall Complex fire suppression.

Figure 4.4. Pinaleño Ecosystem Management project, showing how thinning and prescribed fire reduced fuel loads on the right side of Swift Trail. Compare that to the untreated left side.

Currently, the Coronado National Forest is proposing a large fuel reduction and forest restoration project in the Pinaleños. This project is designed to help reduce the threat of catastrophic wildfire in much of the remaining mixed-conifer zone and will begin to set the forest on a trajectory that will allow a low-intensity fire cycle. Large-diameter trees, snags, and logs of all canopy species will be retained, while select smaller-diameter under- and mid-story trees will be removed to achieve desired forest conditions (species composition, life form structure, landscape matrix of age classes). The mixed-conifer zone currently has the largest block of remaining squirrel habitat. This project should begin implementation after the PEM project has been completed and could take a decade or more to complete.

In the past several years, there has been an emphasis on removing dead and dying hazard trees and insect-killed trees to help abate the pathogenic insect epidemics. Standing dead trees and downed wood, which are important structures to wildlife, are not a limiting resource in the Pinaleños. To help offset future loss of the remaining large Douglas-fir and Engelmann spruce from bark beetle outbreaks, a MCH (3-methyl-2-cyclohexen-1-one) treatment program is in place. MCH is an anti-aggregate pheromone—it keeps beetles from aggregating in large numbers on select trees, thereby reducing tree mortality.

There are efforts underway to curtail the growth and spread of exotic plants. Of the exotic grasses, Lehmann lovegrass (*Eragrostis lehmanniana*) is currently the most problematic. This species was introduced from Africa to stabilize erosion, but has successfully spread across much of the Southwest and is outcompeting native species. It dominates some of the lower elevations around the Pinaleños. Sweet resinbush (*Euryops multifidus*), Canada thistle (*Cirsium arvense*), and other exotics that have the potential to threaten ecosystems in the Pinaleños have been identified as species to be controlled or eradicated (U.S. Forest Service 2002). There is always the concern that new invasive species will reach the Pinaleños, especially after high-intensity wildfires.

Managing for the Mt. Graham Red Squirrel

The Mount Graham Red Squirrel Recovery Plan (U.S. Fish and Wildlife Service 1993) established the basic recommendations for the USFS to manage habitat for the squirrel. Many of the recommendations in the recovery plan were adopted by the Coronado in its Coronado National Forest and Land Resource Management Plan (U.S. Forest Service 1986). The plan established standards and guidelines that determine how we are to manage the Coronado National Forest. The plan specifically addresses conservation needs for the Mt. Graham red squirrel in Management Areas 2 and 2A, which essentially cover the range of the subspecies. Management Area 2 is primarily the mixed-conifer zone and 2A is the subalpine zone, which includes part of the designated critical habitat and is referred to in the plan as the "red squirrel refugium" (U.S. Forest Service 1986:54). Management and public use of forestlands is more restrictive in 2A. The emphasis on red squirrel conservation in the plan is clear. The plan emphasizes that "red squirrel habitat needs will supercede the needs of all other species" (U.S. Forest Service 1986:51,

54–2). Management area standards and guidelines include the closure and rehabilitation of roads; trail and trailhead closures (2A only); the decreased use of motorized vehicles; public education; research on Mt. Graham red squirrel biology; monitoring of Mt. Graham red squirrels; use of silvicultural treatments to improve forest conditions, including thinning and prescribed burning, as well as create wildlife openings; and fuelwood harvest guidelines (U.S. Forest Service 1986:50–54–5).

Although the Coronado National Forest and Land Resource Management Plan and Mount Graham Red Squirrel Recovery Plan were well intentioned, much has changed since these plans were written—and both are currently slated for revision. Habitats within the range of the Mt. Graham red squirrel have been drastically altered in the last decade or so. One catastrophic event was the stand-replacement insect epidemic in the spruce-fir forest in the mid- to late 1990s (Koprowski et al. 2005). About that same time (1996), the Clark Peak fire burned much of the northwestern portion of squirrel habitat, primarily in the mixed-conifer zone. The 2,630 ha fire destroyed about 50 percent of the squirrel middens within its perimeter, and nearly all middens were destroyed within high-severity burn areas (see <http://ag.arizona.edu/research/redsquirrel/fire/html>). Then, in 2004, the Nuttall Complex fire burned along the northeast-facing slopes. While these fires had mixed severity, many areas included large patches of high-severity crown fires. Koprowski et al. (2006) documented direct mortality during an ongoing radiotelemetry study in the fire perimeter. They suggested that 35 percent of the squirrels (7 of 20 individuals) in the fire perimeter were likely direct mortalities, whereas 100 percent (37 individuals) survived in unburned areas during the same period. As of the spring 2005 Mt. Graham red squirrel census, the population was at its lowest level since 1990, with an estimated 214 (plus or minus 12) squirrels (T. Snow, 2005, press release, Arizona Game and Fish Department, Tucson). Of course, not all of the fire effects were negative. The fires have had beneficial effects in areas of low severity, but there are significant short-term effects in areas that experienced high-severity crown fires, and much suitable habitat has been lost in recent years. In all, more than one-half of the limited range of this subspecies has been impacted to some degree by these three events during the past decade.

In addition to these acute catastrophic threats, there exist chronic forest health concerns. The ecosystems are slowly changing because of anthropogenic influences. Forest structure and composition have changed since the arrival of European settlers, and invasive non-natives and exotics are a burgeoning crisis. The proposed fuel-reduction and forest restoration project is designed, in part, to address both the chronic and acute concerns.

Discussion

Given that catastrophic events are likely the primary causes for the recent declines in the Mt. Graham red squirrel and will continue to be the primary threats to the continued existence of that subspecies, the Coronado National Forest feels that reducing the risk of catastrophic wildfire in the remaining squirrel habitat is the most important conservation measure it can undertake. It seems possible that a wildfire originating from the lower slopes on a trajectory into the southwestern portion of its remaining habitat would get into the crown in the mixed-conifer zone (causing high-severity damage) and could reduce the habitat to a point that might decimate the remaining squirrel habitat—and the squirrel. Thus, the Coronado National Forest feels that conservative "baby steps" are more of a risk than a more intensive thinning and prescribed burning regime. At the same time, the Coronado National Forest acknowledges there are uncertainties associated with a more intensive silvicultural treatment in the short-term. To deal with these uncertainties, future management direction needs to incorporate the best science available, monitoring, adaptive management, and interagency and non-government organization collaboration.

Acknowledgments

I thank Randall Smith, George Asmus, Laurie Fenwood, Mike Leonard, and Dick Streeper for comments on an earlier draft of this manuscript, and Randall Smith and Anne Casey for comments on a later draft. The following people provided additional information and assistance: Paul Barrett, Woody Cline, Chuck Duncan, Genice Froehlich, Chris German, Lorean Hennings, David Hodges, Janet Jones, Roger Joos, John Koprowski, Robert Lefevre, Ann Lynch, Chad Munich, Reed Sanderson, Todd Schulke, Tom Skinner, Tim Snow, Sherry Tune, and Craig Wilcox. Carson Coates provided the geographic information system (GIS) illustrations.

Literature Cited

Allen, L. S. 1996. Ecological role of fire in the Madrean province. Pages 5–10 in P. F. Ffolliott, L. F. DeBano, M. B. Baker Jr., G. J. Gottfried, G. Solis-Garza, C. B. Edminster, D. G. Neary, L. S. Allen, and R. H. Hamre, technical coordinators. *Effects of fire on Madrean Province ecosystems: A symposium proceedings.* General Technical Report RM-GTR-289. U.S. Dept. of Agriculture, Forest Service, Rocky Mountain Forest and Range Experiment Station, Fort Collins, Colorado.

Anderson, R. S., and D. S. Shafer. 1991. Holocene biogeography of spruce-fir forests in southeastern Arizona—implications for the endangered Mt. Graham red squirrel. *Madrono* 38:287–295.

Brown, C. K. 1976. Local variation in scale characters of *Sceloporus jarrovi* in the Pinaleño Mountains of Arizona. I. Frequency of head scute polymorphisms. *Herpetologica* 32:189–197.

Fulé, P. Z., and W. W. Covington. 1995. Changes in fire regimes and forest structures of unharvested Petran and Madrean pine forests. Pages 408–415 in L. F. DeBano, G. J. Gottfried, R. H. Hamre, C. B. Edminster, P. F. Ffolliott, and A. Ortega-Rubio, technical coordinators. *Biodiversity and management of the Madrean Archipelago: The sky islands of southwestern United States and northwestern Mexico: September 19–23, 1994, Tucson, Arizona.* General Technical Report RM-GTR-264. U.S. Dept. of Agriculture, Forest Service, Rocky Mountain Forest and Range Experiment Station, Fort Collins, Colorado.

Grissino-Mayer, H. D., C. H. Baisan, and T. W. Swetnam. 1994. *Fire history and age structure analyses in the mixed-conifer and spruce-fir forests of Mt. Graham.* A final report submitted to the Mt. Graham Red Squirrel Study Committee, University of Arizona, Tucson.

——. 1995. Fire history in the Pinaleño Mountains of southeastern Arizona: Effects of human-related disturbances. Pages 399–407 in L. F. DeBano, P. F. Ffolliott, A. Ortega-Rubio, G. J. Gottfried, R. H. Hamre, and C. B. Edminster, technical coordinators. *Biodiversity and management of the Madrean Archipelago: The sky islands of southwestern United States and northwestern New Mexico: September 19–23, 1994, Tucson, Arizona.* General Technical Report RM-GTR-264. U.S. Dept. of Agriculture, Forest Service, Rocky Mountain Forest and Range Experiment Station, Fort Collins, Colorado.

Koprowski, J. L., M. I. Alanen, and A. M. Lynch. 2005. Nowhere to run and nowhere to hide: Response of endemic Mt. Graham red squirrels to catastrophic forest damage. *Biological Conservation* 126:491–498.

Koprowski, J. L., K. M. Leonard, C. A. Zugmeyer, and J. L. Jolley. 2006. Direct effects of fire on endangered Mt. Graham red squirrels. *Southwestern Naturalist* 51:59–63.

McLaughlin, S. P. 1993. Additions to the flora of the Pinaleño Mountains, Arizona. *Journal of the Arizona-Nevada Academy of Science* 27:1–31.

Porath, M. T., and J. L. Nielsen. 2003. Evidence of sexually dimorphic introgression in Pinaleño Mountain Apache trout. *North American Journal of Fisheries Management* 23:172–180.

Rehfeldt, G. E. 1999. Systematics and genetic structure of Ponderosa taxa (Pinaceae) inhabiting the mountain islands of the Southwest. *American Journal of Botany* 86:741–752.

Ruby, D. E., and A. E. Dunham. 1984. A population analysis of the ovoviviparous lizard *Sceloporus jarrovi* in the Pinaleño Mountains of southeastern Arizona. *Herpetologica* 40:425–436.

Sackett, S. S., and S. M. Haase. 1996. Fuel loadings in southwestern ecosystems of the United States. Pages 187–192 in L. F. DeBano, G. J. Gottfried, R. H. Hamre, C. B. Edminster, P. F. Ffolliott, and A. Ortega-Rubio, technical coordinators. *Biodiversity and management of the Madrean Archipelago: The sky islands of southwestern United States and northwestern Mexico: September 19-23, 1994, Tucson, Arizona*. General Technical Report RM-GTR-264. U.S. Dept. of Agriculture, Forest Service, Rocky Mountain Forest and Range Experiment Station, Fort Collins, Colorado.

Sackett, S. S., S. M. Haase, and M. G. Harrington. 1996. Prescribed burning in southwestern ponderosa pine. Page 178–186 in P. F. Ffolliott, L. F. DeBano, M. B. Baker Jr., G. J. Gottfried, G. Solis-Garza, C. B. Edminster, D. G. Neary, L. S. Allen, and R. H. Hamre, technical coordinators. *Effects of fire on Madrean province ecosystems: A symposium proceedings*. General Technical Report RM-GTR-289. U.S. Dept. of Agriculture, Forest Service, Rocky Mountain Forest and Range Experiment Station, Fort Collins, Colorado.

Stromberg, J. C., and D. T. Patten. 1991. Dynamics of the spruce-fir forests on the Pinaleño Mountains, Graham Co., Arizona. *Southwestern Naturalist* 36:37–48.

U.S. Forest Service. 1986. *Coronado National Forest Land and Resource Management Plan*. U.S. Forest Service, Tucson, Arizona.

———. 1999. Regional Forester's Sensitive Species list. Unpublished report, U.S. Forest Service, Albuquerque, New Mexico.

———. 2002. Scoping report: Noxious weed and invasive exotic plant management project, Coronado National Forest. Unpublished report, U.S. Forest Service, Tucson, Arizona.

———. 2005. Wildland fire use. Amendment to U.S. Forest Service, *Coronado National Forest Land and Resource Management Plan*, 1986. Unpublished document, Coronado National Forest, Tucson, Arizona.

U.S. Fish and Wildlife Service. 1993. *Mount Graham red squirrel* (Tamiasciurus hudsonicus grahamensis) *recovery plan*. Albuquerque, New Mexico.

## CHAPTER 5

The Process and Recovery Team Structure for Revising the 1993 Mount Graham Red Squirrel Recovery Plan

Paul J. Barrett

THE MT. GRAHAM RED SQUIRREL was designated a category 2 candidate species by the U.S. Fish and Wildlife Service (USFWS) on 30 December 1982. Although category 2 classifications no longer exist, at the time it meant that endangered or threatened status may be appropriate, but available information was not sufficient to support such a proposal.

On 21 May 1986, the USFWS published in the *Federal Register* a proposed rule to designate the Mt. Graham red squirrel (MGRS) an endangered species, pursuant to the Endangered Species Act of 1973 (Act), as amended (51 FR 18630). On 3 June 1987, the final rule designating the subspecies as endangered was published (52 FR 20994). Critical habitat was designated on 5 January 1990 (55 FR 425) under a separate rule.

The 1987 final rule concluded the MGRS was endangered because its range and habitat had been reduced and its habitat was threatened by a number of factors including the proposed construction of an astrophysical observatory, forest fires, road construction and improvement, and recreational development at high elevations. Furthermore, the rule determined the MGRS might suffer through competition with the introduced Abert's squirrel (*Sciurus aberti*).

Concurrent with the listing of the MGRS, a consortium of research institutions, including the Steward Observatory, the University of Arizona, proposed the construction of several astrophysical observatories in the Pinaleño Mountains. The mountains are federal lands within the Coronado National Forest (Coronado). Therefore, the Coronado entered into consultation with the USFWS as required by section 7 of the Act.

Formal consultation on the proposed astrophysical development and the Coronado National Forest Land and Resource Management Plan was completed on 14 July 1988.

In 1993, the Mount Graham Red Squirrel Recovery Plan (Plan) was finalized (U.S. Fish and Wildlife Service 1993). The MGRS has a recovery priority number of 9c, which means that the taxon is a subspecies with a moderate degree of threat, a high recovery potential, and is in conflict with development projects or other economic activity (Endangered and Threatened Species Listing and Recovery Priority Guidelines [48 CFR 43098 and 48 CFR 52985]).

The 1993 Plan recognized that the primary threat to the MGRS was the loss of habitat in the Pinaleño Mountains. The objective of the Plan was to increase and stabilize the MGRS population by protecting and restoring the mature, closed-canopy conifer forest in the Pinaleños.

Subsequent to the release of the 1993 Plan, conditions in the Pinaleños changed dramatically. In 1996, the Clark Peak fire burned approximately 6,716 acres of pine-oak, mixed-conifer, and spruce-fir vegetation communities on the mountain. In the summer of 2004, the Nuttall Complex fire resulted in a fire perimeter of over 29,000 acres, among the largest ever recorded in the range.

Perhaps a greater threat to the MGRS was the establishment and spread (first noted in the late 1990s) of at least four potentially devastating insect parasites in the high-elevation conifers of the Pinaleños. These species are a looper moth, the spruce bark beetle, the western bark beetle, and the spruce aphid. Details of the extent and ramifications of the infestations are discussed elsewhere in this volume.

Realizing that conditions in the Pinaleños and threats to the MGRS had changed in ways not anticipated in the 1993 Plan, the USFWS decided to revise the Plan. However, not only had conditions on the mountain changed but also our recovery planning process had evolved in the 1990s.

Among the guidelines and policy changes relevant to our planning process, perhaps most pertinent was the USFWS's 1994 Policy on Recovery, Plan Participation, and Implementation under the Endangered Species Act (59 FR 34272) (Policy). Recovery team members advise the USFWS's regional director on the development and implementation of recovery plans. The 1994 Policy mandates that team members "should be selected for their knowledge of the species or for expertise in elements of recovery plan design or implementation (such as local planning, rural sociology, economics, forestry, etc.), rather than their professional or other affiliations." Furthermore, it states that "(t)eams should include representatives of State, Tribal, or Federal agencies, academic institutions, private individuals and organizations, commercial enterprises, and other constituencies with an interest in the species and its recovery, or the economic or social impacts of recovery."

To meet Policy requirements as well as other new mandates, the USFWS formed a much larger and diverse recovery team for the revision than was used to develop the 1993 Plan. Within the USFWS, we developed a list of people with the necessary skills and experience, irrespective of their affiliation, and then contacted them to determine their interest in serving on the recovery team. The new recovery team consists of two subteams and several agency liaisons. The technical subteam is made up of seven experts in squirrel biology, silviculture, fire ecology, forest health, conservation biology, population biology, and entomology. At the request of the subteam, a team leader, with both ecology and editing skills, was added to the team.

Several Native American tribes have unique cultural and historical ties to Mt. Graham. Recognizing this, as well as the trust responsibility of the U.S government to the tribes, the USFWS invited the San Carlos Apache; White Mountain Apache; Hopi Tribes; and the Navajo, Tohono O'odham, Yavapai-Apache, and Zuni Nations to participate in the process. Subsequent to the oral presentation of this paper in May 2003, the four Apache tribes—White Mountain, San Carlos, Payson-Tonto, and Yavapai-Apache Nation—appointed tribal representatives to the technical subteam. We have requested the tribal representative write an appendix to the revised Plan addressing the cultural significance of the MGRS and Mt. Graham.

The technical subteam is charged with identifying threats to the MGRS, establishing recovery goals, and describing objective, measurable criteria used to progress toward recovery goals. The technical subteam addresses the question: "(W)hat do we need to do to recover the Mt. Graham red squirrel?"

The implementation subteam consists of representatives from the

Arizona Game and Fish Department (AGFD), the Coronado, the City of Safford, the City of Thatcher, the Steward Observatory, the Grand Canyon Chapter of the Sierra Club, the Center for Biological Diversity, the Maricopa County Chapter of the National Audubon Society, the Turkey Creek Summer Home Association, and the Graham County Board of Supervisors. Organizations were encouraged to appoint alternates in case a designated representative could not attend a meeting. Other groups that were invited or expressed interest, but have not formally committed to participating, include the Sky Island Alliance and the Columbine Summer Home Association. This subteam will work with the USFWS on issues identified by the technical subteam and help identify the best and most effective way to implement actions identified in the revised Plan. The implementation subteam addresses the question: "(H)ow do we implement Mt. Graham red squirrel recovery goals identified by the technical subteam?"

In addition to full team memberships, several agencies were invited to appoint liaisons to the technical subteam. The purpose of the liaisons is to provide information and guidance that will help the subteam craft the plan. They are not involved in order to advocate their agencies' positions during technical discussions.

The AGFD has a mandate to manage Arizona's wildlife, including the MGRS. In addition, on 26 June 2002, the Southwest Region of the USFWS and the AGFD signed a Memorandum of Agreement (U.S. Fish and Wildlife Service 2002) to facilitate joint participation, communication, coordination, and collaboration regarding implementation of the Act. Accordingly, the Department appointed a liaison to the technical subteam. The AGFD has a separate member on the implementation subteam.

The Coronado has a unique position with regard to the conservation and recovery of the MGRS in that this species occurs solely on national forest lands within the Coronado. Following discussions with the technical subteam, the USFWS asked that the Coronado also appoint a liaison to this subteam. The Coronado's liaison will provide the technical subteam with information on current activities on the mountain, help obtain data and information that might only be found in local district files, and help with background information as needed.

The USFWS also has several liaisons to the recovery team. We provide logistical support, organize meetings, furnish guidance regarding the Act, and perform other duties as necessary. Until the technical subteam decides otherwise, I serve as the head of the recovery team.

None of the liaisons, including the USFWS's, have voting rights during technical subteam meetings. We contribute to discussions and provide feedback, but only subteam members may vote.

The technical subteam met in February 2002 and immediately began to identify threats to the MGRS. Since that time, this team has worked on a draft outline for the Plan and has developed white papers addressing the natural history of the mountain and MGRS, past human uses of the mountain, and legal history of the MGRS. Threats are divided into (1) changes to MGRS habitat, and (2) factors that affect MGRSs directly. Changes to MGRS habitat include the effects of past silvicultural practices, fires and fuel, insects and diseases, non-native plants and animals, and miscellaneous other factors (e.g., construction, recreation, livestock). Direct impacts to the MGRS may be caused by recreation, competition with Abert's squirrels, road mortality, predation, and disease. The technical subteam has divided the issues by expertise, and team members are writing white papers on each topic. After evaluating the "threats papers," subteam members will address the consequences to the species, including demographic and fragmentation concerns.

The implementation subteam also met in February 2002. At that meeting, we explained the recovery planning process to the assembled subteam and additional potential members were identified by subteam members. In March 2003, the implementation subteam reconvened. Progress by the technical subteam, including the draft outline, was discussed and methods of information exchange explained.

Both subteams have secure Web sites, and passwords are required for access. Discussion threads can be started or read, and team members can post information (such as white papers) for review. The technical subteam's site is closed to all but technical subteam members and liaisons. This ensures a free and open exchange of information and opinion. When draft white papers are reviewed and deemed acceptable to the technical subteam, they will be posted on the implementation subteam Web site for further review.

A draft plan will be produced within the next two years. The plan will include an introduction and background information, as well as known and potential threats to the MGRS. It will also identify objective, measurable criteria to develop actions needed to reach our target and achieve recovery. A change from the 1993 Plan will be the requirement that threats and actions be tied to the five criteria identified in the Endangered Species Act that are used to evaluate when a species should be designated as threatened or endangered. These include

1. the present or threatened destruction, modification, or curtailment of its habitat or range;
2. over-utilization for commercial, recreational, scientific, or educational purposes;
3. disease or predation;
4. the inadequacy of existing regulatory mechanisms;
5. other natural or manmade factors affecting its continued existence.

Both professionals and the lay public are often confused by the recovery planning process. I will address several of these misperceptions.

First, recovery plans are intended to be blueprints for recovery. Often, they are used to justify actions and help support funding initiatives and proposals. However, they are advisory documents and do not require any agency, including the USFWS, to undertake any actions or fund any projects.

Second, recovery teams are exempt from the Federal Advisory Committee Act (FACA) for the purposes of "developing and implementing recovery plans." If recovery teams offer us advice for non-recovery-related activities, they may be in violation of FACA. For this reason, we do not ask recovery teams to review specific projects or actions unrelated to the Plan. That is the mandate of the USFWS in cooperation with the AGFD.

Similarly, a recovery team can recommend critical habitat be considered, but it cannot propose or designate critical habitat. Critical habitat designation is a rule-making procedure and requires public notification, possible public hearings, and other actions.

We are still in the early stages of the Plan revision process. After the recovery plan acceptable to the USFWS Regional Director is completed, he will sign it and ask the AGFD's director to also sign the draft Plan. The USFWS will then release the draft recovery plan for peer review and to the public for comment. Comments and reviews will be considered, and the team will reconvene and incorporate appropriate changes. A final Plan will be signed and released to the public.

Literature Cited

U.S. Fish and Wildlife Service. 1993. *Mount Graham Red Squirrel* (Tamiasciurus hudsonicus grahamensis) *Recovery Plan*. Albuquerque, New Mexico.

——. 2002. *Memorandum of Agreement between Region 2 U.S Fish and Wildlife Service and the Arizona Game and Fish Department*. June 26, 2002. 15 pp.

Part III

Mt. Graham Red Squirrel Population Trends

Population, when unchecked, increases in a geometrical ratio.
—Thomas Robert Malthus

MALTHUS FIRST FORMULATED his ideas with respect to human populations; however, Charles Darwin made significant use of these ideas in the derivation of his important ideas on natural selection. Malthus and Darwin viewed the struggle for existence within and between species to be keen, with significant potential for overpopulation to occur. Oh, how many who read this part will long for the prediction of Malthusian growth to be realized for Mt. Graham red squirrels.

In Part III, we provide a review of the status of the Mt. Graham red squirrel population in context for the remainder of the book. First, Kathy Granillo and Genice Froehlich provide a review of the recent history of the Mt. Graham red squirrel and early conservation efforts. Tim Snow and Paul Young independently detail the two long-term methods used to census and enumerate populations of Mt. Graham red squirrels since the 1980s. Both researchers describe considerable fluctuations in squirrel abundance and note recent declines. John Koprowski and Tim Snow follow with an assessment of the level of agreement between the two population monitoring techniques and note a reasonable level of concurrence. Finally, John Koprowski and Bob Steidl review the potential consequences of small population sizes through a literature review and relate these concerns to our state of knowledge of the Mt. Graham red squirrel.

CHAPTER 6

Mt. Graham Red Squirrel Natural History and Pre-observatory Construction Conservation Efforts

Kathleen A. Granillo
and Genice F. Froehlich

THE MT. GRAHAM RED SQUIRREL (MGRS) captures the interest of scientists, land managers, conservationists, and others for many reasons. The squirrel is often considered a surrogate for the old-growth forests of the higher elevations of the Pinaleño Mountains (Gould 1990). This mountain range is one of the most unique ranges in the southwestern United States. Many species of plants and animals are at their southern or northern limit of distribution, and the mountain is home to several endemic species (Jones in this vol.). It is contained within the Safford Ranger District, Coronado National Forest, and managed by the U.S. Forest Service (USFS). The higher-elevation flora and fauna have been isolated from similar communities for at least 11,000 years (U.S. Forest Service 1988). Many consider conserving the red squirrel as synonymous with conserving the unique ecosystems of the Pinaleño Mountains. This chapter provides information on the squirrel's legal status over the years, a brief timeline of events associated with development of the observatory, and background information on natural history and conservation efforts prior to observatory construction in 1990.

Legal Status

In 1976, the Arizona Game and Fish Department (AGFD) listed the MGRS in Group IV of Threatened Wildlife in Arizona. In 1979, the Arizona Natural Heritage Program (ANHP) listed the MGRS as a priority animal for status determination. In 1981, the squirrel was placed on the Regional Forester's list of Sensitive Species for Region 3 of the USFS (Arizona and New Mexico). In December 1982, the squirrel was included in the *Federal Register* Review of Vertebrate Wildlife by the U.S. Fish and Wildlife Service (USFWS) (47 FR 58454–58460) as an animal for which a proposal to list as endangered or threatened might, pending additional biological information, be appropriate. Spicer et al. (1985) provided the first population estimate derived from data collected in a scientific manner. In May 1986, the MGRS was proposed for listing as endangered with critical habitat by the USFWS. The final rule to list the MGRS as endangered was published in the *Federal Register* on June 3, 1987 (52 FR 20994–20999). Critical habitat was designated on January 5, 1990 (55 FR 425–429).

Mt. Graham Astrophysical Area

Concurrent with the increased interest in the MGRS population was increased interest in the mountain as a potential site for an observatory complex. In 1981, the Smithsonian Institution and the University of Arizona (UA), cooperating with the National Optical Astronomy Observatories, began testing the quality of sites for a modern astronomical observatory. In 1982, the Smithsonian Institute requested that the Coronado National Forest, which was then preparing a Land and Resource Management Plan (Forest Plan) for the forest, consider the high peaks of the Pinaleños as a site for a future astronomical facility of major national significance. The first draft of the Forest Plan designated approximately 1,400 ha above 2,926 m elevation in the Pinaleños as a separate management area, called an astrophysical area. In 1984, Steward Observatory at the UA submitted an astrophysical site and facility proposal to the USFS. The USFS was still working on the Forest Plan and incorporated this proposal into the planning effort. High public interest and a lack of site-specific information for the 1,400 ha astrophysical area caused the USFS to withdraw this portion of the forest from the Forest Plan. Instead, the USFS initiated a separate planning effort and associated Environmental Impact Statement (EIS) for the 1,400 ha astrophysical area. The Forest Plan, minus the astrophysical area, was completed in 1986. Planning for the astrophysical area continued. The proposal of the red squirrel as endangered in 1986, the final listing of the red squirrel as endangered in 1987, and then the necessity to consult with the USFWS on the impacts of the proposed astrophysical development on the endangered red squirrel all served to delay the issuance of the final EIS and Forest Plan for the astrophysical area. The USFS completed a Biological Assessment (BA) in 1988 and submitted this to the USFWS for consultation on the proposed project. The consultation process led to a jeopardy determination for the proposed development (U.S. Fish and Wildlife Service 1988). The USFWS issued three reasonable and prudent alternatives to the proposed activity that would not further jeopardize the continued existence of the red squirrel. The University of Arizona found that they could still have a viable observatory by implementing reasonable and prudent alternative #3. The USFS began incorporating this alternative into the EIS and the Forest Plan.

Meanwhile, UA and its partners in the proposed astrophysical development felt that the planning process was proceeding much too slowly. They also knew that any decision by the USFS that allowed for astrophysical development on the mountain was sure to be appealed. Though they felt confident that any such appeal would be denied, they pressed for legislation to establish the facility. On November 18, 1988, the Arizona-Idaho Conservation Act of 1988 (PL 100–696) was signed into law. The law contained a provision (Title IV, Section 601.a) directing the secretary of agriculture to grant a special use permit to UA for the Mt. Graham International Observatory. In January 1989, the final EIS and Record of Decision for the Mt. Graham Astrophysical Area were released to the public. The modified EIS and Record of Decision reflected implementation of PL 100–696.

Construction of the access road to the observatory site began in the fall of 1989, and actual observatory construction began about a year later. The first two telescopes were dedicated in 1993.

Natural History

Description

The red squirrel is a small, grayish-brown arboreal rodent with a rusty to yellowish tinge along the back (Hoffmeister 1986). The tail is fluffy and the ears are slightly tufted in winter (Hoffmeister 1986). In summer, a black lateral line separates the dorsal side from the white underside. The MGRS is slightly smaller than *Tamiasciurus hudsonicus mogollonensis* (the other red squirrel subspecies found in Arizona) in body, hind foot, and skull length, with no sexual dimorphism in measurements of adults (Hoffmeister 1986).

Although Hoffmeister (1986) thought that MGRS was not strongly differentiated from *T. h. mogollonensis*, both Hall (1981) and Hoffmeister (1986) retained the subspecies designation. Recent genetic research using protein electrophoresis indicates that the MGRS population is the most genetically divergent of the red squirrel populations in the Southwest (Sullivan and Yates 1995). The population was estimated to have been isolated for at least 11,000 years. Sullivan and Yates (1995) suspect the genetic divergence of the Mt. Graham population resulted from a founder effect and subsequent genetic drift due to the long isolation of the population. Therefore, the population has probably bottlenecked at least once and may be homozygous (T. Yates, personal communication).

Habitat Requirements

The MGRS is confined to upper-elevation mature and old-growth forest in mixed-conifer and spruce-fir zones above approximately 2,377 m, although it may inhabit lower-elevation drainages that contain mixed-conifer forest. Historically, the MGRS was common above 2,590 m, but as of 1990 was seldom found below 2,804 m (Spicer et al. 1985; U.S. Forest Service unpublished data). Since 1990, the squirrels have been found down to approximately 2,377 m, and more current data show that the mixed-conifer/spruce-fir transition zone contains the bulk of the population (P. Young, personal communication). But during the time period addressed by this chapter, the bulk of the population was thought to inhabit the spruce-fir zone, and a smaller percentage of the population was thought to inhabit the transition zone. As part of the BA associated with the request for an astrophysical area by the UA, the USFS estimated approximately 4,680 ha of suitable habitat remained (U.S. Forest Service 1988). Of this, only 189 ha was considered excellent habitat and 626 ha good habitat (U.S. Forest Service 1988). U.S. Forest Service (1988) provides the specifics of this classification in appendix A of that document.

Suitable habitat for the MGRS depends, in part, on reliable and abundant crops of conifer cones for food, as well as microclimate conditions suitable for storing closed cones. The midden provides moist, cool conditions that help keep cones closed until they are withdrawn for eating (as long as two or three years). Closed-canopy forests (generally greater than 70 percent canopy cover) at higher elevations usually provide these features. Closed-canopy forests also provide conditions hospitable to fungal fruiting bodies, another important component of red squirrel diet. Other key habitat components of suitable habitat are downed logs, snags, and interlocking branch networks (A. Smith 1992). The above elements combined provide red squirrels with adequate food resources; perching, storage, and nesting sites; runways that allow cone retrieval in the winter; and escape routes.

The peaks of the Pinaleños are at about 32 degrees north latitude, which is the southernmost latitude for both the red squirrel and the spruce-fir association. The Pinaleño range also has a low dewpoint isohyte (50°F mean dewpoint), which is an indication of low water vapor density in the atmosphere (U.S. Forest Service 1988). The Pinaleños experience the most direct sunlight of all red squirrel habitats due to their southern latitude. High solar radiation in the Pinaleños might restrict or eliminate red squirrel use of some vegetation types, such as ponderosa pine, that are known to be inhabited year-round in more northerly latitudes (Ferner 1974). It may also increase the need for canopy closures that exceed 60 percent (U.S. Forest Service 1988). To avoid the negative effects of insolation such as higher heat loads and drier forest-floor conditions, MGRS may be more selective about midden placement than squirrels from more northerly latitudes (Froehlich 1990; A. Smith and Mannan 1994).

A. Smith and Mannan (1994) reported that the MGRS midden locations in the spruce-fir and transition associations are found in patches with unusually dense foliage volumes and canopy cover. Canopy cover was >70 percent at 96 percent of all middens measured. Using discriminant function analyses, A. Smith and Mannan (1994) selected foliage volume, canopy closure at the center of each plot, log volume, and density of large snags (>40 cm diameter at breast height) to best distinguish middens from random locations in both associations. The average age of dominant trees at the middens was between 180–215 years (A. Smith 1992). The lower-elevation mixed-conifer and transition zones also had higher numbers of middens than expected on north- and east-facing slopes (Froehlich 1990; A. Smith and Mannan 1994).

Territoriality

The red squirrel is a highly territorial animal (C. Smith 1968) and is considered a central-place forager. Mt. Graham red squirrels scatterhoard at times, but usually each squirrel occupies only one midden (Vahle 1978). Occasionally, conditions arise in which more than one squirrel occupies a midden, or a squirrel uses more than one midden (Froehlich 1990), but these are exceptional cases and occur during extremely abundant or extremely poor food years. In the early years of squirrel midden surveys (1984–1990), the highest densities of middens were in the high-elevation Engelmann spruce–corkbark fir associations. Lower densities of middens were found in mixed-conifer stands dominated by Douglas-fir, with white fir and southwestern white pine as subdominants and little to no spruce. More recent survey data shows that the transition between these two associations usually contains densities equal to or higher than those in the spruce-fir association (Young 1995).

Foods

The foods of the MGRS include (1) conifer seeds from closed cones, (2) aboveground and belowground macro-fungi and rusts, (3) pollen (pistillate cones) and cone buds, (4) conifer twig cambium, (5) bones, and (6) berries and seeds from other broadleaf trees and shrubs. Other food items reported include fledgling birds, bird eggs, mice, young rabbits, carrion, juniper berries, oak acorns, aspen seeds, and ash seeds (Steele 1998). Each food has a seasonal importance: pollen in the spring; bones for lactating females; fungi in the spring and late summer; closed cones low in lipids in the early summer; and closed cones high in lipids for winter storage (C. Smith 1968).

In the Pinaleños, the red squirrel has been observed eating the seeds from the cones of Engelmann spruce, white fir, Douglas-fir, corkbark fir, and southwestern white pine. During the time period covered by this chapter (through 1992), no MGRS was ever recorded eating or caching a ponderosa pine cone, though other red squirrel subspecies do consume ponderosa pine cones in more northerly latitudes (Hatt 1943; Finley 1969; Ferner 1974). Engelmann spruce and Douglas-fir are the most important species supplying food to the MGRS. Douglas-fir, generally a consistent cone producer (Finley 1969), plays a significant role in the diet of MGRS, especially where it co-exists with Engelmann spruce. Douglas-fir may be increasingly important when there are few spruce cones available but there is still an adequate Douglas-fir cone crop (such as 1987, 1988, and 1989). Douglas-fir is more widespread in the Pinaleños than is spruce, but it is more often found in logged and fragmented habitats at lower elevations, which may reduce its overall contribution to the food supply of the red squirrel population.

Miller and Yoder (in this vol.) studied the quality of MGRS food items and determined an average red squirrel required the seeds of approximately 170 cones/day if no other foods were available. Vahle and Patton (1983) stated that at least 9 to 14 cone-producing trees within a red squirrel's home range (about 1 ac in his study) ensured an adequate food supply.

C. Smith (1968) recorded that red squirrel subspecies ate 42 different species of fungi, with a preference for small false truffles (mostly *Rhizopogon* sp.). In two examples, C. Smith (1968) noted that mushrooms and false truffles supplied more than half the squirrel's daily calories. Ferron and Prescott (1977) observed red squirrels in Canada spending up to 20 percent of their foraging time harvesting fungi when fungi were available. By volume, fungi were 77 percent of red squirrel diets in Oregon (Maser et al. 1978). Mushrooms made up at least 10 percent and sometimes 40 percent of MGRS diets during parts of the year (Froehlich 1990). Many of the dry soil species (e.g., earthballs [*Scleroderma* sp.], false truffles, giant puffballs [*Calvatia* sp.]) are symbionts with spruce, fir, or Douglas-fir and cannot produce fruiting bodies without tree rootlets present (Maser et al. 1978). These ectomychorrhizal fungi are a major food source for red squirrels when available. Because they are more dry-tolerant, aboveground (epigeous) species are available to the squirrels longer than belowground (hypogeous) species.

In the Pinaleños, epigeous species occur when there are substantial summer rains, and hypogeous mushrooms appear during the spring snowmelt and after the summer rains begin. The most important seasons for mushrooms appear to be early summer, before cones ripen, and late fall, when squirrels are storing cones and eating the more perishable mushrooms. Observations of MGRS by various biologists confirm mushroom excavation, storage in trees and middens, harvest, and consumption of more than eight species (Froehlich 1990; Young 1995).

Miller and Yoder (in this vol.) analyzed the nutritional content of three hypogeous species of mushroom eaten by Mt. Graham red squirrels. Percent crude protein and percent digestible protein were higher in these mushrooms than for all conifer seeds except Engelmann spruce in summer. C. Smith (1968) found that truffle protein content was also as high as some conifer seed per unit weight.

Population Ecology

Red squirrels generally breed in early spring (February through early April). The sciurids are more specialized than many other rodents in that the female has but one day of fertility during each breeding period (Flyger and Gates 1982). Some populations have begun breeding in January (Layne 1954), and two breeding seasons have been reported in a few populations (Layne 1954; C. Smith 1968; Millar 1970; Lair 1985). It is unknown what percentage of females produce two litters, or how often two breeding seasons occur.

The red squirrel gestation period is 35 to 40 days (Woods 1980). Litter size ranges from two to eight, but most litters number three to five (U.S. Forest Service 1988). Hoffmeister's (1986) analysis of MGRS embryos indicated that these squirrels might have three young per litter. Warshall (1986) observed one MGRS mother with three young. Froehlich (1990) observed eight MGRS litters in 1988 and 1989, ranging from one to five young surviving into the fall. Young (unpublished data) observed six MGRS litters in 1990, with a mean of 2.7 young at emergence from the nest.

The age of first reproduction for female red squirrels occurs after the first winter. The proportion of nulliparous squirrels that breed varies widely from year to year. Rusch and Reeder (1978) and Wood (1967) found "yearling" reproductive rates (number of yearling females producing young) to vary from 24 percent to 88 percent. Yearling female breeding rates were always lower than adult female breeding rates. Red squirrels are considered adults after their second winter. The proportion of adult red squirrel females that breed varies widely from year to year, as does the proportion that produces two litters (Kemp and Keith 1970; Dolbeer 1973; Rusch and Reeder 1978). The proportion of juvenile and adult females that breed each year was unknown for the MGRS as of 1990.

Mortality rates of the MGRS are unknown. For other red squirrel populations, Halvorson and Engeman (1983), Rusch and Reeder (1978), and Kemp and Keith (1970) generally agree that many die between weaning and first reproduction, followed by low adult mortality, ending in increased mortality in older age classes. Survival rates are also unknown for the MGRS, but it is assumed that they fluctuate in response to the amount of closed cones available for storage in the middens (U.S. Forest Service 1988).

The number and quality of stored closed cones influences the length of the breeding season, number of adult females bearing two litters, number of adult and yearling females that breed, longevity of adults born in the year under consideration, eruptive dispersals, diet switches, and perhaps the mean long-term density of the population (M. Smith 1968; Rusch and Reeder 1978; Gurnell 1983). Millar (1970) believes that food availability also influences pre-implantation losses. Although not the only influence on population size and composition, the closed-cone seed crop explained more of the red squirrel demographics that any other single variable pre-1992 (Gurnell 1987; Young 1995).

Population Status—Pre-observatory Construction

The first scientific specimens of red squirrels from the Pinaleños were collected in 1894 when it was formally described as a subspecies, based on pelage differences and its isolation from other red squirrel populations (Allen 1894). There were many anecdotal assessments of the size of the MGRS population made over the next several decades. Hoffmeister (1956) visited the Pinaleños in 1951 and 1952 and reported the squirrel to be uncommon to rare. A note was published in 1968 suggesting the possible extirpation of the red squirrel from the Pinaleños (Minckley 1968). A few years later, Gale Monson (1973) reported that red squirrels persisted on the mountain, but were very difficult to find. The USFWS contracted with ANHP to conduct a status survey for the squirrel. Spicer et al. (1985) conducted this survey of the population, using belt transects. They estimated 300 to 500 squirrels.

Logging operations in the early half of the twentieth century may have affected the MGRS population size (U.S. Forest Service, unpublished data). By 1973, most of the accessible timber had been harvested, thereby altering the age structure and density of much of the red squirrel's forest habitat. Logging operations and road building to accommodate harvest reduced available habitat for the red squirrel further because they created openings that also increased the amount of wind-thrown trees. Additional losses of mature to old-growth coniferous forest and, therefore, red squirrel habitat resulted from both natural and human-caused fires, recreation developments, and the establishment of summer home areas, an administrative site, and a horse pasture. These direct losses not only reduced the amount of suitable habitat but also reduced the quality of remaining habitat by creating more forest edge and reducing the area providing proper microhabitat characteristics for cone storage. Fragmentation of the forest might also have isolated some pockets of the squirrel population and prevented successful dispersal and/or movements between areas, thus reducing genetic flow within the population.

The range and number of the MGRS also may have been reduced by competition with the Abert's squirrel, which was deliberately introduced in the Pinaleños from 1941 to 1943 (Minckley 1968; Hoffmeister 1986; U.S. Forest Service 1988). The tassel-eared squirrel now occupies nearly all coniferous forest on the Pinaleños, and although little is known about the interaction between these two species of squirrel, some authorities have suggested that competition for food resources is occurring and that some competitive exclusion may have occurred (Hoffmeister 1986; U.S. Forest Service 1988). The AGFD halted all hunting of the red squirrel in 1986.

Forest management changed in the Pinaleño Mountains in the 1980s out of concern for the species. In 1988, the USFS stopped all harvesting of fuelwood and Christmas trees and restricted gathering of campfire wood in some areas. Timber harvesting on the mountain had slowed considerably by 1973 because the more accessible timber had been harvested. The area was designated as unregulated (i.e., no scheduled harvest) in 1973. Timber harvesting was completely halted when the squirrel was listed as endangered. The USFS, in conjunction with AGFD, the USFWS, and various volunteer organizations, has closely monitored the

Table 6.1. Population estimates for the Mt. Graham red squirrel,* 1986–1990

Where two estimates are given, the first is a conservative estimate; the second is an optimistic estimate.

Survey Date	Original Estimate	Revised Estimate
Spring 1986	328 ± 55	348± 55
Fall 1987	246 ± 40	235± 40
March 1988	207 ± 62	210± 62
October 1988	178 ± 62	194± 62
	226 ± 62	258± 62
Jan/Feb 1989	197 ± 63	210± 63
June 1989	116 ± 29	146± 29
	167 ± 32	221± 32
October 1989	162 ± 15	191± 15
	185 ± 15	204± 15
May 1990	132 ± 15	152± 15
	146 ± 16	169± 16
October 1990	265 ± 7	260± 7

*Pinaleño Mountains, Graham County, Arizona.

red squirrel population since 1986, and began monitoring the conifer cone crops on the Pinaleños in 1989.

Most biologists agreed that the main cause for the decline seen in the mid-twentieth century was a cumulative loss of habitat. In the 1980s, only about one-half of the original upper-elevation coniferous forests were still considered suitable habitat (U.S. Forest Service 1988).

From 1986 through 1990, the USFS, along with cooperating agencies and individuals, conducted nine population surveys to estimate numbers of the MGRS. The first survey, based on strip transects, estimated 328 ± 5 squirrels (U.S. Forest Service 1988). During that five-year time period, estimates fluctuated (with a low of 146 individuals in June 1989 and a high of 348 in spring 1986). The October 1990 estimate was 272, and this coincides with the start of construction of the observatory site. Population estimates after 1990 are addressed by Snow (in this vol.) and Young (in this vol.).

Two methods were used to estimate populations. Originally, surveys estimated the average occupancy rate for all middens surveyed, then multiplied by the estimated number of middens on the mountain, which was 444 (U.S. Forest Service 1988). Beginning in the fall of 1990, different occupancy rates for each vegetation association were calculated. The estimated number of middens within each vegetation association was multiplied by the percent of occupied middens in that association (table 6.1). Previous population estimates were adjusted, using this new information (table 6.1). Assumptions for both methods are (1) squirrel occupancy can be determined from signs of recent caching, digging, and the condition of midden material, even when a squirrel is not directly observed, and (2) one squirrel occupies only one active midden at a time.

Population Vulnerability and Viability

From the time the Pinaleños were isolated as a sky island some 10,000 years ago, the small MGRS population has undoubtedly been subjected to many threats. Presently, the greatest management concern is to maintain enough habitat to continually insure a viable population. At the time of the BA, the USFS thought that the squirrels might be subject to a "habitat/population" bottleneck before the full recovery of the upper-elevation forests. The USFS estimated that a minimum of 100 to 200 years was needed to recover significant amounts of habitat (U.S. Forest Service 1988). Meanwhile, the population was viewed as vulnerable to natural catastrophes and possible loss of genetic fitness. The team that wrote the BA put together the following list of population concerns (U.S. Forest Service 1988:89–90):

1. The population is inherently vulnerable from a long-term evolutionary historical perspective because it is on an isolated mountain range and is dependent on a small Pleistocene-relict forest that, like the subspecies, is at the southern edge of its distribution.
2. It is an isolated population with no recolonization possible.
3. It is a rodent population dependent on booms in bumper-crop years to provide "cushioning" for the bust years. But, population booms cannot be very large on the Pinaleños because it is an isolated mountain range with limited and degraded forest habitat.
4. The population dynamics of this squirrel are largely unknown.

These uncertainties could mean that the squirrel is even more vulnerable than believed at this time.
5. The potential habitat of the subspecies on the Pinaleños has been fragmented and reduced in a piecemeal fashion over the last several decades and these effects have become cumulative.
6. The habitat of the MGRS is also relatively unstudied.

While many aspects of these concerns have been studied during the last fifteen years (other chapters in this volume detail the results of that research), the concerns remain highly applicable.

The effective genetic population size of a population without loss of genetic fitness for a few centuries has been estimated at 500 for polygenic and much higher for single-locus genetic characters (Lande and Barrowclough 1987). This translates to an even larger actual population size when age structure, sex ratios, reproductive status, kinship constraints, and fluctuating population size are considered. If this number is even close to that necessary to maintain a long-term population of red squirrels, then the MGRS was a candidate for high concern because, as of 1990, there has been no population estimate of more than about 400 individuals.

Models

To address these concerns, the USFS used two models to look at the probability of persistence of the red squirrel. These were the USFS Habitat Capability Model (HCM) (R. Wadleigh, unpublished USFS document) and the Population Dynamics Simulation Model (POPDYN) (Perez-Trejo 1986; R. Wadleigh, unpublished USFS document). At the time, the USFS lacked many pieces of information to develop a complete population viability assessment, so they combined the use of models with a qualitative best guess from evaluating the taxon and ecological status of the population on purely biological grounds (U.S. Forest Service 1988).

Habitat Capability Model. The HCM estimated the maximum population capability of the Pinaleños and estimated reductions in carrying capacity due to human actions (loss of habitat). Mixed-conifer and spruce-fir habitats were rated for two variables: food and cover. These values were derived from current biological opinion, and the values were input into the model. The model added all the acreage units of similar conifer types and structural stages and multiplied it by maximum density and habitat effectiveness scores. The most effective habitat, old-growth spruce-fir with closed canopy, was estimated to be able to support 0.23 squirrels per acre (U.S. Forest Service 1988).

The HCM projected a maximum current carrying capacity of 502 breeding red squirrels (U.S. Forest Service 1988). The USFS prepared age tables for seral stages of spruce-fir and mixed-conifer, and then constructed projected vegetative conditions 200 years into the future. The maximum carrying capacity was projected to go from 502 to 725 squirrels. This modeling effort was based on the assumption that no timber would be harvested, that currently degraded areas would recover, and that there would be no major catastrophic events. It was assumed that some small events (small-scale wind throw, fires, disease pockets) would occur and set back seral stages on 17 to 20 percent of the old-growth acreage (U.S. Forest Service 1988).

Population Dynamics Simulation Model. Samson et al. (1984) developed this interactive, Leslie Matrix dynamics model. It evaluates the effects of management strategies on the general trends of population sizes of endangered species (Samson et al. 1985).

The Biological Assessment team used this model to predict population trends of the red squirrel over a 30-year period with and without astrophysical development. The POPDYN model uses demographic data and is not dependent upon habitat-carrying capacity or other environmental indicators. The demographic values used for the MGRS were derived from literature on red squirrels. Therefore, this was considered a preliminary population viability assessment (U.S. Forest Service 1988).

The probability of extinction observed from the model for the fall 1987 population size of 246 was 29.2 ± 5.7 percent within the 30-year period. Under a future carrying capacity of 725, the observed extinction rate dropped to 14 ± 10 percent. A population level of 704 was considered a simulation of the effects of losing the maximum number of squirrels under astrophysical development and resulted in the same probability of extinction as under 725 squirrels (U.S. Forest Service 1988).

Conclusions

The HCM, using 1980s assumptions about suitable habitat, estimated available habitat would support 502 squirrels. It projected future habitat capability to support 725 squirrels. This modeling effort needs to be reassessed given current knowledge of suitable habitat.

The POPDYN model projected a 71 percent probability of survival for 30 years or an 8 percent probability of survival for 100 years. It was largely based on demographic data from other red squirrel populations. A new look at population dynamics using MGRS demographic data is certainly warranted.

Reviewers of the modeling and other data generally agreed that the population had already passed through any genetic bottlenecks and was safely homozygous (U.S. Forest Service 1988). Questions still remain about what constitutes a minimum viable population for this subspecies, and this is certainly a topic for future investigation.

The data gathered in the 1980s painted a dim picture of the future of the MGRS. Given what was known about the status of the MGRS population and its habitat at the start of observatory construction, the Mt. Graham Red Squirrel Recovery Team recommended aggressive habitat protection and restoration to achieve relative security in the near-term (U.S. Fish and Wildlife Service 1993). While more recent information has expanded our concept of suitable habitat, and squirrel population numbers are greater than they were at the end of the 1980s, many recent events have threatened squirrel habitat, including major wildfires, and forest pest and disease outbreaks. This subspecies is by no means out of the woods in terms of extinction probability, and efforts need to continue to conserve habitat and to learn more about this unique representative of a unique sky island.

Literature Cited

Allen, J. A. 1894. Descriptions of ten new North American mammals, and remarks on others. *Bulletin of the American Museum of Natural History* 6:320–321.

Dolbeer, R. A. 1973. Reproduction in the red squirrel (*Tamiasciurus hudsonicus*) in Colorado. *Journal of Mammalogy* 54:536–540.

Ferner, J. W. 1974. Habitat relationships of *Tamiasciurus hudsonicus* and *Sciurus aberti* in the Rocky Mountains. *Southwest Naturalist* 18:470–473.

Ferron, J., and J. Prescott. 1977. Gestation, litter size, and number of litters of the red squirrel (*Tamiasciurus hudsonicus*) in Quebec. *Canadian Field Naturalists* 91:83–84.

Finley, R. B., Jr. 1969. Cone caches and middens of *Tamiasciurus* in the Rocky Mountain region. Pages 233–273 in J. K. Jones Jr., ed. *Contributions in mammalogy.* University of Kansas Museum of Natural History Miscellaneous Publication 51.

Flyger, V., and J. E. Gates. 1982. Pine squirrels: *Tamiasciurus hudsonicus* and *T. douglasii.* Pages 230–238 in J. A. Chapman and G. A. Feldhamer, eds. *Wild mammals of North America.* John Hopkins University Press, Baltimore, Maryland.

Froehlich, G. F. 1990. Habitat use and life history of the Mt. Graham red squirrel. Master's thesis, University of Arizona, Tucson.

Gould, S. J. 1990. The golden rule—a proper scale for our environmental crisis. *Natural History* September:24–30.

Gurnell, J. 1983. Squirrel numbers and the abundance of tree seeds. *Mammalogy Review* 13:133–148.

——. 1987. *The natural history of squirrels.* Facts on File, New York.

Hall, E. R. 1981. *The mammals of North America*, vol. I, 2nd ed. John Wiley and Sons, New York.

Halvorson, C. H., and R. M. Engeman. 1983. Survival analyses for red squirrel population. *Journal of Mammalogy* 64:332–336.

Hatt, R. T. 1943. The pine squirrel in Colorado. *Journal of Mammalogy* 24:311–345.

Hoffmeister, D. F. 1956. Mammals of the Graham (Pinaleño) Mountains, Arizona. *American Midland Naturalist* 55:257–288.

——. 1986. *Mammals of Arizona.* University of Arizona Press and Arizona Game and Fish Department, Tucson.

Kemp, G. A., and L. G. Keith. 1970. Dynamics and regulation of red squirrel (*Tamiasciurus hudsonicus*) populations. *Ecology* 51:763–779.

Lair, H. 1985. Mating seasons and fertility of red squirrels in southern Quebec. *Canadian Journal of Zoology* 63:2323–2327.

Lande, R., and G. F. Barrowclough. 1987. Effective population size, genetic variation, and their use in population management. Pages 87–123 in M. E. Soulé, ed. *Viable populations for conservation.* Cambridge University Press, Cambridge.

Layne, J. N. 1954. The biology of the red squirrel, *Tamiasciurus hudsonicus logua* (Bangs), in central New York. *Ecological Monographs* 24:227–267.

Maser, C., J. M. Trappe, and D. C. Ure. 1978. Implications of small mammal mycophagy to the management of western coniferous forest. *Transactions of the North American Wildlife and Natural Resources Conference* 43:78–88.

Millar, J. S. 1970. The breeding season and reproductive cycle of the western red squirrel. *Canadian Journal of Zoology* 48:471–473.

Minckley, W. L. 1968. Possible extirpation of the spruce squirrel from the Pinaleño (Graham) Mountains, south central Arizona. *Journal of the Arizona Academy of Science* 5:110.

Monson, G. 1973. Mount Graham red squirrel. Page 23 in Unique birds and mammals of the Coronado National Forest. Unpublished report, U.S. Forest Service, Coronado National Forest, Tucson, Arizona.

Perez-Trejo, F. 1986. User manual, POPDYN, population simulation model. Unpublished manuscript. Department of Range Science, Colorado State University, Fort Collins.

Rusch, D. A., and W. G. Reeder. 1978. Population ecology of Alberta red squirrels. *Ecology* 59:400–420.

Samson, F. B., L. A. Bennett, F. Perez-Trejo, R. P. Concotta, and H. Salwasser. 1984. A simplified model for determining minimum viable population sizes and a diagnostic horizon for the grizzly bear. Third Predator Symposium, University of Montana, Missoula, March 23, 1984.

Samson, F. B., F. Perez-Trejo, H. Salwasser, L. F. Ruggiero, and M. L. Shaffer. 1985. On determining and managing minimum population size. *Wildlife Society Bulletin* 13:425–433.

Smith, A. A. 1992. Identification of distinguishing characteristics around middens of Mount Graham red squirrels. Master's thesis, University of Arizona, Tucson.

Smith, A. A., and R. W. Mannan. 1994. Distinguishing characteristics of Mount Graham red squirrel midden sites. *Journal of Wildlife Management* 58(3):437–445.

Smith, C. C. 1968. The adaptive nature of social organization in the genus of three [*sic*] squirrels *Tamiasciurus*. *Ecological Monographs* 38:31–63.

Smith, M. C. 1968. Red squirrel responses to spruce cone failure in interior Alaska. *Journal of Wildlife Management* 32:305–317.

Spicer, R. B., J. C. deVos Jr., and R. L. Glinski. 1985. Status of the Mount Graham red squirrel, *Tamiasciurus hudsonicus grahamensis* (Allen), of southeastern Arizona. Unpublished report, Arizona Game and Fish Department to U.S. Fish and Wildlife Service, Office of Endangered Species, Albuquerque, New Mexico.

Steele, M. A. 1998. *Tamiasciurus hudsonicus*. *Mammalian Species* 586:1–9.

Sullivan, R. M., and T. C. Yates. 1995. Population genetics and conservation of relict populations of red squirrel. Pages 193–208 in C. A. Istock and R. S. Hoffmann, eds. *Storm over a mountain island: Conservation biology and the Mt. Graham affair.* University of Arizona Press, Tucson.

U.S. Fish and Wildlife Service. 1988. *Biological Opinion on Coronado National Forest Plan and Mt. Graham astrophysical area plan.* Provided to Regional Forester, U.S. Forest Service, July 14, 1988.

———. 1993. *Mount Graham Red squirrel* (Tamiasciurus hudsonicus grahamensis) *Recovery Plan.* Albuquerque, New Mexico.

U.S. Forest Service. 1988. *Mt. Graham red squirrel: An expanded biological assessment.* Coronado National Forest, Tucson, Arizona.

Vahle, J. R. 1978. Red squirrel use of southwestern mixed coniferous habitat. Master's thesis, Arizona State University, Tempe.

Vahle, J. R., and D. R. Patton. 1983. Red squirrel cover requirements in Arizona mixed conifer forests. *Journal of Forestry* 81:14–15, 22.

Warshall, P. 1986. Mt. Graham squirrel evaluation. Unpublished report. Coronado National Forest, Tucson, Arizona.

Wood, T. J. 1967. Ecology and population dynamics of the red squirrel (*Tamiasciurus hudsonicus*) in Wood Buffalo National Park. Master's thesis, University of Saskatchewan, Saskatoon, Canada.

Woods, S. E., Jr. 1980. *The squirrels of Canada.* National Museum of Canada, Ottawa, Ontario, Canada.

Young, P. 1995. Monitoring the Mt. Graham red squirrel. Pages 226–246 in C. A. Istock and R. S. Hoffmann, eds. *Storm over a mountain island: Conservation biology and the Mt. Graham affair.* University of Arizona Press, Tucson.

CHAPTER 7

Mt. Graham Red Squirrel Interagency Midden Surveys

1991–2007

TIM K. SNOW

THE MOUNT GRAHAM RED SQUIRREL RECOVERY PLAN (U.S. Fish and Wildlife Service 1993) calls for estimating red squirrel abundance once each year during spring. Abundance estimates are derived from the number of active red squirrel middens observed during each survey. Because red squirrels are territorial (Smith 1968) and defend areas centered on their middens, midden activity rates may provide a reliable index of the overall squirrel population (Vahle 1978). Because red squirrels exclude other squirrels from their midden territories, the general assumption is one adult red squirrel per active midden. Therefore, midden activity rates may serve as a reliable index for determining the abundance of red squirrels. Employees and volunteers from the Arizona Game and Fish Department (AGFD), the U.S. Forest Service (USFS), the University of Arizona (UA), and the U.S. Fish and Wildlife Service (USFWS) participate in each red squirrel survey.

This chapter summarizes the interagency Mt. Graham red squirrel population surveys conducted from 1991–2007. The intent is to (1) describe Mt. Graham red squirrels and the importance of red squirrel middens in estimating squirrel abundance, (2) provide a synopsis of past survey efforts, (3) describe current survey methodology used to estimate abundance of red squirrels on Mt. Graham, and (4) present population estimates from 1991–2007.

History of Mt. Graham Red Squirrel Surveys

Prior to 1986, the USFS Safford Ranger District organized and conducted Mt. Graham red squirrel survey efforts. Belt transects were used to locate red squirrels and their middens. During this time, the USFS, along with UA Wildlife and Fisheries Resources, and Arid Lands programs, used aerial photography to assess red squirrel habitat in the upper elevations of the Pinaleño Mountains. Habitat was classified into three vegetation types, including spruce-fir, ecotone or transition, and mixed-conifer. Because red squirrels are considered to be spruce-fir obligates, the vegetation types were weighted for their value to red squirrels, with the spruce-fir type having the highest rating.

In 1986, the first interagency survey effort was conducted with the primary objective to identify red squirrels and their middens. A small number of representatives from each agency (AGFD, USFS, USFWS, and UA) participated in these efforts, concentrating in the spruce-fir and ecotone vegetation types.

The first attempt at estimating abundance of the Mt. Graham red squirrel population was made in 1988, based on a simple random sample of known red squirrel middens. The interagency survey effort was increased to include fall surveys in addition to the established spring survey and added sampling in the mixed-conifer vegetation association.

In 1989, the current method of randomly selecting middens was stratified by vegetation type. Initial surveys using this method were complex efforts involving more than 30 people. From 1987–1991, midden locations were marked on a 7.5-minute topographical map and then manually transferred onto a base map. However, with advances in global positioning system (GPS) technology that improved locating, marking, and tracking middens, survey efforts required fewer participants working in smaller crews and covering larger areas. From 1993–1995, the AGFD and USFS mapped the majority of known middens on the mountain using GPS units and incorporated these locations into a geographic information system (GIS).

In 1994, survey responsibility was transferred to the AGFD Nongame Branch. Since 1996, the AGFD Region V (Tucson) Office has coordinated the interagency semi-annual survey efforts and managed the Mt. Graham red squirrel survey databases.

Intensive area searches began in 1997 to find new middens and update the status of disappeared (abandoned or non-apparent) middens. The entire Mt. Graham red squirrel survey area was divided into 21 smaller units based on topography and accessibility (fig. 7.1). Survey participants

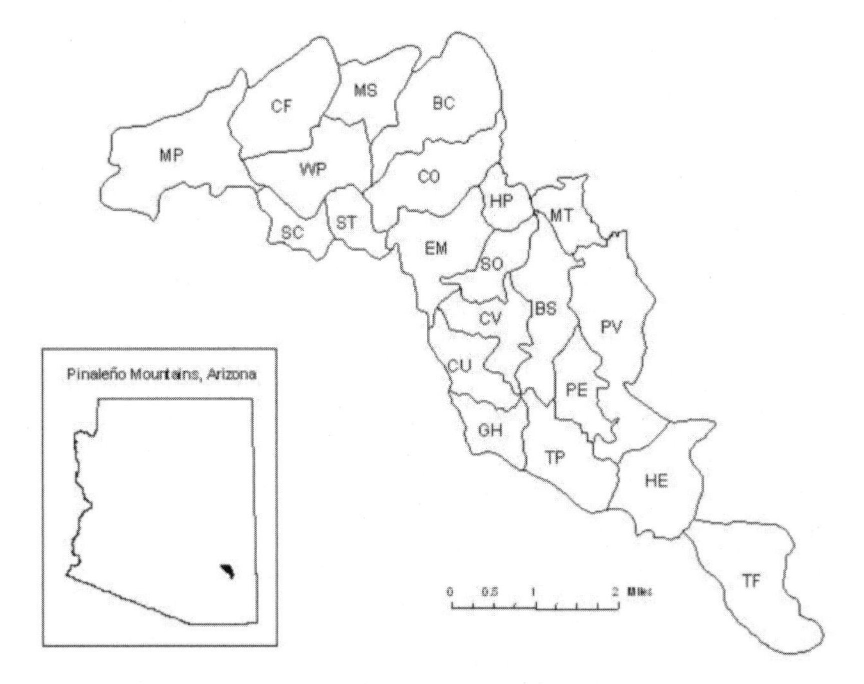

Figure 7.1. The 21 interagency subdivisions of the overall Mt. Graham red squirrel survey area in the Pinaleño Mountains.

were spaced at 10–15 m intervals and walked in a line through each area, marking or recording GPS locations for all middens encountered. In 1998, four of the smaller survey units were intensely searched (WP, SO, PE, and ST). In 2001, MP and WP units were searched in conjunction with USFS fire reduction planning.

In 2002, AGFD began randomly selecting units for complete midden searches during the spring survey period. Searches were conducted in four units: MP, PE, GH, and TP. For these efforts, all known middens (current and disappeared) in each unit were visited.

Survey Sampling Design and Methodology

The current survey method estimates abundance of Mt. Graham red squirrels based on activity rates of randomly selected middens. Thus, abundance is the product of the number of available territories (middens in the database) and the number of occupied middens (number of active middens found during the survey). Because cone production differs year to year between the mixed-conifer, spruce-fir, and ecotone or transition vegetation types, which results in varying red squirrel activity within vegetation types, the number of middens visited during the survey is proportionally selected based on the number of known middens found in each vegetation type. As a result, 30–40 percent of all known middens in each vegetation type are sampled each spring and fall.

Since 1990, abundance estimates for Mt. Graham red squirrels have been reported with 95 percent confidence intervals. The confidence level and/or the number of known middens visited during the survey can change the degree of precision. Because it is generally conventional to report 95 percent confidence intervals, and because this was the confidence interval used in previous Mt. Graham red squirrel estimates, sampling effort (number of known middens visited) is adjusted for each survey so that we may achieve as close as possible to a 95 percent confidence interval. Additional middens are added to the sample to compensate for middens that could not be found or might not be visited (inadvertently missed).

Surveys begin with survey participants assessing whether each red squirrel midden visited is currently "active," "inactive," or "uncertain." The assessment is based on three factors: the presence of a red squirrel at the midden; signs of recent activity such as digging, feeding, or caching cones in the midden; and distribution and consistency of cone scales in the midden area.

Survey participants also assess whether there is Abert's squirrel (*Sciurus aberti*) activity in the red squirrel midden area. The Abert's squirrel is an introduced species in the Pinaleño Mountains and may compete with Mt. Graham red squirrels. It is vital that survey participants be able to distinguish red squirrel and Abert's squirrel activity so that red squirrel activity at middens is accurately assessed. Abert's squirrels tend to scatter their debris much more than red squirrels because they consume their food where they find it, although debris concentrations can occur surrounding Abert's squirrel nest trees. Cone cobs deposited by Abert's squirrels usually are tattered in appearance, leaving more than 6 mm of the cone scales attached to the cob. In contrast, cone cobs deposited by red squirrels usually have a neatly cut appearance and scales are cut close to the cob. Red squirrel debris is concentrated around the midden and nearby feeding perches.

Finally, any shifts in midden activity or updates to midden diagrams also are recorded. All undocumented middens that are found during the survey are noted for future confirmation and for possible inclusion in the database should these meet the criteria of a new midden. To be considered a new midden for inclusion in the database: (1) the newly found midden must not be a simple shift of activity by the resident squirrel, (2) it must be an adequate distance (greater than 50 m) away from an existing midden, and/or (3) it must be defended by a different red squirrel than the one occupying the marked site (two squirrels must be observed).

Estimating Abundance

Two different red squirrel abundance estimates are calculated after completion of the survey. The first estimate is conservative, based only on middens classified as "active." The second estimate is optimistic, based on the number of "uncertain" middens added to the number of active middens. The proportion of active middens in each vegetation type is calculated and extrapolated for the number of known middens in that type. The number of red squirrels (one active midden = one squirrel) in each vegetation type is then summed to determine the population estimate. A confidence interval also is calculated for both estimates.

The formulas used to estimate abundance are as follows:

$$\text{Conservative Estimate:} \quad A_i/n_i\,(N_i) = CP_i$$

$$\text{Optimistic Estimate:} \quad [(A_i + U_i)/n_i]\,(N_i) = OP_i$$

Where:

A_i = number of middens determined active in *i* vegetation type

U_i = number of middens where activity is uncertain in *i* vegetation type

n_i = number of middens visited in *i* vegetation type = T–NF

N_i = total number of known middens in *i* vegetation type (database)

CP_i = estimated number of active middens in *i* vegetation type (conservative)

= estimated number of Mt. Graham red squirrels in *i* vegetation type (conservative)

OP_i = estimated number of active middens in *i* vegetation type (optimistic)

= estimated number of Mt. Graham red squirrels in *i* vegetation type (optimistic)

The formulas used to calculate 95 percent confidence intervals for abundance are as follows:

Determine the percentage of middens that are active for all vegetative types combined.

Conservative:

$$P = \text{number of active middens/number visited}$$

$$= \Sigma A_i / \Sigma n_i$$

Optimistic:

$$P = (\text{active middens} + \text{uncertain middens})/\text{number visited}$$

$$= (\Sigma A_i + \Sigma U) / \Sigma n_i$$

Calculate the confidence interval as follows:

$$CI = n(1.96) \times [P(1-P)/n]^{1/2} \times [(N-n)/N]^{1/2}$$

Where:

n = original sample size before extra middens are added

P = percentage of active middens calculated in Step 1

N = potential population size

CI = number of middens in the database or 650 if the number in the database is less than 650 (per recovery plan)

Survey Results: 1991–2007

Mt. Graham red squirrel population estimates have ranged from 562 (±11) individuals in the spring of 1999 to 199 (±12) individuals in the spring of 2006 (fig. 7.2). Although none of the population estimates statistically

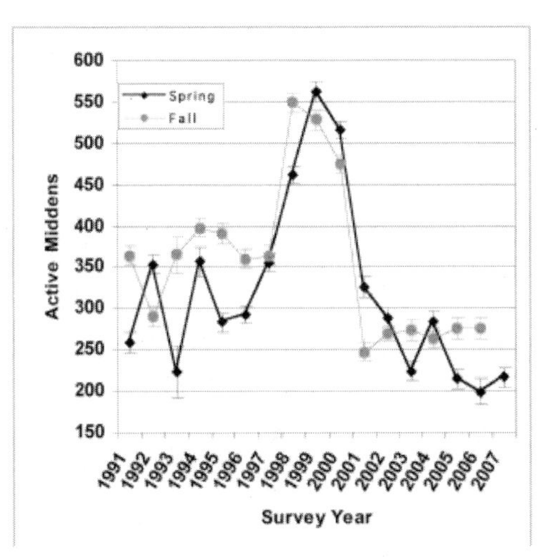

Figure 7.2. Mt. Graham red squirrel conservative population estimates and 95% confidence intervals: 1991–2007. Note the grouped results before and after large increase in the population during 1998–2000.

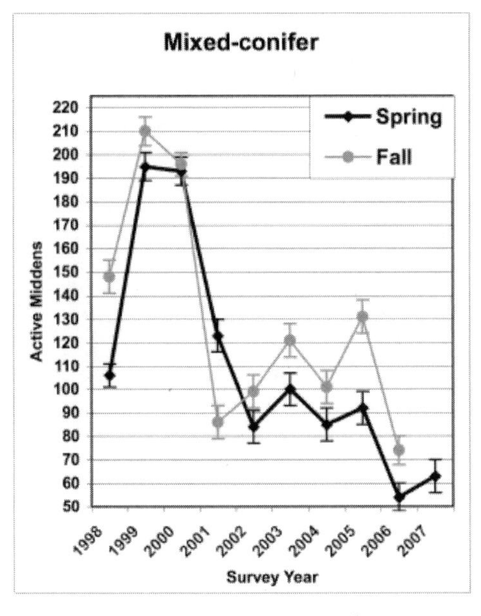

Figure 7.3. Conservative population estimates and 95% confidence intervals for mixed-conifer habitat, 1998–2007.

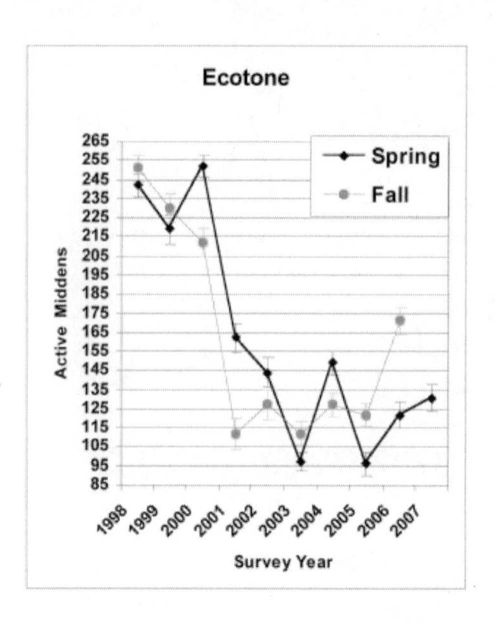

Figure 7.4. Conservative population estimates and 95% confidence intervals for ecotone habitat, 1998–2007.

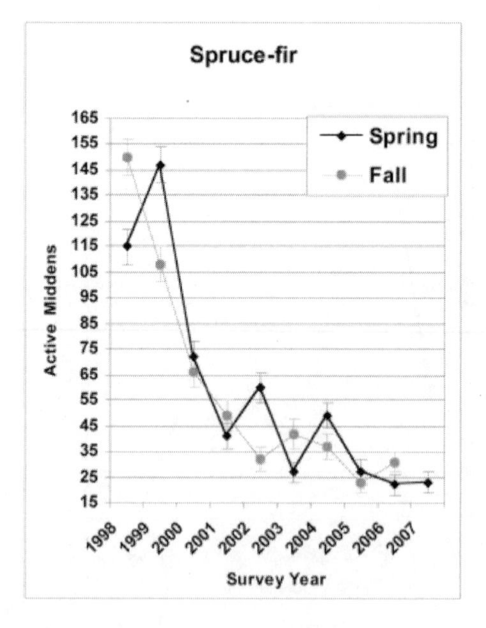

Figure 7.5. Conservative population estimates and 95% confidence intervals for spruce-fir habitat, 1998–2007.

indicate a trend in the population, our estimates show two groups of figures with a spike during 1998–2000. It is apparent that Mount Graham red squirrels had a population of around 350 individuals from 1991–1997, and over the last seven years (2001–2007), the number has fallen to somewhere around 250 individuals.

In addition, the conservative estimates of Mt. Graham red squirrels generally have declined across all vegetation types from 1998 to 2007 (figs. 7.3, 7.4, and 7.5). Estimates of red squirrels within the spruce-fir vegetation type (fig. 7.5) exhibited the greatest decrease (85 percent), from 150 (±7) squirrels in fall of 1998 to 23 (±4) squirrels in fall of 2006.

Conclusions

Surveys are important for managing and monitoring Mt. Graham red squirrels. Not only do they provide information about population size, they help determine how close we are to achieving recovery goals by providing information about red squirrel habitat and history of use at each midden. The interagency group will continue to monitor the red squirrel population in the Pinaleño Mountains, identify population trends, and work toward recovery of Mt. Graham red squirrels.

Literature Cited

Smith, C. C. 1968. The adaptive nature of social organization in the genus of three [*sic*] squirrels *Tamiasciurus*. *Ecological Monographs* 88(1):31–63.

U.S. Fish and Wildlife Service. 1993. *Mount Graham Red Squirrel* (Tamiasciurus hudsonicus grahamensis) *Recovery Plan*. Albuquerque, New Mexico.

Vahle, J. R. 1978. Red squirrel use of southwestern mixed-coniferous habitat. Master's thesis, Arizona State University, Tempe.

CHAPTER 8

Mt. Graham Red Squirrel Populations on the University of Arizona's Mt. Graham Red Squirrel Monitoring Program Area

1989–2003

PAUL J. YOUNG

A Brief History

DRS. RUSSELL DAVIS AND CONRAD ISTOCK (Department of Ecology and Evolutionary Biology, the University of Arizona) organized the University of Arizona's Mt. Graham Red Squirrel Monitoring Program (MGRSMP) in 1988. Their proposed monitoring program was included in the Mt. Graham Red Squirrel Management Plan as required under the special use permit issued to the University of Arizona (UA) for construction and operation of the Mt. Graham International Observatory (MGIO). The permit issued for the observatory included many conditions and mitigating actions because it was to be constructed in the habitat of the Mt. Graham red squirrel, which was listed as a federally endangered species in 1987. The permit mandated that the observatory fund a program to monitor the red squirrel population around the observatory site "for the life of the astrophysical complex" (Coronado National Forest 1989:30). The monitoring program was to collect data on the potential impact of the construction and operation of the observatory on the red squirrel population and to report directly to the U.S. Forest Service (USFS). The findings of the monitoring program were to be used to determine whether or not the effects of construction and operation of MGIO on the red squirrel exceeded those predicted and allowed by the Biological Opinion that was issued by the U.S. Fish and Wildlife Service (USFWS) and that set the conditions for building the telescopes. Additionally, the data and findings of the monitoring program were to be used to help in future decisions on the possible construction of up to four more telescopes.

Beginning in May 1989, Dr. Davis, with the help of two field assistants, surveyed the areas to be monitored for active and inactive red squirrel territories, and began collecting data on the location and number of squirrels and on some aspects of the behavior of the squirrels. The MGRSMP was expanded in August 1989 with the hiring of two additional full-time field assistants and me as the program supervisor. Pre-construction monitoring officially began in August 1989 with a concerted effort to trap and tag squirrels and to collect data from behavior observations. The monitoring plan was modified over time, with the approval of USFS wildlife biologists, to improve the data collection and to mediate problems that became apparent as the data were analyzed.

During the first three years, much of the monitoring effort was directed at collecting data on the behavior of red squirrels relative to their distance from construction activity. This was terminated after 1991 because it was apparent that there were no discernible differences in behavior between squirrels near the MGIO site and those farther away. The program then changed focus to collecting data on the numbers and distribution of red squirrels on the four monitoring program study areas. From August 1989 to May 1996, a full census of the squirrels on the monitored areas was made each month during the snow-free period of the year and bi-monthly when there was snow cover. Since May 1996, we have conducted a full census four times each year (March, June, September, and December).

The census was conducted by visiting each known midden site within the study areas, looking for signs of red squirrel activity, and remaining at a midden with activity for up to three hours to confirm its occupancy. In most instances, the sex of the occupying squirrel could be determined through observation with binoculars. New midden locations were added to the census list as they were discovered, and a thorough search of the entire study area was made each autumn for new middens.

The mandate, requiring the UA to fund the monitoring program in virtual perpetuity, has provided a unique opportunity to conduct population studies on a small mammal species for an extremely long period of time. For this chapter, I have focused on the changes in the red squirrel populations on the areas of Mt. Graham monitored by the UA's

MGRSMP from August 1989 through March 2003. I have refrained from discussing whether or not the construction and operation of MGIO has had any negative effect on the red squirrel population, because that issue is covered in detail in the many reports prepared by the MGRSMP for the USFS.

Study Areas

The size of the monitored areas has changed over the years to accommodate changes in the monitoring program and the configuration of the MGIO footprint. The accuracy of our mapping techniques was greatly enhanced, starting in 1994–1995, when satellite-based geographic positioning system (GPS) methods became available and allowed us to accurately map the location of each red squirrel midden site. Each midden site was located using a handheld GPS receiver, collecting a minimum of 180 position points. These points were subjected to differential correction equations, using data from the USFS base station in Tucson, and then averaged to improve accuracy to approximately 2 m. Many of the trails, roads, and other physical features were also mapped using GPS methods, and the entire mapping data were converted to geographic information system (GIS) format for use in ArcView software (ESRI Corp., Dublin, CA). The GIS data were used to determine the size of the study areas and GIS densities of middens and squirrels, as well as various calculations such as distance to nearest neighbor and distance to construction activity.

From May through August 1989, the MGRSMP monitored an area approximately 173.5 ha in size surrounding the construction area to a distance of 300 m. The boundary of this area was defined by the USFWS in the Biological Opinion for the Mt. Graham red squirrel as the area within 1,000 ft (304.8 m) of the road(s) and telescopes. For convenience, the distance was rounded down to 300 m in the MGIO management plan (Coronado National Forest 1989). This area was divided into upper and lower study areas along the 3,048 m (10,000 ft) elevation contour, which also approximates the lower boundary of the spruce-fir forest on the Pinaleño Mountains. The upper area (88.3 ha) is referred to as the Spruce-Fir Construction Area (SFC), while the lower area (85.2 ha) is referred to as the Transitional Forest Construction Area (TRC).

In 1989, a comparison area of approximately 104.8 ha was also established in spruce-fir habitat. This area, referred to as the Spruce-Fir Non-construction Area (SFN), is located approximately 750 m to the east of the boundary of the SFC area and extends from the top of Hawk Peak eastward to encompass High Peak (aka Mt. Graham). A smaller comparison area (21.0 ha) was added in mixed-conifer or transition forest (Transition Non-construction Area [TRN]) in September 1989, bringing the total size of the monitored areas to approximately 306 ha. In 1993, the decision to relocate the Large Binocular Telescope site to the east of the Vatican Advanced Technology Telescope and the Submillimeter Telescope required the inclusion of an additional 12.2 ha to the SFC, bringing that area up to 100.5 ha.

In April 1996, the Clark Peak fire started from undetermined causes on the northwest end of the Pinaleños near Riggs Flat Lake, approximately six miles from the telescopes and the MGRSMP study areas. Before the fire was contained, it burned through an area encompassing more than 2,625 ha, including 34.0 ha on the TRC and 24.5 ha on the SFC areas. A small spot fire on the TRN area caused some damage to about 0.01 ha, but the SFN area was untouched. While only about half of the area within the entire Clark Peak fire perimeter was catastrophically burned, almost all of the forest on the burned portions of the monitored areas was destroyed.

Habitat Quality

The quality of existing and potential red squirrel habitat on the study areas was examined by using the habitat management zone (HMZ) classifications described in the Mount Graham Red Squirrel Recovery Plan (U.S. Fish and Wildlife Service 1993) as a basis for determining how much of each of the study areas was deemed to be within each of the classifications. The recovery plan described seven categories of forested areas of the mountain based on their perceived potential to support red squirrels now and into the future. The relative quality of zones 1, 2, and 3 can generally be described as high, moderate, and low, with zones 4–7 having very low to no potential for supporting red squirrels.

The use of HMZ classification for rating red squirrel habitat has several drawbacks that are unavoidable. The most important with respect to this report is that the resolution of the habitat characteristics within each zone is very coarse; each zone consists of a mosaic of patches of varying quality that cannot be represented. In 1993–1994, the Safford Ranger District contracted with a private consultant to conduct a forest stand assessment to better determine the structural characteristics of the forested areas on the Pinaleños on a stand-by-stand basis. Individual stands were identified from aerial photographs, and each stand was examined on the ground using either standard transect or plot assessment techniques. The district supplied the MGRSMP with the raw data from these assessments in 1996. We used these data in an attempt to develop a more fine-grained map of red squirrel habitat quality; however, our attempts were not successful. The stand analyses were not specifically designed to measure structural characteristics relevant to red squirrel preferences and have proven to be less helpful than was hoped. Additionally, many of the stands are missing data for some key variables and must be excluded from the models and analyses. Missing data create gaps in the map, many of which are in areas known to contain relatively large numbers of red squirrel middens.

The AGFD is working on a new method for defining and classifying MGRS habitat based on remote sensing and GIS technology. Although this method promises to provide a more realistic and better-resolved classification of the habitat, it was not completed in time for inclusion in this report.

Squirrel Census Methods

A census was conducted every month during the snow-free months and bi-monthly during winter, weather permitting, from 1989 through 1995. In 1996, the Clark Peak fire disrupted the schedule, and at that time, we changed to conducting the census on a quarterly basis (March, June, September, and December).

Each census was conducted by visiting each midden site, examining it for evidence of red squirrel activity, and, if thought to be occupied, observing the site for up to three hours to verify the presence of a resident squirrel. Multiple visits were often made to midden sites to confirm the residency of a squirrel, and to attempt to determine its sex, age, and any natural or human-made marks. While this method may have missed counting some non-territorial, or transient, individuals, it is generally agreed that red squirrels are solitary and highly territorial. A midden site is almost always occupied by only one adult squirrel and is aggressively defended from all intruders. The results of each census were used as an estimate of the minimum number of red squirrels known to be alive on the monitored areas at the time of the census.

The census data were used to compare changes in the size and spatial distribution of the squirrel populations between construction and non-construction areas.

Results and Discussion

Habitat Quality on the Monitored Areas

Only three of the seven HMZ classifications (1, 2, and 4) occur on the monitored areas (fig. 8.1). The largest portions of all the areas are classified as being high quality (HMZ-1). While a large portion of both construction areas (TRC and SFC) is classified as moderate quality (HMZ-2), only the TRC area contains a significant amount of area classified as low quality (HMZ-4). Based on the USFS GIS maps of the HMZ, the monitored areas contain 31 percent of the high-quality, 4.7 percent of the moderate-quality, and 2.5 percent of the low-quality habitat on the whole mountain.

The Clark Peak fire destroyed a large amount of habitat on the TRC and SFC areas in 1996 (table 8.1). The TRC area was most affected: 40 percent of the area was engulfed in a catastrophic fire. About one-third of the TRC high-quality habitat and one-half of the moderate-quality habitat were destroyed. In contrast, only about 3 percent of the high-quality SFC habitat was destroyed, while over one-half of the moderate- and low-quality SFC habitat was catastrophically burned. The most important effect of the fire, other than the outright loss of squirrel habitat and midden sites, was the change in the relative amount of habitat quality on the SFC area. The fire did not appreciably alter the proportion of each habitat type on the TRN area. However, the moderate- and low-quality SFC habitat was extensively decreased and the high-quality habitat was considerably increased. The single most important result of the fire was the effective destruction of almost all suitable red squirrel habitat within

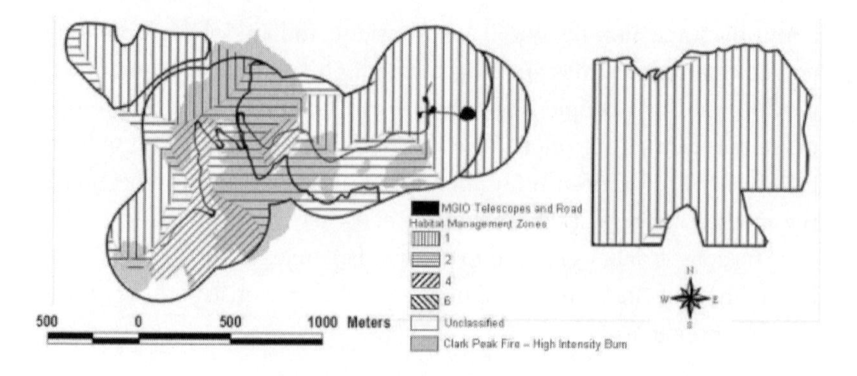

Figure 8.1. Map showing overlay of habitat management zones on the monitored areas along with the area intensely burned in the Clark Peak fire in 1996.

100 m of the lower two-thirds of the access road. This has effectively obliterated whatever effect the road may have had on the adjacent habitat or the nearby squirrel population.

Most of the middens on the monitored areas are found within areas designated as high-quality habitat (table 8.2). While the number and proportion of middens found in moderate-quality habitat has increased since 1989, there were no middens in low-quality habitat in 1989, and the single attempt by a squirrel to establish one failed. Although many midden sites have been added on the study areas and some have been lost due to inactivity, the greatest single loss of middens occurred during the Clark Peak fire. Middens that are inactive for three years are temporarily removed from the census database and no longer checked during the quarterly census. This allows us to calculate a more realistic value for the proportion of the middens occupied during the census. Some of the removed middens were re-added to the database when they were re-occupied after more than three years' inactivity.

The Clark Peak fire resulted in the permanent loss of a large number of midden sites on the TRC and SFC areas. While most of the middens lost were in high-quality habitat portions of the TRC, most of the middens lost on the SFC were located in moderate-quality habitat.

Table 8.1. Amount of area (ha) in each habitat management zone within each study area destroyed in the Clark Peak fire

Habitat management zone classes range from 1 (high) to 7 (very poor). Forest Types: Transition (TR), Spruce-Fir (SF); Construction (C), Non-construction (N)

	Habitat Management Zone Class								
AREA	1	2	3	4	5	6	7	None	All Areas
TRC	7.8	15.6	-	8.5	-	-	-	2.2	34.1
TRN	0.0	0.0	-	0.0	-	-	-	0.0	0.0
SFC	1.7	21.0	-	1.8	-	-	-	0.0	24.5
SFN	0.0	0.0	-	0.0	-	-	-	0.0	0.0
Total	9.5	36.6		10.3				2.2	58.6

Red Squirrel Populations 1989–2003

When initially surveyed in May 1989, the areas to be monitored by MGRSMP were occupied by only 18 red squirrels, all but 2 of which were located in spruce-fir (SF) habitat. The red squirrel population on the monitored areas was still very small when monitoring actually started in August 1989; of the 83 midden sites known on the area at that time, only 27 were occupied. Most of the squirrels were located in the TRC area at that time. Due to striking differences in the behavior of squirrels on the TRC area compared to the two SF areas, and the increase in number of squirrels in the transitional (TR) habitat, we added an additional non-construction site in TR habitat (TRN) to the monitored areas.

The population increased in the TR habitat but decreased in the SF during autumn 1989, and several new midden sites were discovered on the monitored areas. By November, the population had reached a high of 33 squirrels. Over winter, the populations declined in both habitats, on construction and non-construction areas alike. By summer 1990, the overall population reached a low of 27 squirrels. Concurrent with an abundant Engelmann spruce cone crop in autumn 1990, the population increased greatly (84), far exceeding the population of the previous year (fig. 8.2). This was followed by an over-winter decline on all but the SFC area, and

Table 8.2. Distribution of midden sites (n) by habitat management zones (HMZ) within the MGRSMP study areas

	TRC HMZ			TRN HMZ			SFC HMZ			SFN HMZ		
	1	2	4	1	2	4	1	2	4	1	2	4
Sep 1989	10	1	0	6	0	0	21	2	0	49	0	0
# added	38	12	1	29	4	0	84	13	0	72	1	0
# removed	21	5	1	6	1	0	7	9	0	24	1	0
Dec 1998	27	8	0	29	3	0	98	6	0	97	0	0

Habitat management zone classes from 1 (high) to 7 (very poor).
Forest Types: Transition (TR), Spruce-Fir (SF); Construction (C), Non-construction (N).

another larger increase in all populations in autumn 1991. The populations then declined again through 1992 and until autumn 1993 when, with the occurrence of another good cone crop, they recovered to nearly the same numbers as in 1991. All of the populations on the monitored areas remained fairly stable with normal seasonal increases and decreases until autumn 1995. In 1995, the populations in TR habitat showed modest increases; however, the populations in SF habitat increased significantly. The total population of the monitored areas reached its greatest number to date (228) in November and December 1995 (fig. 8.2).

The Clark Peak fire, which started on 24 April 1996 from unknown cause, greatly reduced the amount of available squirrel habitat on the TRC and SFC areas but did not affect either the TRN or SFN areas. Approximately 15 red squirrels disappeared from the TRC and 8 from the SFC areas during the fire. The fire and modest cone crops contributed to a general decline in the squirrel populations from autumn 1995 through summer 1998 when they were reduced to numbers similar to those seen in 1993 and 1994. Also, beginning in 1996, but particularly in 1997 and 1998, the populations on the SF areas had begun to be affected by an outbreak of defoliating insects (looper moth larvae) that was encompassing large areas of SF habitat on the High Peak and Hawk Peak areas. The looper outbreak was followed by an outbreak of two species of bark beetles in the stressed trees, which continues to affect the spruce and corkbark fir. In the winter of 1999–2000, the Pinaleños were invaded by yet another insect pest, an

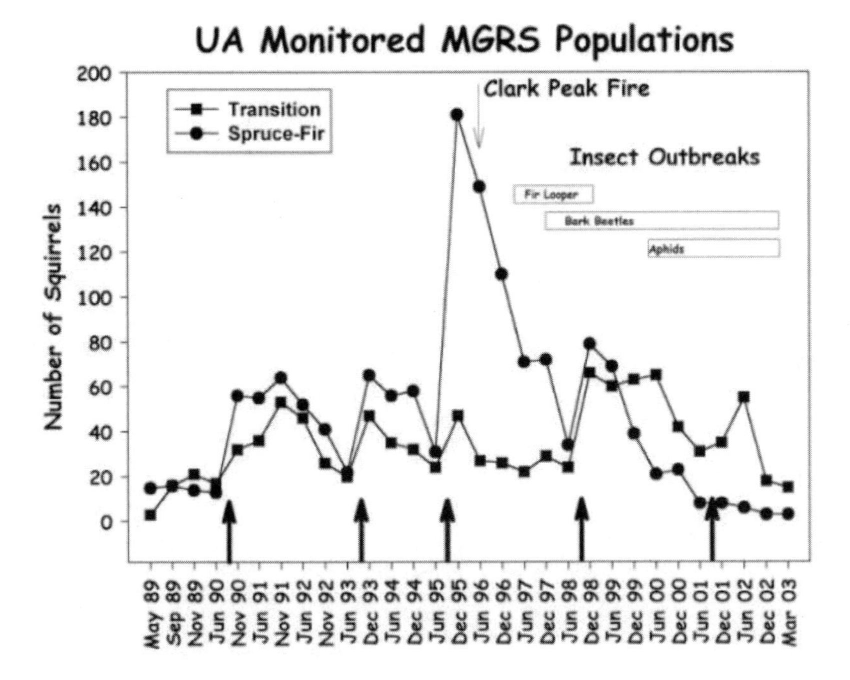

Figure 8.2. Seasonal changes in the Mt. Graham red squirrel populations on the monitored areas, May 1989 through March 2003. Populations are separated by habitat type (transition and spruce-fir); arrows along x-axis indicate years with good conifer cone crops.

exotic species of aphid, which has contributed further to the decline of the spruce-fir forest, as well as the transition-zone forest. These events are most likely responsible for the noticeable decline in the populations of the SF areas from December 1997 through June 1998, during which time the populations on the TR areas remained stable. Populations on all areas increased again in autumn 1998 in conjunction with another moderately good conifer cone crop. However, the recovery was short-lived in the SF areas. By March of 2003, the population of squirrels in these areas had virtually disappeared. And the squirrel populations on the monitored areas had come full circle with a total of 18 squirrels present, as was the case in May 1989. The distribution of squirrels was significantly changed, however, with most of the squirrels now living within TR habitat, and the SF habitat being almost totally unoccupied (fig. 8.2).

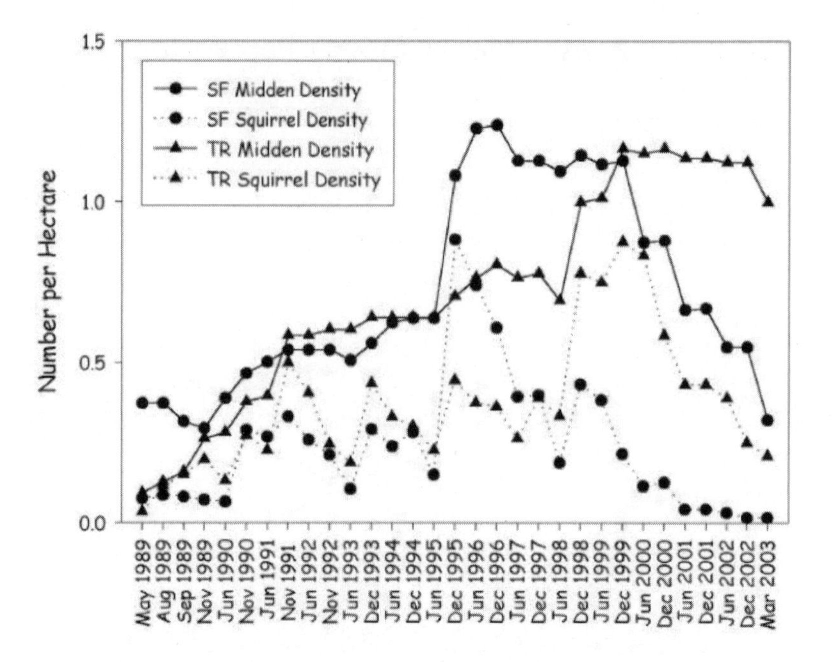

Figure 8.3. Densities (#/ha) of midden locations and red squirrels by habitat type on the monitored areas, May 1989 through March 2003. Transition forest (TR), spruce-fir (SP).

Densities of middens and squirrels on all the areas were initially, and remain, considerably lower than that reported for other red squirrel populations (Hamilton 1939; Layne 1954; Obbard 1987). In 1989, the density of middens thought to be regularly occupied was around 0.10/ha in TR habitat and approaching 0.40/ha in SF habitat. The density of squirrels was below 0.10/ha in both habitats (fig. 8.3). We recorded a steady increase in the density of midden sites through 1995 as the squirrel population increased and more middens sites were established. There were sharp increases in midden density in TR habitat in the fall of 1991 and again in 1998 and also in SF habitat in 1995, all corresponding with large conifer seed crops in those habitats and a concurrent large increase in the squirrel populations. The density of middens in SF habitat has been in decline since 1996 as the habitat has been degraded by a series of insect infestations and the middens have become abandoned. The density of middens in TR habitat has remained above 1.00/ha since 1999 but has also been declining in recent years. The densities of squirrels in both habitats has fluctuated widely, reaching a peak of 0.88/ha in SF habitat in the fall of 1995, and 0.87/ha in TR habitat in the fall of 1999.

Summary

It is obvious that the Mt. Graham red squirrel continues to experience a precarious existence at best. Yet, in the face of the seemingly endless series of natural disasters of fire and insects, they continue to persist. Bear in mind that we are witness to only a tiny fraction of the history of this species. Both the squirrel, and the forest ecosystem it lives in, are relics of an ice age that ended some 12,000 years ago, and this latest disaster is likely just one of many that have occurred over the millennia. The squirrel's future is ultimately tied to the future of the forest. The recent insect outbreaks in the spruce-fir forest have laid the foundation for a stand-replacing fire. In spite of the boreal appearance of the forest, this sky island environment is far different from the cooler and wetter forests of the north. And what grows out of any stand-replacing event is anyone's guess at this time. Will the spruce-fir forest recover? Or has the climate changed enough since the last stand replacement that it will now be replaced by mixed conifers or aspen?

It is apparent that the squirrel population is tied closely to the conifer seed crops, and ultimately to the health of the forest ecosystem on the Pinaleños. We initially identified the spruce-fir forest as a "boom-and-bust" habitat for the squirrel; the squirrels responded quickly in the presence of good cone crops and declined in the interval. The mixed-conifer (or transition habitat), on the other hand, appears to support a more stable population. The current state of the spruce-fir habitat, decimated by insects, has interrupted the boom and bust cycle of the spruce-fir and has virtually restricted the squirrel to the lower elevations of the mountain. While this may be cause for concern, or even pessimistic forecasts of doom, it also provides an opportunity to study some of the more pressing concerns of modern conservation biology. We should take advantage of the situation to learn how this fragmentation of the squirrel's habitat affects it. Can they breach the barrier of a regenerating forest? How will the surviving small subpopulations surrounding the spruce-fir fare? Will they go locally extinct, or can enough migration occur to sustain them?

I hope that the future still holds a place for the Mt. Graham red squirrel. A species with the tenacity of this one certainly deserves the right to survival. And I hope that as observers of this grand natural experiment, we have the wisdom to learn more about how our ecosystem works.

Literature Cited

Coronado National Forest. 1989. *Management Plan for the Mount Graham International Observatory.* Coronado National Forest, Tucson, Arizona.

Hamilton, W. J. 1939. Observations on the life history of the red squirrel in New York. *American Midland Naturalist* 22:732–745.

Layne, J. N. 1954. The biology of the red squirrel, *Tamiasciurus hudsonicus loquax* (Bangs) in central New York. *Ecological Monographs* 24:227–267.

Obbard, M. E. 1987. Red squirrel. Pages 265–281 in M. Novak, J. Baker, M. Obbard, and B. Mallock, eds. *Furbearer managements and conservation in North America.* Ontario Ministry of Natural Resources, Toronto, Ontario, Canada.

U.S. Fish and Wildlife Service. 1988. *Biological Opinion for the Mt. Graham red squirrel.*

———. 1993. *Mount Graham Red Squirrel* (Tamiasciurus hudsonicus grahamensis) *Recovery Plan.* Albuquerque, New Mexico.

CHAPTER 9

A Comparison of Two Sampling Techniques to Assess Population Trends in Endangered Mt. Graham Red Squirrels

John L. Koprowski and Tim K. Snow

The abundance of animals in an area is a fundamental parameter of a population of interest (Caughley 1977). Population viability models, harvest strategies, landscape management, and conservation status require accurate and precise assessments of abundance (Caughley 1994; Munzbergova and Ehrlen 2005). The conservation of a species of precarious conservation status is likely to be more effective with, if not dependent upon, precise and accurate estimates of population size.

A number of techniques can be applied to assess populations of mammals involving indices, estimates, and counts (Wilson et al. 1996; Schwarz and Seber 1999; Braun 2005). In tree squirrels, methods of assessing population trends or abundance include live trapping (Nixon et al. 1967), point-area and transect methods (Bouffard and Hein 1978; Don 1985; Hein 1997; Gurnell et al. 2001), feeding sign (Brown 1982; Bryce et al. 1997; Gurnell et al. 2001), hair tubes (Gurnell et al. 2001, 2004), and track plates and counts (Carey and Witt 1991; Mattson and Reinhart 1996; Drennan et al. 1998).

Mt. Graham red squirrels are found only in the highest-elevation forests of the Pinaleño Mountains of southeastern Arizona (Hoffmeister 1986). The subspecies has been protected as an endangered species since 1987, and numbers appear to have fluctuated below 600 animals since this time (Snow in this vol.). Red squirrels in the coniferous forests of North America larderhoard numerous cones in conspicuous piles of cone scales known as middens (Finley 1969; A. Smith and Mannan 1994; Steele 1998). Signs of feeding or caching activity at these midden sites occupied by a single squirrel appear to be useful indicators of abundance

(Mattson and Reinhart 1996). Two methods have been applied to census Mt. Graham red squirrels. Herein, we compare the census results of the mountain-wide interagency census (Snow in this vol.) and the complete direct counts of a smaller study area monitored by the University of Arizona (UA) (Young in this vol.).

Methods

A mountain-wide census of a random stratified sample of known middens occurred in late spring (May–June) and fall (late September–October) and was conducted with volunteers and personnel from a variety of state and federal natural resource agencies (Snow in this vol.). In addition, a quarterly complete direct count of middens was undertaken typically in March, June, September, and December on a smaller 311.3 ha area by the UA since 1989 (Young in this vol.). Middens were scored as active based upon signs of activity (Rasmussen et al. 1975; Mattson and Reinhart 1996) during a single visit to the midden in the interagency census (Snow in this vol.) or multiple visits by biologists and a consensus decision in the UA count (details in Young 1994; Young in this vol.).

We used the abundance estimates of each mountain-wide and local census technique during the spring and fall in our analyses. Seasonality of estimates was assessed using t-tests and ANOVA. Correlation and simple linear regression analyses were used to assess the potential relationships between the two techniques.

Results and Discussion

The two census techniques applied to the Mt. Graham red squirrel population yielded similar trends in population numbers. Mountain-wide census values did not differ between spring (337.46 ± 32.80) and fall (360.00 ± 27.95: $t = 0.59$, df = 24, $P = 0.59$). Similarly, local census values did not vary across seasons (spring: 87.85 ± 15.12, summer: 80.77 ± 11.17, fall: 80.36 ± 9.19, winter: 94.50 ± 13.74; $F_{5,30} = 0.30$, $P = 0.83$). Spring and fall estimates were highly correlated for the mountain-wide census ($r = 0.81$, df = 12, $P < 0.005$), while the correlation between these seasons was positive but less strong for the local census ($r = 0.33$, df = 11, $P < 0.20$). The local census was a very effective predictor of the mountain-wide census

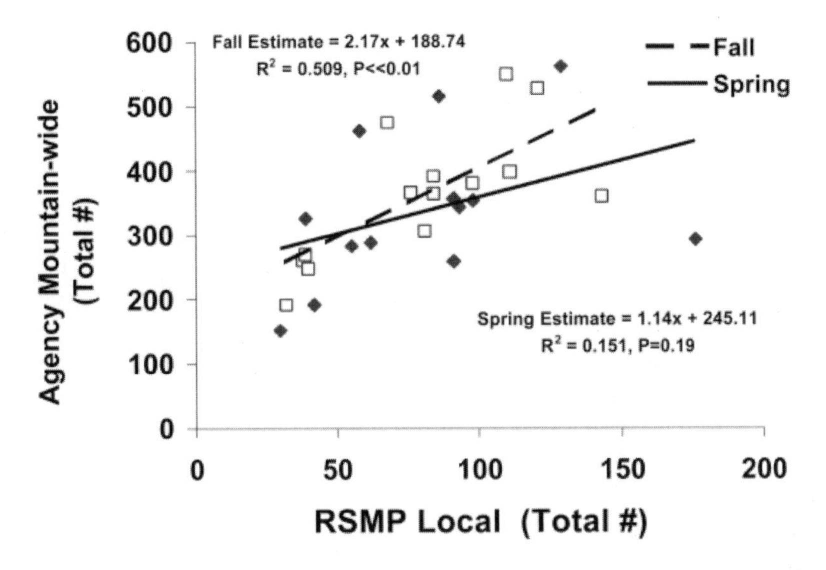

Figure 9.1. Relationship between mountain-wide interagency estimates and local counts of the University of Arizona. Solid diamonds represent spring abundances; open squares signify fall abundances. Simple regression lines are presented to elucidate trends.

(fig. 9.1) in fall ($F_{1,12}$ = 12.42, R^2 = 50.85 percent, P <0.001), but not in spring ($F_{1,12}$ = 1.96, R^2 = 15.14 percent, P = 0.189).

The general agreement between the mountain-wide census and the smaller sample used by the UA's Mt. Graham Red Squirrel Monitoring Program suggests that both are tracking similar population level changes throughout the mountain; however, local events such as ongoing insect outbreaks (Koprowski et al. 2005) or fire could change this agreement (Koprowski et al. 2006). Fall censuses demonstrate the greatest agreement, and this is no surprise. Territoriality is particularly strong in fall (M. Smith 1968; Steele 1998), when squirrels are caching green conifer cones from the current year's crop (Gurnell 1984; Hurly and Lourie 1997). As a result, fresh caches, digging, and squirrel presence greatly facilitate accurate assessment of midden occupancy (Mattson and Reinhart 1996). Strong agreement exists between the two methods of assessing population trends on Mt. Graham for past censuses, which suggests that the same trends in abundance are being tracked with each technique.

Acknowledgments

I appreciate the comments of P. Lurz and C. Salsbury on an earlier draft of this manuscript. The Arizona Game and Fish Department, U.S. Forest Service, U.S. Fish and Wildlife Service, the University of Arizona's Office of the Vice President for Research, and Steward Observatory administered or funded the census efforts on which this analysis is based.

Literature Cited

Bouffard, S. H., and D. Hein. 1978. Census methods for eastern gray squirrels. *Journal of Wildlife Management* 42:550–557.

Braun, C. E. 2005. *Techniques for wildlife investigations and management*, 6th ed. The Wildlife Society, Bethesda, Maryland.

Brown, D. E. 1982. The use of clippings to index tassel-eared squirrel population levels. *Journal of Wildlife Management* 46:520–525.

Bryce, J., J. S. Pritchard, N. K. Waran, and R. J. Young. 1997. Comparison of methods of obtaining population estimates for red squirrels in relation to damage due to bark stripping. *Mammal Review* 27:165–170.

Carey, A. B., and J. W. Witt. 1991. Track counts as indices to abundances of arboreal rodents. *Journal of Mammalogy* 72:192–194.

Caughley, G. 1977. *Analysis of vertebrate populations*. John Wiley and Sons, Chichester, England.

——. 1994. Directions in conservation biology. *Journal of Animal Ecology* 63:215–244.

Don, B. A. C. 1985. The use of drey counts to estimate grey squirrel populations. *Journal of Zoology* 206:282–286.

Drennan, J. E., P. Beier, and N. L. Dodd. 1998. Use of track stations to index abundance of sciurids. *Journal of Mammalogy* 79:352–359.

Finley, R. B., Jr. 1969. Cone caches and middens of *Tamiasciurus* in the Rocky Mountain region. Pages 233–273 in J. K. Jones Jr., ed. *Contributions in mammalogy*. University of Kansas Museum of Natural History Miscellaneous Publication 51.

Gurnell, J. 1984. Home range, territoriality, caching behaviour, and food supply of the red squirrel (*Tamiasciurus hudsonicus fremonti*) in a subalpine lodgepole pine forest. *Animal Behaviour* 32:1119–1131.

Gurnell, J., P. W. W. Lurz, and H. Pepper. 2001. Practical techniques for surveying and monitoring squirrels. *Forestry Commission Practice Note* 11:1–12. Forestry Commission, Edinburgh, Scotland.

Gurnell, J., P. W. W. Lurz, M. D. F. Shirley, S. Cartmel, P. J. Garson, L. Magris, and J. Steele. 2004. Monitoring red squirrels *Sciurus vulgaris* and grey squirrels *Sciurus carolinensis* in Britain. *Mammal Review* 34:51–74.

Hein, E. W. 1997. Demonstration of line transect methodologies to estimate urban gray squirrel density. *Environmental Management* 21:943–947.

Hoffmeister, D. F. 1986. *Mammals of Arizona*. University of Arizona, Tucson.

Hurly, T. A., and S. A. Lourie. 1997. Scatterhoarding and larderhoarding by red squirrels: Size, dispersion, and allocation of hoards. *Journal of Mammalogy* 78:529–537.

Koprowski, J. L., M. I. Alanen, and A. M. Lynch. 2005. Nowhere to run and nowhere to hide: Response of endemic Mt. Graham red squirrels to catastrophic forest damage. *Biological Conservation* 126:491–498.

Koprowski, J. L., K. M. Leonard, C. A. Zugmeyer, and J. L. Jolley. 2006. Direct effects of fire on endangered Mt. Graham red squirrels. *Southwestern Naturalist* 51:59–63.

Mattson, D. J., and D. P. Reinhart. 1996. Indicators of red squirrel (*Tamiasciurus hudsonicus*) abundance in the whitebark pine zone. *Great Basin Naturalist* 56:272–275.

Munzbergova, Z., and J. Ehrlen. 2005. How best to collect demographic data for population viability analysis models. *Journal of Applied Ecology* 42:1115–1120.

Nixon, C. M., W. R. Edwards, L. Eberhardt. 1967. Estimating squirrel abundance from live-trapping data. *Journal of Wildlife Management* 31:96–101.

Rasmussen, D. I., D. E. Brown, and D. Jones. 1975. Use of ponderosa pine by tassel-eared squirrels and a key to determine evidence of their use from that of red squirrels and porcupines. *Arizona Game and Fish Dept. Wildlife Digest* 10:1–12.

Schwarz, C. J., and G. A. F. Seber. 1999. Estimating animal abundance: Review III. *Statistical Science* 14:427–456.

Smith, A. A., and R. W. Mannan. 1994. Distinguishing characteristics of Mount Graham red squirrel midden sites. *Journal of Wildlife Management* 58:437–445.

Smith, M. C. 1968. Red squirrel responses to spruce cone failure. *Journal of Wildlife Management* 32:305–316.

Steele, M. A. 1998. *Tamiasciurus hudsonicus. Mammalian Species* 586:1–9.

Wilson, D. E., F. R. Cole, J. D. Nichols, R. Rudran, and M. S. Foster, eds. 1996. *Measuring and monitoring biological diversity: Standard methods for mammals.* Smithsonian Institution Press, Washington, D.C.

Young, P. J. 1994. Monitoring the Mount Graham red squirrel. Pages 227–246 in C. A. Istock and R. S. Hoffman, eds. *Storm over a mountain island: Conservation biology and the Mt. Graham affair.* University of Arizona Press, Tucson.

## CHAPTER 10

Consequences of Small Populations and Their Impacts on Mt. Graham Red Squirrels

JOHN L. KOPROWSKI AND ROBERT J. STEIDL

MT. GRAHAM RED SQUIRRELS are a single, apparently panmictic population of 200–300 individuals (Snow in this vol.), inhabiting high-elevation forests in the Pinaleño Mountains, Graham County, Arizona. Small, narrowly distributed populations, such as that of Mt. Graham red squirrels, are increasingly vulnerable to factors affecting extinction (Soulé 1987; Belovsky et al. 1994). Given unprecedented reductions in habitat and habitat quality due to factors such as wildfire, fire suppression, insect damage, development, and forest succession, the likelihood of reductions in population size and connectivity among habitat patches has increased concerns about persistence of this squirrel population. Consequently, we review issues affecting persistence of small populations and focus on the principal interrelated concerns of genetic and demographic effects with reference to the ecology of red squirrels.

Genetics of Red Squirrels

The most recent extensive surveys of genetic variation in *Tamiasciurus* did not include *Tamiasciurus hudsonicus grahamensis* (Arbogast et al. 2001; Wilson et al. 2005). A number of recent studies have incorporated molecular methods into assessments of genetics in other portions of the range of red squirrels (Arbogast et al. 2001; Doyon et al. 2003; Wilson et al. 2005), and a number of polymorphic loci and primers are available (Gunn et al. 2005). Although no fixed differences were found in allozymes among populations studied, evidence indicates three unique lineages, including one from Arizona and New Mexico (Arbogast et al. 2001). The only other relevant genetic information using protein electrophoresis indicated that *T. h. grahamensis* was monomorphic at all loci, with no heterozygotes detected; one allozyme was found almost exclusively in the Mt. Graham population, except for one individual in the White–Mogollon Mountains that also possessed the allele (Sullivan and Yates 1994).

Genetic Consequences of Small Population Size

The likelihood of genetic problems emerging in populations increases as population size decreases (Lande 1988, 1994, 1995; Mills and Smouse 1994; Frankham 1995a, 1995b, 1995c; Lynch et al. 1995). Most adverse genetic consequences stem from reductions in available genetic variation caused by reductions in population size. Populations typically have lower genetic variation after undergoing large reductions in population size, known as population bottlenecks (Maruyama and Kimura 1980; Hedrick 1995, 1996; Conner and White 1999). Smaller populations also have a higher probability of genetic drift, which is the result of changes in allele frequencies due to random chance (Hedrick 1983). Adverse effects of genetic drift are more likely for small populations where stochastic variation in rates of fecundity and survival can cause substantial changes in allele frequencies.

If population numbers remain low, probability of matings between individuals with a high degree of genetic similarity increases, which could reduce rates of survival and fecundity (Hedrick 1983; Lande 1988). Adverse consequences of matings between closely related individuals are known as inbreeding depression, which is a result of increased genetic homozygosity that results in the expression of deleterious recessive alleles or decreased heterozygosity where heterozygote advantage is present (Crow 1948). Three requirements are necessary for inbreeding depression to reduce population persistence: (1) inbreeding occurs, (2) inbreeding depression occurs, and (3) affected traits must reduce population viability (Allendorf and Ryman 2002). Inbreeding depression is likely to occur in all species (Hedrick and Kalinowski 2000), and some evidence of inbreeding depression has been reported for mammals (Ralls et al. 1986, 1988; Lacy 1997; Crnokrak and Roff 1999).

The ability of a population to express a phenotype in response to future environmental changes depends in part on the genetic diversity inherent in that population, as measured by the number of alleles (Slade

1992; Burger and Lynch 1995; Edwards and Potts 1996). Phenotypic variation in life history traits that has a heritable component such as timing of reproduction, age at first reproduction, offspring size, and litter size can increase population persistence (Conner and White 1999; Allendorf and Ryman 2002). Therefore, loss of genetic variation brings an associated lost potential for a population to respond to environmental changes through a shift in genotypic frequency, which has been called the loss of evolutionary potential (Allendorf and Ryman 2002).

Accumulation of deleterious mutations in small populations can result in a decline in population viability and likelihood of persistence (Lynch and Gabriel 1990; Gabriel and Burger 1994; Lande 1995; Lynch et al. 1999). Mitochondrial mutations historically were thought to be of little consequence because mtDNA is maternally inherited in the cytoplasm and thus not likely subject to natural selection (Allendorf and Ryman 2002); however, mutations in mtDNA may impact fertility of males because sperm are powered by mitochondria (Gemmell and Allendorf 2001).

Environmental Effects

As populations decrease in size, their vulnerability to environmental changes increases, including slow, long-term trends (deterministic) and brief, high-intensity (stochastic) events. These environmental changes can influence the demographics of a population directly, through changes in survival and recruitment rates, and indirectly, through changes in characteristics of the physical environment that affect the quantity and quality of habitat. Herein, we briefly explore only a few examples of environmental effects, as others will be explored in detail elsewhere in this volume.

Deterministic environmental effects are due to long-term ecological or physical processes including succession (Peterson 2000), global warming (Flannigan et al. 2000; Hughes 2000; Peterson 2000; Reale et al. 2003), and habitat alterations by humans that increase edge effects and fragmentation (Lovejoy et al. 1986; Wilcove et al. 1986; Irland 2000; Peterson 2000).

Stochastic environmental effects include fire (J. K. Brown and Smith 2000; Flannigan et al. 2000; J. Smith 2000), insect infestations (Ayres and Lombardero 2000; Logan et al. 2003), epizootics (Yuill 1987; Gerber and Hilborn 2001), and weather-induced changes to the physical environment that affect habitat (Flannigan et al. 2000; Hughes 2000; Irland 2000; Peterson 2000). A commonly suggested consequence of conservation of populations in geographic isolation is increased susceptibility to stochastic events that increase the probability of extinction (Lande 1998; Rushton et al. 2000; Alvarez 2001), although such cases rarely are documented (Mangel and Tier 1994; Akcakaya and Baur 1996; Crawford et al. 2001). Theoretical models suggest that stochastic events that remove a significant proportion of the population can be countered by the potential response of a species to adapt via natural selection under shorter time scales than are traditionally proposed; however, the risk of extinction remains high (Gomulkiewicz and Holt 1995; Boulding and Hay 2001). Catastrophic events in mammals are usually associated with disease, anthropogenic overharvest, or starvation due to harsh climatic conditions (Yuill 1987; Gerber and Hilborn 2001; Koprowski et al. 2005).

Demographic Consequences of Small Population Sizes

Demographics of small populations can be adversely affected by a number of factors, all of which increase risk of extinction (Shaffer 1981; Gilpin and Soulé 1986; Goodman 1987; Lande 1993). Demographic stochasticity is the random change in characteristics of populations, such as sex ratio and rates of fecundity and mortality, and is thought to have an important influence on risk of extinction (Goodman 1987; Gabriel and Burger 1992; Lande 1993). Population viability analyses and modeling efforts suggest that red squirrel populations, including those on Mt. Graham, may be sensitive to changes in adult survival (Buenau and Gerber 2004; Rushton et al. 2006).

Although demographic processes can vary randomly, especially within small populations, deterministic effects can alter population parameters. In addition to differential reproduction and survival based on age or dominance status (reviewed for squirrels in Gurnell 1987; Steele 1998), sex ratios can vary deterministically. Because sex in mammals is determined by chromosomal mechanisms (females are the homogametic sex, e.g., XX; males are the heterogametic sex, e.g., XY), deviations in sex ratios from an expected ratio of 50:50 are often considered to be stochastic. Recent examples of adaptive variation in sex ratios are known for a number of mammals, including rodents (Verme 1983; Clutton-Brock et al. 1986; Armitage 1987; B. Smith et al. 1996; Kohlmann 1999; Allaine et al. 2000).

Cooperative behaviors such as foraging, offspring rearing, predator detection and defense, and mating often increase with population size (Allee et al. 1949) and decline as population size decreases (Odum and Allee 1954; Philip 1957). Allee effects describe instances where the probability of finding a mate or other cooperative behavior is reduced because of low population sizes (Lande 1988). Presence of adult males is necessary for subadult female eastern gray squirrels to become reproductively active (Webley and Johnson 1983). Because red squirrels are only in estrus for <1 day/breeding season, and instances of females producing a second litter during a year are rare (Lair 1985; Steele 1998), Allee effects could become increasingly consequential for Mt. Graham red squirrels at low densities.

Red squirrels are territorial in coniferous forests of western North America (Koprowski 1998; Steele 1998); therefore, cooperative foraging and parental care are not likely important. However, red squirrels call in response to predators, to which other squirrels respond (Gurnell 1987; Steele 1998). Further, communal nesting has been documented rarely (Ackerman and Weigl 1970), including during winter on Mt. Graham (Munroe et al. in this vol.).

Dispersal patterns also can influence population persistence. For small populations that function as a component of a larger metapopulation, emigrants provide the source of individuals for demographic rescue effects (J. H. Brown and Kodric-Brown 1977; Lande 1993). Mammals typically have low levels of natal philopatry and high levels of natal dispersal (Greenwood 1980; Waser and Jones 1983; Johnson and Gaines 1990), which is true of most tree squirrels including red squirrels (Koprowski 1998; Larsen and Boutin 1998). Red squirrels are capable of dispersing distances of more than 2 km (Larsen and Boutin 1994; Steele 1998; Goheen et al. 2003), and juveniles have dispersed similar distances on Mt. Graham (Kreighbaum and Van Pelt 1996; J. Koprowski, personal observation).

Immigration of red squirrels to the Pinaleños from other montane environments is highly unlikely given the expanse of uninhabitable desert and grassland environments surrounding Mt. Graham. Consequently, maximizing the availability and maintaining the health of forests, as well as a high degree of connectivity among forested patches in and near Mt. Graham, appears an appropriate strategy for facilitating movements of dispersing individuals, and minimizing environmental and demographic effects to which small populations are vulnerable. Our lack of knowledge of the genetic composition and demography of Mt. Graham red squirrels makes assessments of the risk of extinction difficult; however, all attempts to model population persistence suggest that the challenges are great (Buenau and Gerber 2004; Rushton et al. 2006; Harding et al. in this vol.). Solutions and strategies must be developed that consider the short- and long-term consequences of actions on the genetic and demographic composition of the isolated Mt. Graham population in order to conserve this highly endangered subspecies.

Literature Cited

Ackerman, R., and P. D. Weigl. 1970. Dominance relations of red and gray squirrels. *Ecology* 51:332–338.

Akcakaya, H. R., and B. Baur. 1996. Effects of population subdivision and catastrophes on the persistence of a land snail metapopulation. *Oecologia* 105:475–483.

Allaine, D., F. Brondex, L. Graziani, J. Coulon, and I. Till-Bottraud. 2000. Male-biased sex ratio in litters of Alpine marmots supports the helper repayment hypothesis. *Behavioral Ecology* 11:507–514.

Allee, W. C., A. E. Emerson, O. Park, T. Park, and K. P. Schmidt. 1949. *Principles of animal ecology*. W. B. Saunders, Philadelphia.

Allendorf, F. W., and N. Ryman. 2002. The role of genetics in population viability analysis. Pages 50–85 in S. R. Beissinger and D. R. McCullogh, eds. *Population viability analysis: Assessing models for recovering endangered populations*. University of Chicago Press, Chicago.

Alvarez, L. H. R. 2001. Does increased stochasticity speed up extinction? *Journal of Mathematical Biology* 43:534–544.

Arbogast, B. S., R. A. Browne, and P. D. Weigl. 2001. Evolutionary genetics and Pleistocene biogeography of North American tree squirrels (*Tamiasciurus*). *Journal of Mammalogy* 82:302–319.

Armitage, K. B. 1987. Do female yellow bellied marmots adjust the sex ratios of their offspring? *American Naturalist* 129:501–519.

Ayres, M. P., and M. J. Lombardero. 2000. Assessing the consequences of global change for forest disturbance from herbivores and pathogens. *Science of the Total Environment* 262:263–286.

Belovsky, G. E., J. A. Bissonette, R. D. Dueser, T. C. Edwards, C. M. Luecke, M. E. Ritchie, J. B. Slade, and F. H. Wagner. 1994. Management of small populations:

Concepts affecting the recovery of endangered species. *Wildlife Society Bulletin* 22:307–316.

Boulding, E. G., and T. Hay. 2001. Genetic and demographic parameters determining population persistence after a discrete change in the environment. *Heredity* 86:313–324.

Brown, J. H., and A. Kodric-Brown. 1977. Turnover rates in insular biogeography: Effect of immigration on extinction. *Ecology* 58:445–449.

Brown, J. K., and J. K. Smith, eds. 2000. *Wildland fire in ecosystems: Effects of fire on flora*, vol. 2. General Technical Report RMRS-GTR-42. U.S. Dept. of Agriculture, Forest Service, Rocky Mountain Research Station, Ogden, Utah.

Buenau, K. E., and L. R. Gerber. 2004. Developing recovery and monitoring strategies for the endemic Mount Graham red squirrels (*Tamiasciurus hudsonicus grahamensis*) in Arizona. *Animal Conservation* 7:1–6.

Burger, R., and M. Lynch. 1995. Evolution and extinction in a changing environment: A quantitative genetic analysis. *Evolution* 49:151–163.

Clutton-Brock, T. H., S. D. Albon, and F. E. Guinness. 1986. Great expectations: Dominance, breeding success and offspring sex ratios in red deer. *Animal Behaviour* 34:460–471.

Conner, M. M., and G. C. White. 1999. Effects of individual heterogeneity in estimating the persistence of small populations. *Natural Resources Modeling* 12:109–127.

Crawford, R. J. M., J. H. M. David, L. J. Shannon, J. Kemper, N. T. W. Klages, J. P. Roux, L. G. Underhill, V. L. Ward, A. J. Williams, and A. C. Wolfaardt. 2001. African penguins as predators and prey—coping (or not) with change. *South African Journal of Marine Science* 23:435–447.

Crnokrak, P., and D. A. Roff. 1999. Inbreeding depression in the wild. *Heredity* 83:260–270.

Crow, J. F. 1948. A consequence of the dominance hypothesis of hybrid vigor. *Genetics* 33:101–102.

Doyon, C., V. L. Trudeau, B. M. Hibbert, L. A. Howes, and T. W. Moon. 2003. mRNA analysis in flattened fauna: Obtaining gene-sequence information from road-kill and game-hunting samples. *Canadian Journal of Zoology* 81:692–698.

Edwards, S. V., and W. K. Potts. 1996. Polymorphism of genes in the major histocompatibility complex (MHC): Implications for conservation genetics of vertebrates. Pages 3–21 in T. B. Smith and R. K. Wayne, eds. *Molecular genetic approaches in conservation*. Oxford University Press, New York.

Flannigan, M. D., B. J. Stocks, and B. M. Wotton. 2000. Climate change and forest fires. *Science of the Total Environment* 262:221–229.

Frankham, R. 1995a. Conservation genetics. *Annual Review of Genetics* 29:305–327.

——. 1995b. Effective population size—adult population size ratios in wildlife: A review. *Genetical Research* 66:95–107.

——. 1995c. Inbreeding and extinction: A threshold effect. *Conservation Biology* 9:792–799.

Gabriel, W., and R. Burger. 1992. Survival of small populations under demographic stochasticity. *Theoretical Population Biology* 41:44–71.

——. 1994. Extinction risk by mutational meltdown: Synergistic effects between population regulation and genetic drift. Pages 69–86 in V. Loeschcke, J. Tomiuk, and S. K. Jain, eds. *Conservation genetics*. Birkhauser Verlag, Basil, Switzerland.

Gemmell, N. J., and F. W. Allendorf. 2001. Mitochondrial mutations may decrease population viability. *Trends in Ecology and Evolution* 16:115–117.

Gerber, L. R., and R. Hilborn. 2001. Catastrophic events and recovery from low densities in populations of otariids: Implications for risk of extinction. *Mammal Review* 31:131–150.

Gilpin, M. E., and M. E. Soulé. 1986. Minimum viable populations: Processes of species extinction. Pages 19–34 in M. E. Soulé, ed. *Conservation biology: The science of scarcity and diversity*. Sinauer Associates, Sunderland, Massachusetts.

Goheen, J. R., R. K. Swihart, T. M. Gehring, and M. S. Miller. 2003. Forces structuring tree squirrel communities in landscapes fragmented by agriculture: Species differences in perceptions of forest connectivity and carrying capacity. *Oikos* 102:95–103.

Gomulkiewicz, R., and R. D. Holt. 1995. When does evolution by natural selection prevent extinction? *Evolution* 49:201–207.

Goodman, D. 1987. The demography of chance extinction. Pages 11–34 in M. E. Soulé, ed. *Viable populations for conservation*. Cambridge University Press, Cambridge.

Greenwood, P. J. 1980. Mating systems, philopatry, and dispersal in birds and mammals. *Animal Behaviour* 28:1140–1162.

Gunn, M. R., D. A. Dawson, A. Leviston, K. Hartnup, C. S. Davis, C. Strobeck, J. Slate, and D. W. Coltman. 2005. Isolation of 18 polymorphic microsatellite loci from the North American red squirrel, *Tamiasciurus hudsonicus* (Sciuridae, Rodentia), and their cross-utility in other species. *Molecular Ecology Notes* 5:650–653.

Gurnell, J. C. 1987. *The natural history of squirrels*. Facts on File, New York.

Hedrick, P. W. 1983. *Genetics of populations*. Science Books International, Boston.

——. 1995. Elephant seals and the estimation of a population bottleneck. *Journal of Heredity* 86:232–235.

——. 1996. Bottleneck(s) or metapopulations in cheetahs. *Conservation Biology* 10:897–899.

Hedrick, P. W., and S. T. Kalinowski. 2000. Inbreeding depression in conservation biology. *Annual Review of Ecology and Systematics* 31:139–162.

Hughes, L. 2000. Biological consequences of global warming: Is the signal already apparent? *Trends in Ecology and Evolution* 15:56–61.

Irland, L. C. 2000. Ice storms and forest impacts. *Science of the Total Environment* 262:231–242.

Johnson, M. L., and M. S. Gaines. 1990. Evolution of dispersal: Theoretical models and empirical tests using birds and mammals. *Annual Review of Ecology and Systematics* 21:449–480.

Kohlmann, S. G. 1999. Adaptive fetal sex allocation in elk: Evidence and implications. *Journal of Wildlife Management* 63:1109–1117.

Koprowski, J. L. 1998. Conflict between the sexes: A review of social and mating systems of the tree squirrels. Pages 33–41 in M. A. Steele, J. F. Merritt, and D. A. Zegers, eds. *Ecology and evolutionary biology of tree squirrels.* Virginia Museum of Natural History, Martinsville.

Koprowski, J. L., M. I. Alanen, and A. M. Lynch. 2005. Nowhere to run and nowhere to hide: Response of endemic Mt. Graham red squirrels to catastrophic forest damage. *Biological Conservation* 126:491–498.

Kreighbaum, M. E., and W. E. Van Pelt. 1996. *Mount Graham red squirrel juvenile dispersal telemetry study.* Nongame and Endangered Wildlife Program Technical Report 89. Arizona Game and Fish Department, Phoenix.

Lacy, R. C. 1997. Importance of genetic variation to the viability of mammalian populations. *Journal of Mammalogy* 78:320–335.

Lair, H. 1985. Length of gestation in the red squirrel (*Tamiasciurus hudsonicus*). *Journal of Mammalogy* 66:809–810.

Lande, R. 1988. Genetics and demography in biological conservation. *Science* 241:1455–1460.

——. 1993. Risks of population extinction from demographic and environmental stochasticity and random catastrophes. *American Naturalist* 142:911–927.

——. 1994. Risk of population extinction from fixation of new deleterious mutations. *Evolution* 48:1460–1469.

——. 1995. Mutation and conservation. *Conservation Biology* 9:782–791.

——. 1998. Anthropogenic, ecological, and genetic factors in extinction and conservation. *Researches on Population Ecology* 40:259–269.

Larsen, K. W., and S. Boutin. 1994. Movements, survival, and settlement of red squirrel (*Tamiasciurus hudsonicus*) offspring. *Ecology* 75:214–223.

——. 1998. Sex-unbiased philopatry in the North American red squirrel: (*Tamiasciurus hudsonicus*). Pages 21–32 in M. A. Steele, J. F. Merritt, and D. A. Zegers, eds. *Ecology and evolutionary biology of tree squirrels.* Special Publication no. 6. Virginia Museum of Natural History, Martinsville.

Logan, J. A., J. Regniere, and J. A. Powell. 2003. Assessing the impacts of global warming on forest pest dynamics. *Frontiers in Ecology and the Environment* 1:130–137.

Lovejoy, T. E., R. O. Bierregaard Jr., A. B. Rylands, J. R. Malcolm, C. E. Quintela, L. H. Harper, K. S. Brown Jr., A. H. Powell, G. V. N. Powell, H. O. R. Schubart, and M. Hays. 1986. Edge and other effects of isolation on Amazon forest fragments. Pages 257–285 in M. E. Soulé, ed. *Conservation biology: The science of scarcity and diversity.* Sinauer Associates, Sunderland, Massachusetts.

Lynch, M., J. Conery, and R. Burger. 1995. Mutation accumulation and the extinction of small populations. *American Naturalist* 146:489–518.

Lynch, M., and W. Gabriel. 1990. Mutation load and the survival of small populations. *Evolution* 44:1725–1737.

Lynch, M., M. Pfrender, K. Spitze, N. Lehman, D. Allen, J. Hicks, L. Latta, M. Ottene, F. Bogue, and J. Colbourne. 1999. The quantitative and molecular genetic architecture of subdivided species. *Evolution* 53:100–110.

Mangel, M., and C. Tier. 1994. Four facts every conservation biologist should know about persistence. *Ecology* 75:607–614.

Maruyama, T., and M. Kimura. 1980. Genetic variability and effective population size when local extinction and recolonization of subpopulations are frequent. *Proceedings of the National Academy of Sciences of the United States of America* 77:6710–6714.

Mills, L. S., and P. E. Smouse. 1994. Demographic consequences of inbreeding in remnant populations. *American Naturalist* 144:412–431.

Odum, H. T., and W. C. Allee. 1954. A note on the stable point of populations showing both intraspecific cooperation and disoperation. *Ecology* 35:95–97.

Peterson, C. J. 2000. Catastrophic wind damage to North American forests and the potential impact of climate change. *Science of the Total Environment* 262:287–311.

Philip, J. R. 1957. Sociality and sparse populations. *Ecology* 38:107–111.

Ralls, K., K. Brugger, J. Ballou, and A. Templeton. 1988. Estimates of lethal equivalents and the cost of inbreeding in mammals. *Conservation Biology* 2:185–193.

Ralls, K., P. H. Harvey, and A. M. Lyles. 1986. Inbreeding in natural populations of birds and mammals. Pages 35–56 in M. E. Soulé, ed. *Conservation biology: The science of scarcity and diversity.* Sinauer Associates, Sunderland, Massachusetts.

Reale, D., A. G. McAdam, S. Boutin, and D. Berteaux. 2003. Genetic and plastic responses of a northern mammal to climate change. *Proceedings of the Royal Society of London B, Series B* 270:591–596.

Rushton, S. P., P. W. W. Lurz, J. Gurnell, and R. Fuller. 2000. Modelling the spatial dynamics of parapoxvirus disease in red and grey squirrels: A possible cause of the decline in the red squirrel in the UK? *Journal of Applied Ecology* 37:997–1012.

Rushton, S. P., D. J. A. Wood, P. W. W. Lurz, and J. L. Koprowski. 2006. Modelling the population dynamics of the Mt. Graham red squirrel: Can we predict its future in a changing environment with multiple threats? *Biological Conservation* 131:121–131.

Shaffer, M. L. 1981. Minimum population sizes for species conservation. *Bioscience* 31:131–134.

Slade, R. W. 1992. Limited MHC polymorphism in the southern elephant seal: Implications for MHC evolution and marine mammal population biology. *Proceedings of the Royal Society of London Series B, Biological Science* 249:163–171.

Smith, B. L., R. L. Robbins, and S. H. Anderson. 1996. Adaptive sex ratios: Another example? *Journal of Mammalogy* 77:818–825.

Smith, J. K., ed. 2000. *Wildland fire in ecosystems: Effects of fire on fauna,* vol. 1. General Technical Report RMRS-GTR-42. U.S. Dept. of Agriculture, Forest Service, Rocky Mountain Research Station, Ogden, Utah.

Soulé, M. E., ed. 1987. *Viable populations for conservation.* Cambridge University Press, Cambridge.

Steele, M. A. 1998. *Tamiasciurus hudsonicus. Mammalian Species* 586:1–9.

Sullivan, R. M., and T. L. Yates. 1994. Population genetics and conservation biology of relict populations of red squirrels. Pages 193–208 in C. A. Istock and R. S. Hoffmann, eds. *Storm over a mountain island: Conservation biology and the Mt. Graham affair.* University of Arizona Press, Tucson.

Verme, L. J. 1983. Sex ratio variation in *Odocoileus*: A critical review. *Journal of Wildlife Management* 47:573–582.

Waser, P. M., and W. T. Jones. 1983. Natal philopatry among solitary mammals. *Quarterly Review of Biology* 58:355–390.

Webley, G. E., and E. Johnson. 1983. Reproductive physiology of the grey squirrel (*Sciurus carolinensis*). *Mammal Review* 13:149–154.

Wilcove, D. S., C. H. McLellan, and A. P. Dobson. 1986. Habitat fragmentation in the temperate zone. Pages 237–256 in M. E. Soulé, ed. *Conservation biology: The science of scarcity and diversity.* Sinauer Associates, Sunderland, Massachusetts.

Wilson, G. M., R. A. Den Bussche, K. McBee, L. A. Johnson, and C. A. Jones. 2005. Intraspecific phylogeography of red squirrels (*Tamiasciurus hudsonicus*) in the central rocky mountain region of North America. *Genetica* 125:141–154.

Yuill, T. M. 1987. Diseases as components of mammalian ecosystems: Mayhem and subtlety. *Canadian Journal of Zoology* 65:1061–1066.

Part IV

Mt. Graham Red Squirrel Habitat

> "Nature" is what we see—
> The Hill, the Afternoon—
> Squirrel—
> —Emily Dickinson

MANY AUTHORS AND NATURALISTS have referred to the Mt. Graham red squirrel as the "spruce squirrel" in apparent reference to a predilection for spruce cones and spruce forests. More recently, biologists realized that the species could be found at lower elevations, and many middens in the initial surveys conducted for the species were found at relatively low elevations. The delineation of distribution and range of the species in the Pinaleño Mountains has not been an easy task, in part due to rugged and varied terrain; however, it is a most important one.

In Part IV, we address the state of Mt. Graham and the Pinaleños as habitat for the endangered Mt. Graham red squirrel. Scott Anderson and Susan Smith assess the historical changes in the high-elevation forests of Mt. Graham and echo the sentiments of Swetnam et al. in Part I that recount considerable fluctuation in the Pinaleño forest ecosystems through time. Jim Hatten details the status and potential availability of habitat for Mt. Graham red squirrels throughout the Pinaleños using spatial imagery; he notes the significant decline in habitat in recent years due to disturbances such as fire and insect damage. Our final collection of chapters examines the location of red squirrel middens relative to random locations in the forests of Mt. Graham. Sadie Bertelsen and

John Koprowski demonstrate that new middens are initiated in more open areas than traditional middens are. Marit Alanen and others found little evidence for differential selection and use of middens by males and females, suggesting that, at the broad habitat scale, such sex differences may not be important. Collectively, the chapters in Part IV provide the reader with an excellent sense of Mt. Graham red squirrel habitat and its past, present, and future.

CHAPTER 11

Vegetation Changes within the Subalpine and Mixed-Conifer Forest on Mt. Graham, Arizona

Proxy Evidence for Mt. Graham Red Squirrel Habitat

R. SCOTT ANDERSON AND SUSAN J. SMITH

THOUGH IMPORTANT AS HABITAT for the endangered Mt. Graham red squirrel, and increasingly important as a recreational resource, little is known about the antiquity of the spruce-fir forests on the Pinaleño Mountains (but see Swetnam et al. in this vol.). In this chapter we address further aspects of the vegetation history of the range, first investigated by Anderson and Shafer (1991). Our goals are threefold. First, we seek to clarify the Late Holocene vegetation history of the mixed-conifer and spruce-fir forests of the range. Second, we explore the history and importance of fire within the spruce-fir forest over the course of the post-glacial period. Third, we speculate on the antiquity of the habitat of the endangered Mt. Graham red squirrel, one of two subspecies of red squirrel found in Arizona (Hall 1981). Currently, the Pinaleño Mountains are the only habitat of the Mt. Graham red squirrel (Hoffmeister 1986).

Because the Mt. Graham red squirrel has an incomplete fossil history, a realistic approach to determining the antiquity of the squirrel's habitat is to examine proxy evidence of its occurrence. Using this paradigm, we have obtained sediment cores from cienegas (small marshes) on Mt. Graham and have analyzed pollen and plant remains from the sediments to determine long-term forest history, and charcoal particles to reconstruct forest fires.

Considerable change in vegetation distribution within the southwestern United States has occurred in the last 15,000 years (Betancourt et al. 1990). During the last glacial period, elevational limits of most high-elevation conifers were depressed as much as 1,000 m (Cole 1982; Anderson et al 2000). In southeastern Arizona, areas that now support desert grassland, desert scrub, or Sonoran Desert vegetation previously supported juniper (*Juniperus* sp.) woodland (Van Devender 1990a, 1990b). The transition to modern communities occurred during the Early Holocene in most parts of the region, with less substantial changes occurring over the last ca. 8,500 years. Our records from the Pinaleños do not include the late glacial period, but do include most of the Holocene.

Study Sites

Sediments from four sites in the Pinaleño Mountains—Emerald Springs Cienega, High Peak Cienega, Soldier Creek Meadow, and Hospital Flat Meadow—were analyzed. The sites are located within a ca. 6 km radius of each other, approximately 20 km southwest of Safford, Arizona (fig. 11.1; table 11.1).

Emerald Springs Cienega is a wet meadow located in a small, freeze-thaw– generated nivation hollow. The vegetation surrounding the cienega is dominated by Engelmann spruce, with minor amounts of corkbark fir. Understory growth is sparse and consists of currant (*Ribes wolfii*), orange gooseberry (*Ribes pinetorum*), cranesbill (*Geranium richardsonii*), and blueberry (*Vaccinium myrtillus* (nomenclature follows Johnson 1988). Grasses (Poaceae) and sedges (Cyperaceae) cover the cienega proper.

High Peak Cienega is also a wet meadow located in a nivation hollow. Upland vegetation around this site is dominated by Engelmann spruce, though corkbark fir is more abundant here than at Emerald Springs Cienega. Bearberry honeysuckle (*Lonicera involucrata*), currant, strawberry (*Fragaria* sp.), raspberry (*Rubus* sp.), wintergreen (*Pyrola picta*), cow parsnip (*Heracleum lanatum*), and members of the sunflower family (Asteraceae) grow in open areas of the understory. Corn lily (*Veratrum californicum*), violet (*Viola* sp.), cranesbill, grasses, and sedges grow on the marsh surface.

Soldier Creek Meadow is a small, open meadow within the mixed-conifer forest of the range. Trees are dominated by variable amounts of Engelmann spruce, white fir, and Douglas-fir, as well as southwestern white and ponderosa pines.

Hospital Flat Meadow is the lowest-elevation site in this study and is presently being incised by an intermittent stream. The coring site was

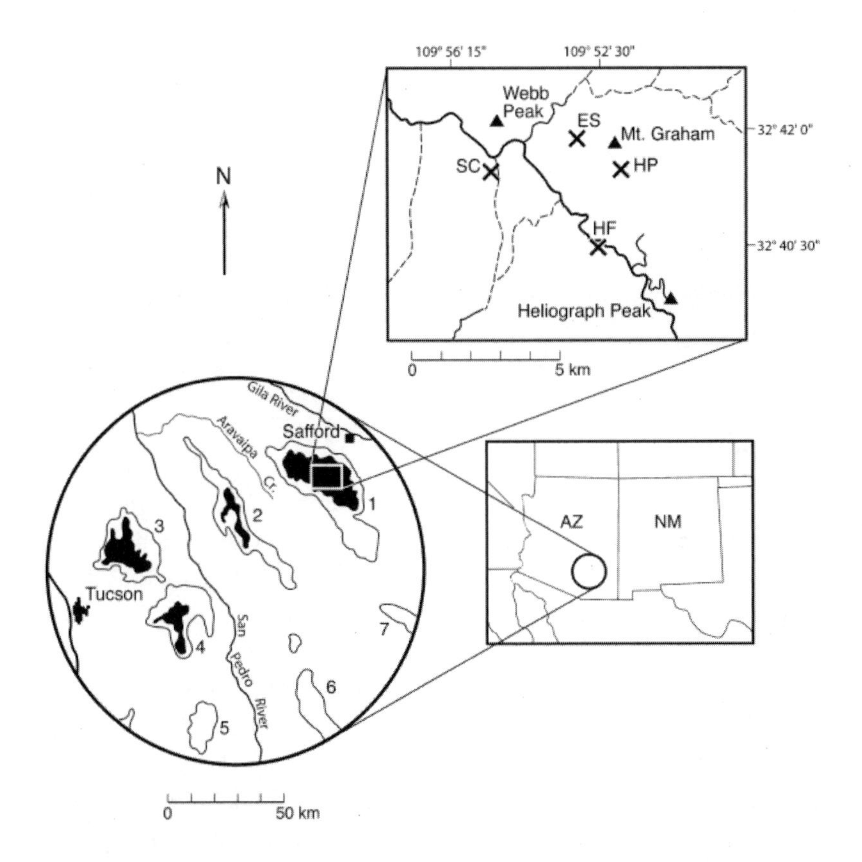

Figure 11.1. Location of sites on Mt. Graham reported in the text. ES = Emerald Springs Cienega; HF = Hospital Flat Meadow; HP = High Peak Cienega; SC = Soldier Creek Meadow. Blackened area on regional map corresponds to mixed conifer and spruce–fir forests (after McLaughlin 1995). Sky Islands of the region are 1 = Pinaleño Mtns.; 2 = Galiuro Mtns.; 3 = Santa Catalina Mtns.; 4 = Rincon Mtns.; 5 = Whetstone Mtns.; 6 = Dragoon Mtns.; and 7 = Dos Cabezas Mtns.

located on the west side of the meadow, near the upland forest, which is composed of ponderosa pine, with minor amounts of Douglas-fir, Engelmann spruce, southwestern white pine, white fir, and quaking aspen (*Populus tremuloides*). Herbs include yarrow (*Achillea lanulosa*), thistle (*Cirsium* sp.), goldenrod (*Solidago* sp.), and grasses. Blue wild rye (*Elymus glaucus*) grows on the moist meadow surface.

Table 11.1. Location of the four Mt. Graham study sites

Site Name	Elevation m	Latitude	Longitude	USGS Quad
Emerald Springs Cienega	3,143	32° 42' N	109° 53' 30" W	Webb Peak 7.5'
High Peak Cienega	3,118	32° 41' 30"N	109° 52' 00" W	Mt. Graham 7.5'
Soldier Creek Meadow	2,860	32° 41' 30"N	109° 55' 00" W	Webb Peak 7.5'
Hospital Flat Meadow	2,750	32° 40' 30"N	109° 52' 32" W	Webb Peak 7.5'

Methods

At least one sediment core was collected from each site. Cores were taken from Emerald Springs Cienega in October 1986 and 1987; cores were taken from the High Peak, Soldier Creek, and Hospital Flat sites in November 1990. Cores were collected with either a Livingstone piston corer (Wright 1980) or a modified Dachnowsky corer.

Sediments of each of the records were described in the laboratory. Depending upon sediment type, pollen was concentrated from raw sediment by using modifications of standard techniques (Fægri and Iversen 1989). Sediments from Emerald Springs core #1 included treatments with dilute potassium hydroxide (KOH), hydrochloric acid (HCl), hydrofluoric acid (HF), and acetolysis solution, with suspension in silicone oil. *Lycopodium* spore tracers were added for calculation of pollen concentration. KOH treatment was eliminated and acetolysis was limited for core #2 because of poor pollen recovery. Samples from the other three sites were processed with additional steps, including pollen separation using a heavy liquid (zinc bromide). The pollen assemblages were counted to ca. 300 terrestrial grains at 400X. Grains were identified using the reference collection at the Laboratory of Paleoecology (LOP), Northern Arizona University.

Plant macrofossils were isolated by water sieving larger portions of core sediments through soil sieves after suspension in water. Macrofossils were identified under a stereoscope at 10 to 70X, with reference to plant materials deposited at the LOP.

Sedimentary charcoal was identified from High Water core #1 only. The technique consisted of embedding each continuous 5 cm long core section with epoxy resin, mounting the resulting section on a microscope slide, and grinding the preparation to a thin section (Smith and Anderson 1995). The thin-section slide was scanned for charcoal at an interval of 1 mm, providing a near-continuous record of sedimentary charcoal for large sections of the core. However, due to difficulties in preparing sandy sediment, several gaps occur in the record.

Chronology of the resulting record was provided by bulk radiocarbon dating, performed by Beta Analytic, Inc. (Miami, FL).

Results

Chronology and Sedimentology

Seven radiocarbon dates were obtained for four cores (table 11.2). Only the record from Emerald Springs Cienega extends back to the Early Holocene, with a bottom date exceeding 9,200 years ago (Anderson and Shafer 1991).

Emerald Springs Cienega. Three radiocarbon dates were obtained from the two Emerald Springs Cienega cores. In addition, we assign the age of AD 1685 to a large charcoal peak at 21.2 cm depth in the core, corresponding to the most recently documented fire in the vicinity of Emerald Springs, as determined by stand origination data (Grissino-Mayer et al. 1995). The 134 cm long core #1 consists mostly of homogeneous organic silt, with varying amounts of sands, to a depth of 128 cm. Sediments below 128 cm are tan, oxidized, sandy silts.

High Peak Cienega. One radiocarbon date was obtained from the single sediment core. The top 3 cm of this 42 cm core consisted of dark brown peat. Dark-brown silty clays, with little visible organics, extended from ca. 3 to 40 cm. The bottom 2 cm consisted of light brown sands.

Soldier Creek Meadow. Two radiocarbon dates were obtained from this core. We also found a charcoal layer at a depth of 18–19 cm, which may correspond to the last major fire in the vicinity of Soldier Camp Meadow at ca. AD 1871 (Grissino-Mayer et al. 1995). The 95 cm long record consists largely of a homogeneous dark-brown peaty colluvium

Table 11.2. Radiocarbon dates for the Mt. Graham sediment cores

Site Name	Core #	Lab #	Depth (cm)	^{14}C Age (yr BP)	Median Calibrated Years (2 σ)
Emerald Springs Cienega	1	Beta-54267	54.0 – 60.0	3,630 ± 110	3,954
		Beta-18365	121.0 – 129.0	8,250 ± 160	9,211
Emerald Springs Cienega	2	Beta-32263	107.0 – 112.0	6,010 ± 150	6,869
High Peak Cienega	1	Beta-40691	33.0 – 40.0	3,750 ± 90	4,120
Soldier Creek Meadow	1	Beta-57985	51.5 – 57.5	890 ± 70	815
		Beta-40692	84.0 – 92.0	1,370 ± 70	1,289
Hospital Flat Meadow	1	Beta-44180	22.0 – 32.0	650 ± 70	613

to 53 cm depth. Sandier sections occur from depths of ca. 7–21 cm and 30–35 cm. Below 53 cm the color becomes darker, and the clay content increases.

Hospital Flat Meadow. One radiocarbon date occurs near the bottom of this 61 cm core. Although Hospital Flat Meadow does not lie within the study area of Grissino-Mayer et al. (1995), we assign the age of AD 1871 to a charcoal layer that occurred at a depth of 8.5 cm depth. The core top is silty peat with large organic fragments from its top to a depth of 30 cm. From ca. 30 to ca. 49 cm, the core is a gray, sandy unit. Finer sands and colluvium occurs from 49 to 61 cm.

Pollen and Plant Macrofossils

Emerald Springs Cienega. Anderson and Shafer (1991) reported on this record, though additional data are added here. The Emerald Springs Cienega record is the only one from the Pinaleños that extends back to the Early Holocene, with a bottom age exceeding 9,200 years old. Pollen was analyzed from 13 samples from core #1, with a minimum of 29 pollen and spore types identified. The concentration and preservation of pollen was excellent in the top 10 cm of the core, but declined rapidly with depth. This was true for all major pollen types (spruce, fir, pine, oak, aster family, etc.) except pollen of the grasses, which remained abundant throughout. Pollen recovery from core #2 was similar to core #1.

The dominant, identifiable plant remains found were Engelmann spruce needles (Anderson and Shafer 1991). Needle fragments of corkbark or subalpine fir and achenes of sedge were also found. All fragments had been burned, which probably enhanced their preservation.

High Peak Cienega. This record extends back to ca. 4,200 years ago. Twenty pollen samples were analyzed from the single core, but only the top (surface) sample contained enough well-preserved pollen to yield a 300-grain count. That sample contained 73 percent spruce, 10 percent pine, 5 percent fir, 4 percent ragweed (*Ambrosia* sp.), and 2 percent each of grass and goosefoot family (Chenopodiaceae) pollen, with smaller amounts of other types. Wetland pollen included 8 percent lily family (Liliaceae) and 5 percent sedge. Such percentages of pollen are characteristic of the spruce-fir forest in the Southwest (R. S. Anderson et al., Northern Arizona University, unpublished data).

The macrofossil data complements the pollen data. Spruce remains occur in four of the five macrofossil samples analyzed, while fir needle fragments were found in only one of the samples. Moreover, the concentration of spruce remains increased toward the top of the core, indicating either (1) an actual increase in the local abundance of spruce, or (2) an increase in preservation toward the surface.

Soldier Creek Meadow. The record from Soldier Creek extends back to over 1,300 calendar years ago and, along with that from Hospital Flat, provides the most complete record of vegetational change for the Late Holocene in that mountain range. Eighteen pollen samples were analyzed from this core, with good pollen preservation in 17 samples.

Changes in the pollen assemblage during this period reflect changes in both the upland forest community as well as in the meadow environment itself. Periods of little change in the pollen assemblages, perhaps corresponding to periods of relative stability, occur from ca. 750 to 1,300 years ago and from ca. 100 to 375 years ago. Periods of rapid change in the record occur ca. 375 to 750 yr BP, and within the last 100 years.

Prior to ca. 750 years ago, pine, grass, and goosefoot (Cheno-Am) pollen types dominate the pollen assemblage (fig. 11.2). Other conifers, including spruce and fir, are a minor component of the assemblage, although Douglas-fir pollen increases by ca. 1,000 years ago. Pollen characteristic of the piñon-juniper woodland at lower elevations (piñon pine, *Pinus edulis*; Cupressaceae, mostly junipers; oaks [*Quercus*]; ragweed; and sagebrush [*Artemisia* sp.]) is present in small but consistent quantities. The dominant riparian pollen type is alder (probably *Alnus incana*), while sedge pollen is also abundant.

The period 750 to 375 years ago shows variable pollen assemblages, perhaps indicating a period of adjustment in the local vegetation. Pine remains the dominant conifer; spruce, juniper/cypress, and oak pollen increase; fir pollen reaches its maximum then declines slightly; and Douglas-fir pollen percentages remain steady. However, the pollen of dwarf mistletoe (*Arceuthobium vaginatum*) begins a steady increase. Since this parasitic plant produces little pollen and is poorly dispersed, its increase probably signifies its abundance within host pine trees surrounding the meadow. Alternatively, this may represent a period of tree encroachment upon the meadow, nearer the coring site. A decline in percentages of grass pollen, with an increase in thistle, provides support for this hypothesis. Declines in alder, meadow rue (*Thalictrum*), and lily family pollen may also indicate a drying of the meadow.

Sediments deposited ca. 375 to 100 years ago show a steady increase in spruce and grass, and a decline in Douglas-fir. Towards the end of this period, riparian and wet meadow pollen indicators, including alder, walnut (*Juglans*), dock (*Rumex*), and meadow rue, increase, while sedge pollen remains prominent. These changes suggest that water tables in the meadow were higher than previously, perhaps corresponding to increased effective precipitation.

Changes in the pollen assemblages in the top portion of the core are undoubtedly associated with local modification of the forests on Mt. Graham and vicinity. Most prominent among these changes is a further decline in pine, associated with local logging operations, and its replacement in the assemblage by spruce pollen. Oak pollen also reaches its highest percentage in the record at this time.

Hospital Flat Meadow. The record from Hospital Flat covers the last 1,200 years. Although the core was 60 cm long, pollen was not recovered below 50 cm depth. Eleven pollen samples were analyzed, and due to

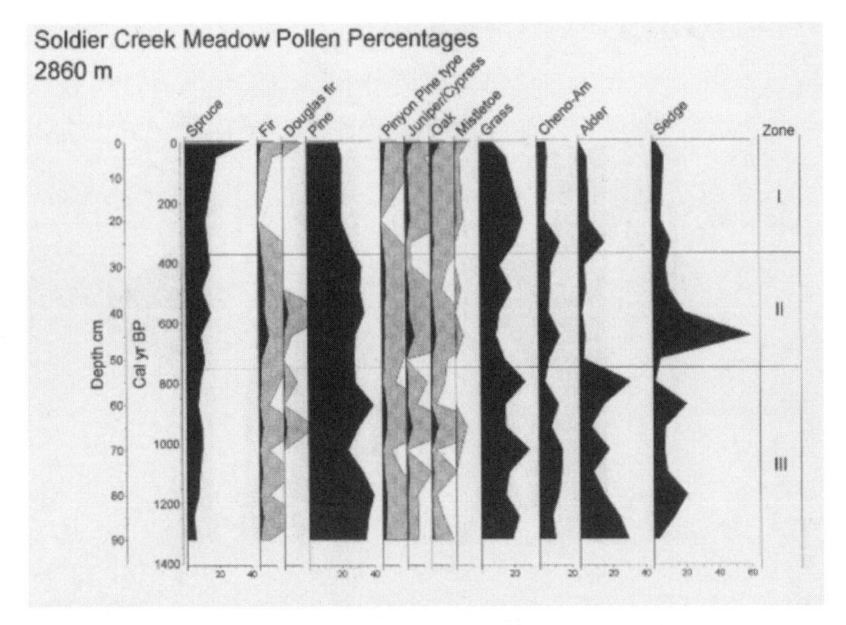

Figure 11.2. Summary pollen percentage diagram for Soldier Creek Meadow.

poorer preservation than at the Soldier Creek site, the average pollen sum was less (ca. 225 grains) than for the higher site.

Changes in the pollen assemblages suggest a somewhat simpler story here than for Soldier Creek. Pollen assemblages older than ca. 725 years are dominated by pine, which varies from 40 percent to over 60 percent, contrasting with samples from Soldier Creek that never exceed 40 percent. Small amounts of other conifer pollen were also found (fig. 11.3), with the most abundant of these being the piñon type. Dominant shrub and herb species include members of the sunflower, grasses, and goosefoot families, as well as alder.

A transition in the pollen assemblage at ca. 725 years ago is shown by declines in herbs (e.g., sunflower family, grasses), alder, and the carrot family (Apiaceae). Pollen from sedge increases to over 80 percent of the pollen sum. Pine pollen also increases, reaching a maximum of 75 percent between ca. 500 and 350 years ago. Dwarf mistletoe is common at this time and, as at Soldier Creek, suggests either a heavy infestation of the pines surrounding the meadow or tree encroachment on the meadow

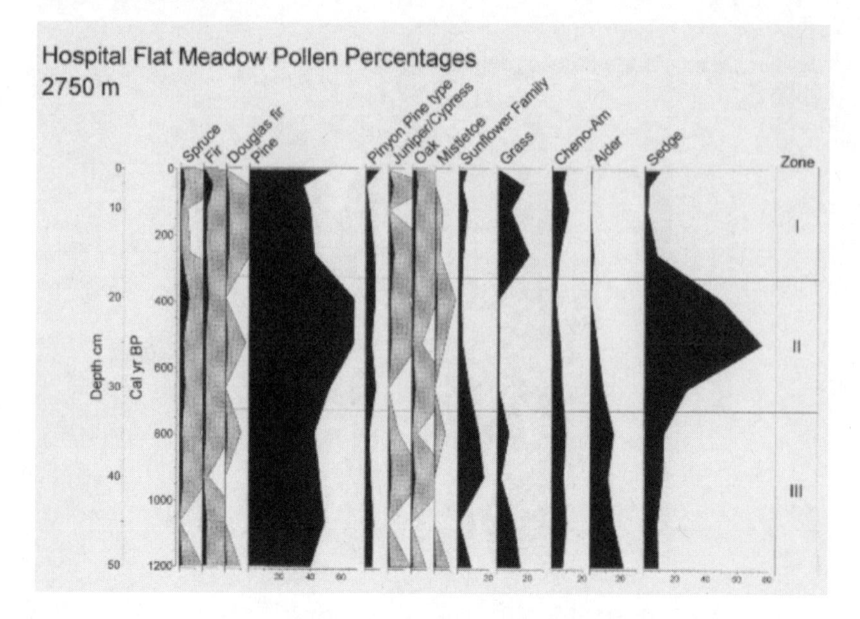

Figure 11.3. Summary pollen percentage diagram for Hospital Flat Meadow.

itself. By ca. 300 years ago, grasses had replaced sedges as the dominant herbs of the meadow. The dominant grass at the location today is wild blue rye, which prefers moist soils (Kearney and Peebles 1951).

Pine dominates at this site, and other conifers such as spruce, Douglas-fir, and fir are not as important in the fossil pollen record as they are at Soldier Creek (figs. 11.2 and 11.3). The macrofossil record complements the pollen evidence. The dominant plant macrofossil is ponderosa pine, and macrofossils in general are most abundant during the time of maximum pine-pollen percentages. This also suggests tree encroachment upon the meadow, and if so, zone II-time should have been the driest period of the sequence.

Sedimentary Charcoal

The occurrence of sedimentary charcoal in cores is a proxy record of forest fire. Our technique (Smith and Anderson 1995) has allowed us to identify the amount of charcoal in each successive centimeter of sediment of a core. Charcoal data are available only for the Emerald Springs site (fig. 11.4).

Many factors are responsible for the abundance of charcoal in sediment cores (Whitlock and Larsen 2001; Whitlock and Anderson 2003). Previous studies on charcoal sedimentation suggested that the smallest charcoal sizes are probably produced by sub-continental to global sources (Clark 1988). This "background" charcoal can be found at nearly all levels, and probably does not represent local fires (those burning at or very near the site). Large increases in charcoal above background levels within the core probably represent nearby fires, or even fires that burned across the meadow surface (Whitlock and Anderson 2003).

In the Emerald Springs record, distinct periods of high charcoal abundance alternate with periods of low charcoal abundance. We believe, however, that our chronology is inadequate for calculation of fire-event frequency (e.g., Millspaugh and Whitlock 1995; Long et al. 1998; Brunelle and Anderson 2003). Nevertheless, charcoal deposited in three core segments far exceeds background values in this record—91.2 to 91.3 cm, 73.2 to 76.2 cm, and 19.4 to 21.6 cm. Based on interpolation between radiocarbon dates, the first and second periods correspond to ca. 5,825 to 5,850 and 4,850 to 5,000 median calibrated years BP. Given the origination age of the modern spruce forest there, we believe the uppermost charcoal peak originated from the AD 1685 fire that was widespread across the range (Grissino-Mayer et al. 1995). Though four core sections were too sandy to permit our sediment epoxy technique (the largest being from ca. 27.0 to 55.0 cm depth), sedimentary charcoal was identified from nearly all other levels. These data suggest that major fires, such as the AD 1685 conflagration, have probably been rather rare near Emerald Springs during the Holocene, but fire has been a regular factor in the disturbance of local forests over the period of record.

Discussion and Conclusions

Perhaps less is known regarding the history of the spruce-fir forest type than any other forest association in the Southwest. Much of our knowledge of vegetation history comes from packrat (*Neotoma* sp.) middens, preserving a record of change in lowland vegetation. During much of the last glacial period, until ca. 11,000 to 10,000 yr BP, spruce grew at eleva-

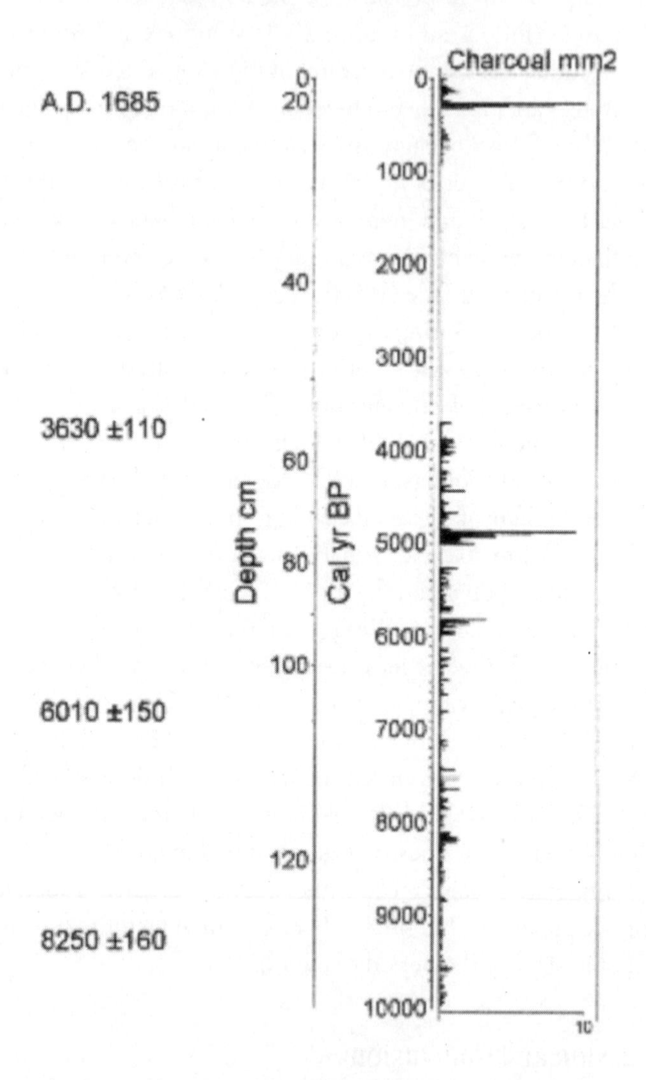

Figure 11.4. Charcoal area (mm²) for individual levels in a core from Emerald Springs Cienega.

tions lower than today. Engelmann spruce grew as low as 2,222 m near Flagstaff, Arizona (Anderson 1993), and at 2,000 m in the Guadalupe Mountains, Texas (Van Devender 1990a). Spruce was also found at 2,285 m within the White Mountains of Arizona (Jacobs 1983), and at 2,780 m within the Chuska Mountains of New Mexico (Wright et al. 1973). With the increasing temperatures of the Early Holocene, spruce retreated to its present elevational range.

Even fewer data exist for the region closer to Mt. Graham. Pluvial Lake Cochise sediments from Willcox Playa show higher percentages of spruce pollen during the late glacial period than today (Martin 1963). Packrat middens have not been identified from the area, so we are unsure of the species of spruce present at that time. However, results from Emerald Springs Cienega led Anderson and Shafer (1991) to conclude that Engelmann spruce has grown on Mt. Graham for at least the last 9,200 calendar years.

The records from the Mt. Graham sites do not provide us with a continuous vegetational history for the Holocene. However, it is clear that Engelmann spruce dominated at the highest elevations for much of that time. This contrasts with the record from Soldier Creek and Hospital Flat, where mixed-conifer dominated, at least for the last 1,200 to 1,400 years. Pines, both ponderosa and southwestern white, were dominant at Hospital Flat, while spruce was more important at the slightly higher Soldier Creek site. Pollen data also suggest that vegetational changes were largely synchronous at these two sites, even though the individual characteristics of the mixed-conifer forest were different, due to elevational differences. Vegetational changes occur at ca. 750 and 375 years ago, and within the last century. Moist conditions are inferred for the first period, followed by drier climates, succeeded by wetter conditions again. A general cooling trend is noted with a long-term increase in spruce in the Soldier Creek record and an increase in fir and Douglas-fir in the Hospital Flat record.

Based on these data, it is likely that the habitat of the Mt. Graham red squirrel existed essentially in variants of its present form for much of the Holocene (also see Swetnam et al. in this vol.). This gives us confidence in concluding that habitat requirements were present for the squirrel during that period.

Acknowledgments

We thank the following for their help with this project: administrative support, L. Fitzpatrick (U.S. Fish and Wildlife Service); fieldwork, P. Anderson, P. Chouinard, N. Giggy, P. Koehler, D. Shafer, and R Sweeney; and logistical assistance, T. Newman (U.S. Forest Service) and C. Duncan (U.S. Forest Service). Funding was provided by the University of Arizona through the Mt. Graham Red Squirrel Study Committee, which is comprised of representatives from the Arizona Game and Fish Department, the University of Arizona, the U.S. Forest Service, and the U.S. Fish and Wildlife Service (USFS contract 40–8197–0–0499). Laboratory of Paleoecology Contribution No. 80.

Literature Cited

Anderson, R. S. 1993. A 35,000 year vegetation and climate history from Potato Lake, Mogollon Rim, Arizona. *Quaternary Research* 40:351–359.

Anderson, R. S., J. L. Betancourt, J. I. Mead, R. H. Hevly, and D. P. Adam. 2000. Middle- and late-Wisconsin paleobotanic and paleoclimatic records from the southern Colorado Plateau, USA. *Palaeogeography, Palaeoclimatology, Palaeoecology* 155:31–57.

Anderson, R. S., and D. S. Shafer. 1991. Holocene biogeography of spruce-fir forests in southeastern Arizona—implications for the endangered Mt. Graham red squirrel. *Madroño* 38:287–295.

Betancourt, J. L., T. R. Van Devender, and P. S. Martin, eds. 1990. *Packrat middens: The last 40,000 years of biotic change.* University of Arizona Press, Tucson.

Brunelle, A., and R. S. Anderson. 2003. Sedimentary charcoal as an indicator of late-Holocene drought in the Sierra Nevada, California, and its relevance to the future. *The Holocene* 13:21–28.

Clark, J. S. 1988. Particle motion and the theory of charcoal analysis: Source area, transport, deposition, and sampling. *Quaternary Research* 30:67–80.

Cole, K. L. 1982. Late Quaternary zonation of vegetation in the Grand Canyon. *Science* 217:1142–1145.

Fægri, K. and J. Iversen. 1989. *Textbook of pollen analysis*, 4th ed. John Wiley and Sons, Chichester, England.

Grissino-Mayer, H. D., C. H. Baisan, and T. W. Swetnam. 1995. Fire history in the Pinaleño Mountains of southeastern Arizona: Effects of human-related disturbances. Pages 399–407 in L. F. DeBano, P. F. Ffolliott, A. Ortega-Rubio, G. J. Gottfried, R. H. Hamre, and C. B. Edminster, technical coordinators. *Biodiversity and management of the Madrean Archipelago: The sky islands of southwestern United States and northwestern New Mexico: September 19-23, 1994, Tucson, Arizona.* General Technical Report RM-GTR-264. U.S. Dept. of Agriculture, Forest Service, Rocky Mountain Forest and Range Experiment Station, Fort Collins, Colorado.

Hall, E. R. 1981. *The mammals of North America*. John Wiley and Sons, New York.
Hoffmeister, D. F. 1986. *Mammals of Arizona*. University of Arizona Press, Tucson.
Jacobs, B. F. 1983. Past vegetation and climate of the Mogollon Rim area, Arizona. PhD diss., University of Arizona, Tucson.
Johnson, W. T. 1988. Flora of the Pinaleño Mountains, Graham County, Arizona. *Desert Plants* 8:147–191.
Kearney, T. H., and Peebles, R. H. 1951. *Arizona flora*. University of Arizona Press, Tucson.
Long, C. J., C. Whitlock, P. J. Bartlein, and S. H. Millspaugh. 1998. A 9000–year fire history from the Oregon Coast Range, based on a high-resolution charcoal study. *Canadian Journal of Forest Research* 28:774–787.
Martin, P. S. 1963. *The last 10,000 years*. University of Arizona Press, Tucson.
McLaughlin, S. P. 1995. An overview of the flora of the sky islands, southeastern Arizona: Diversity, affinities and insularity. Pages 160–170 in L. F. DeBano, P. F. Ffolliott, A. Ortega-Rubio, G. J. Gottfried, R. H. Hamre, and C. B. Edminster, technical coordinators. *Biodiversity and management of the Madrean Archipelago: The sky islands of southwestern United States and northwestern Mexico: September 19-23, 1994, Tucson, Arizona*. General Technical Report RM-GRT-264. U.S. Dept. of Agriculture, Forest Service, Rocky Mountain Forest and Range Experiment Station, Fort Collins, Colorado.
Millspaugh, S. H., and C. Whitlock. 1995. A 750–year fire history based on lake sediment records in central Yellowstone National Park, USA. *The Holocene* 5:283–292.
Smith, S. J., and R. S. Anderson. 1995. A method for impregnating soft sediment cores for thin-section microscopy. *Journal of Sedimentary Research* A65:576–577.
Van Devender, T. R. 1990a. Late Quaternary vegetation and climate in the Chihuahuan Desert, United States and Mexico. Pages 104–133 in J. L. Betancourt, T. R. Van Devender, and P. S. Martin, eds. *Packrat middens: The last 40,000 years of biotic change*. University of Arizona Press, Tucson.
———. 1990b. Late Quaternary vegetation and climate in the Sonoran Desert, United States and Mexico. Pages 134–165 in J. L. Betancourt, T. R. Van Devender, and P. S. Martin, eds. *Packrat middens: The last 40,000 years of biotic change*. University of Arizona Press, Tucson.
Whitlock, C., and R. S. Anderson. 2003. Fire history reconstructions based on sediment records from lakes and wetlands. Pages 3–31 in T. T. Veblen, W. L. Baker, G. Montenegro, and T. W. Swetnam, eds. *Fire and climatic change in temperate ecosystems of the western Americas*. Ecological Studies, vol. 160. Springer, New York.
Whitlock, C., and C. Larsen. 2001. Charcoal as a fire proxy. Pages 75–98 in J. P. Smol, H. J. B. Birks, and W. M. Last, eds. *Tracking environmental change using lake sediments*. Vol. 3, *Terrestrial, algal, and siliceous indicators*. Kluwer, Dordrecht, Netherlands.
Wright, H. E., Jr. 1980. Cores of soft lake sediments. *Boreas* 9:107–114.
Wright, H. E., Jr., A. M. Bent, B. S. Hansen, and L. J. Maher Jr. 1973. Present and past vegetation of the Chuska Mountains, northwestern New Mexico. *Geological Society of America Bulletin* 84:1150–1180.

## CHAPTER 12

Mapping and Monitoring Mt. Graham Red Squirrel Habitat with GIS and Thematic Mapper Imagery

JAMES R. HATTEN

TO ESTIMATE THE MT. GRAHAM RED SQUIRREL (MGRS) population, personnel visit a proportion of middens each year to determine their occupancy (Snow in this vol.). The method results in very tight confidence intervals (high precision), but the accuracy of the population estimate is dependent upon knowing where all the middens are located. I hypothesized that there might be areas outside the survey boundary that contained Mt. Graham red squirrel middens, but the ruggedness of the Pinaleño Mountains made mountain-wide surveys difficult. Therefore, I started exploring development of a spatially explicit (geographic information system [GIS]–based) habitat model in 1998 that could identify MGRS habitat remotely with satellite imagery and a GIS. A GIS-based model would also allow us to assess changes in MGRS habitat between two time periods because Landsat passes over the same location every 16 days, imaging the earth in 185 km swaths (Aronoff 1989). Specifically, the objectives of this analysis were to (1) develop a pattern recognition model for MGRS habitat, (2) map potential (predicted/modeled) MGRS habitat, (3) identify changes in potential MGRS habitat between 1993 and 2003, and (4) evaluate the current location of the MGRS survey boundary.

Study Area

The Pinaleño Mountains trend northwest to southeast for approximately 35 km, are less than 20 km wide, and have an extensive high-elevation plateau, reaching a height of 3,268 m. There are 4,097 ha of terrain above the 2,744 m contour, and 538 ha above the 3,049 m contour, supporting one of the southernmost spruce-fir forests in North America. The northwest/southeast orientation of the Pinaleño Mountains creates aspects that generally face northeast or southwest, creating temperature differences that influence the distribution of plants and animals. The topography inside the MGRS survey boundary is gentle compared to the steep slopes that fall sharply away from the upper plateau.

Methods

Modeling Overview

I used the Arizona Game and Fish Department (AGFD) MGRS midden database for model development because it contained >1,000 midden locations collected over a 15-year period. To increase spatial and model accuracies, I used only sites that were spatially referenced with a global positioning system (GPS), with horizontal accuracy varying between 5 and 20 m. The AGFD midden database also contained habitat information collected at hundreds of sites (e.g., mixed-conifer, ecotone, spruce-fir). I extracted all other variables used in the characterization of MGRS habitat from TM (Thematic Mapper) imagery and a digital elevation model (DEM).

I completed seven steps to develop and test a spatially explicit habitat suitability model. First, I created a boundary for the GIS analysis by masking (excluding) vegetation communities (e.g., oak grasslands, upper Sonoran Desert) that do not provide the structural characteristics necessary to support MGRS habitat. Second, I classified a Landsat TM 1993 image of the unmasked portion of the Pinaleño Mountains. Third, I identified spectrally suitable and unsuitable areas by overlaying 50 percent of the MGRS midden locations and the classified TM image with a GIS. Fourth, I conducted an accuracy assessment inside the MGRS survey boundary to determine the accuracy of the classified image by overlaying the remaining midden locations not used in model development. Fifth, I identified potential MGRS habitat outside the survey boundary by overlaying the MGRS survey boundary on predicted MGRS habitat. Sixth, MGRS biologists inspected randomly selected sites to determine the suitability of predicted MGRS habitat outside the survey boundary. Seventh, I conducted change detection by comparing classified TM imagery from 1993, 1997, and 2003.

Topographic Analysis

I created a continuous elevation surface of the Pinaleño Mountains with U.S. Geological Survey DEMs (30 m resolution). I extracted slope, elevation, and aspect data from the DEMs with GRID functions (ESRI 1992) and aggregated them into discrete classes. Elevation data were aggregated into 13, 76 m (~250 ft) classes; slope data were aggregated into 4 classes (0–10°, 11–20°, 21–30°, and >30°); and aspect data were aggregated into 4 classes (north [315–45°], east [46–135°], south [136–225°], and west [225–314°]). I generated midden frequencies with a GIS by overlaying topographic classes and midden data.

Spectral Analysis

I characterized the spectral properties of the Pinaleño Mountains with a TM image acquired on June 19, 1993. The TM image had a pixel resolution (ground sample distance) of 28.5 m, contained seven spectral bands, and had ~30 m horizontal accuracy. I used bands 1–5, which correspond to the blue, green, red, near-infrared (IR), and mid-IR portions of the electromagnetic spectrum (Avery and Berlin 1992). ERDAS IMAGINE software (ERDAS Inc., Atlanta, GA) was used for all image-processing tasks, and ArcInfo software (ESRI Inc., Redlands, CA) was used for all GIS analysis. I created an additional composite band for image classification by calculating the Normalized Difference Vegetation Index (NDVI = [band 4 – band 3 / band 4 + band 3]) because of its proven utility in discriminating differences in vegetation density and biomass (Jensen 1983) and minimizing shadow effects.

I characterized the spectral properties of the forest canopy with pattern recognition, a clustering algorithm that finds patterns in spectral data that can be extracted through classification (Schrader and Pouncey 1997). Before conducting the classification, I created a vegetation-density grid of the Pinaleño Mountains by calculating NDVI from TM imagery. The NDVI ranged from –0.50 to 0.75, with smaller NDVI values having less density and/or biomass than higher values (Avery and Berlin 1992). I identified an approximate NDVI cutpoint between coniferous forests and desert scrub vegetation at 0.45, as determined from the vegetation-density grid, and masked areas <0.45 (NDVI) because they did not contain MGRS middens. I then used the spectral information contained in TM bands 1–5, plus NDVI, to divide the unmasked conifer forests into 12 spectral classes with the ISODATA (iterative, self-organizing data analysis) algorithm (Tou and Gonzalez 1974).

To characterize spectral and structural properties of the forest canopy occupied by MGRS, I randomly selected 50 percent of the midden locations (511). The remaining middens (507) were used later in accuracy assessment. Identifying spectral areas occupied by MGRS was an iterative process. Midden data were overlaid on the 12 spectral classes, and the resultant frequencies were examined. Spectral classes that contained relatively few middens (<5 percent) were collapsed into a single unsuitable class, while the remaining classes were considered spectrally suitable. There were no criteria to guide this process of spectral class clumping, just careful examination of the midden data overlaid on the imagery. Next, I used the habitat data that had been collected at the midden locations to characterize the forest composition within each spectral class. If the spectral classes contained two or more habitat types, they were candidates for collapsing (merging) with other mixed classes in an effort to create the most effective, simple model. I used ancillary topographic data (slope, aspect, and elevation) in conjunction with pattern recognition to provide insight into the distribution of MGRS middens.

Accuracy Assessment

I conducted accuracy assessment in 3 areas: (1) predicted MGRS habitat within the survey boundary, (2) predicted (modeled) unsuitable areas within the survey boundary, and (3) predicted MGRS habitat outside the survey boundary. While I had a great deal of validation data within the survey boundary (middens), virtually no data existed outside the survey boundary. To assess model accuracy, I overlaid 507 randomly selected middens—the middens not used in model development—on the final classified image. Accuracy was calculated by errors of omission or the number of middens that fell outside of predicted MGRS habitat (Story and Congalton 1986). Using middens to determine model accuracy was convenient because it eliminated the difficulty of identifying MGRS habitat, for which I had no proven set of criteria. I examined the accuracy of the unsuitable class, as determined from the model, by visiting 18 randomly selected sites located within meadows, burn areas, rock outcrops, or pine/oak/aspen thickets. Field notes were collected to aid in interpreting classification error.

To determine whether predicted MGRS habitat outside the survey boundary was actually suitable, MGRS biologists visited 17 randomly selected locations. Data were collected at each random point on elevation, slope, aspect, seral stage, site potential, evidence of squirrel presence, and tree species. Seral stage was denoted as pole (young trees), mature, old growth, and mixed ages. A qualitative habitat suitability ranking was developed by MGRS personnel and assigned to each site visited: (1) low = little to no potential, (2) moderate = habitat did not look too unsuitable and probably could support squirrels, and (3) high = very good habitat or squirrels seen or heard. Qualitative habitat criteria included presence or absence of standing snags or downed logs, canopy density, site lushness, presence of large cone-bearing trees, slope, and aspect.

Lively discussions ensued at each random site between MGRS biologists as to whether it constituted MGRS habitat or not. In the strictest sense, any area that was mixed-conifer, ecotone, or spruce-fir qualified as suitable because those habitat types contained MGRS middens within the survey boundary. However, a lack of fine-scaled habitat criteria made ranking habitat potential outside the survey boundary difficult and somewhat qualitative. Sites where MGRS were seen or heard were considered suitable, but sites without evidence of MGRS required a judgment call on habitat suitability.

Change Detection

To examine changes in MGRS habitat in the Pinaleño Mountains, I acquired TM imagery from 1993, 1997, and 2003. Because I was interested in changes to MGRS habitat, I used the 1993 TM image as a baseline image to which I compared the other two images. While all three images were acquired during the summer months (June–August), small differences in the solar illumination angle could interfere with the change detection, so I adjusted the tonal qualities of the 1997 and 2003 images to match the 1993 image with histogram matching (Schrader and Pouncey 1997). Once the images were tonally matched, I calculated NDVI for each time period (12 classes) and used pixel (digital number) subtraction to identify areas where NDVI had decreased. To minimize change-detection error, I only considered pixels that had increased or decreased by at least two NDVI classes. A field reconnaissance in 2000 into burn areas from the 1996 Clark Peak fire found that NDVI was an effective metric to monitor changes in the forest canopy.

Survey Boundary Analysis

To ascertain whether the MGRS survey boundary was accurately placed, I overlaid midden and topographic data (slope, aspect, elevation) to characterize surrounding terrain features occupied by MGRS. Aspect and elevation were examined together because they both regulate vegetation and microclimatic variables such as temperature, relative humidity, and tree species. I also examined whether the northwest/southeast trend in the Pinaleño Mountains had an impact on midden distribution within similar elevation and aspect classes, but on opposite sides of the Pinaleño backbone.

Results

Topographic Analysis

Middens were sparse between 2,286 m and 2,743 m (fig. 12.1A), with no middens observed on southward slopes below 2,743 m, and none observed on westward slopes below 2,670 m. The lowest elevation at which a midden was observed was 2,353 m, in the Turkey Flat survey unit, found on a north aspect with a gentle slope (<10°). Midden concentrations increased above 2,743 m and extended all the way to the top of Mt. Graham (3,268 m). Concerning aspect (fig. 12.1B), the north slopes of the mountain contained the greatest number of middens, east and west slopes contained similar numbers of middens, and southerly aspects contained the fewest. Regarding slope (fig. 12.1C), classes 1–2 (0–20°) had the most middens, with a rapid drop in midden frequency in slope classes 3 and 4. Very few middens were observed over 30° and none over 40°. To refine the GIS analysis, I created a GIS layer that divided the Pinaleños into two zones: (1) within the survey boundary (zone 1), and (2) outside the survey boundary above 2,353 m elevation (zone 2).

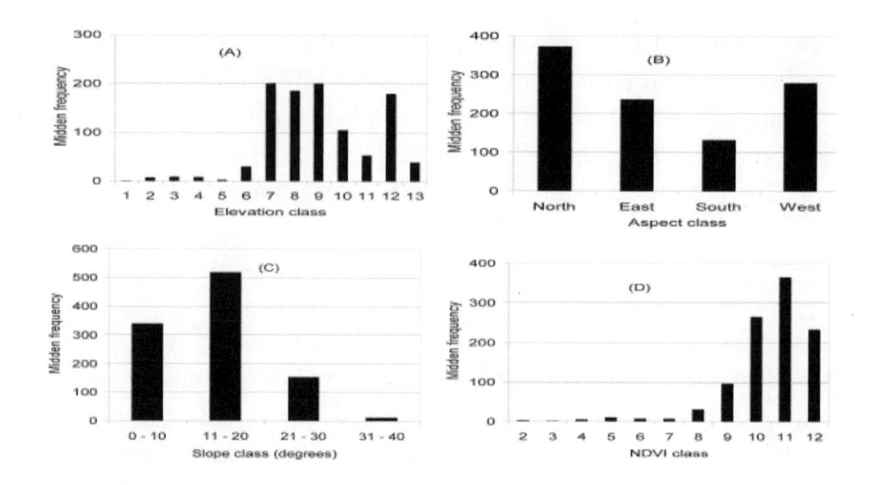

Figure 12.1. Relationships among the topographic variables (elevation, aspect, and slope), NDVI, and middens. Elevations (Figure 12.1A) are divided into 13 consecutive 76 m (250 ft) elevation bands; class 1 starts at 2,286 m, and class 13 starts at 3,200 m.

Spectral Analysis

Relatively few middens (8.3 percent) were observed below NDVI class 8 (NDVI raw value <0.45), and several field trips helped qualitatively define the contents of the NDVI classes. Classes 1–3 contained rock outcrops, semi-desert grassland, meadows, water features, and bare soils. In contrast, classes 4–7 contained oak woodlands, piñon-juniper, and pine-oak communities, while classes 8–12 corresponded with mixed-conifer, ecotone, and spruce-fir habitats commonly associated with MGRS. Because NDVI classes <8 were unsuitable for MGRS, I masked them out of all subsequent image analyses (fig. 12.1D).

Spectral classes 1–6 contained 91 percent of the middens (457), and classes 7–12 contained 9 percent (50). A close inspection of the middens, when overlaid on the unclassified TM image, revealed that the majority of middens in classes 7–12 were found along the edges of features, such as roads and meadows (spectrally confused areas). Thus, spectral classes 7–12 were aggregated into a single class and labeled unsuitable for MGRS.

Class 1 was the only spectral class comprised of a relatively pure habitat type (88.5 percent spruce-fir). The other five spectral classes had substantial mixing of two or more habitat types. Due to habitat mixing, I simplified the model by aggregating spectral classes 1–6 into a single spectral class that I referred to as potential or predicted MGRS habitat.

Accuracy Assessment

Classification accuracy of the suitable class (as determined from errors of omission) was 93 percent inside the survey boundary, and 83 percent for the unsuitable class. Close examination of the unclassified TM imagery revealed that most classification errors appeared related to spatial (positional) error because their locations were less than 1 pixel (28.5 m) from a feature edge, such as a meadow or forest boundary. Such areas often had two or more features represented (covered) by a single 28.5 × 28.5 m (0.08 ha) pixel and were spectrally mixed.

All but 1 of the 17 random points MGRS biologists visited outside the survey boundary, but within the suitable class, contained Douglas-fir. The other site was located at an elevation of 3,085 m, had a north aspect, and contained subalpine fir and Engelmann spruce. Thus, the GIS-based model worked inasmuch as it identified potential MGRS habitat, but seral stage and aspect reduced the habitat potential at some sites. Four of the 17 sites (2 northward and 2 eastward sites) contained good or moderate habitat, with MGRS seen or heard at 2 sites. The ground slopes of the moderate-to-good habitats were between 20–45° and contained old-growth or mixed-age forest. Two of the sites were located at a relatively low elevation (2,515 m), and adult red squirrels were observed clipping cones from Engelmann spruce. While no middens were observed outside the survey boundary, MGRS feeding and foraging suggested that middens were probably in the immediate vicinity.

To determine the suitability of predicted MGRS habitat outside the survey boundary, MGRS staff inspected eight sites outside the survey boundary with southward or westward aspects. All eight sites had low-quality MGRS habitat and no MGRS were seen or heard. There was Douglas-fir or white fir at every site, but the overall quality of the habitat appeared low. Generally, the sites tended to be quite open, steep, hot, and had few quality snags or large downed logs. Also, the south and west slopes appeared to be less lush compared to the randomly selected sites on the north and east

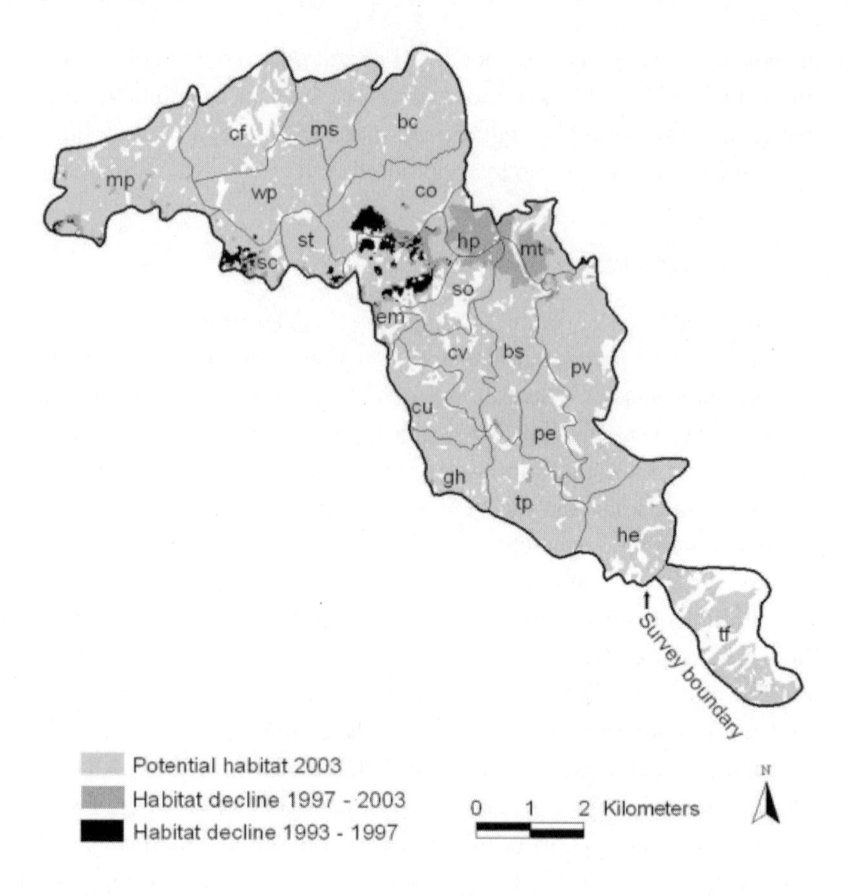

Figure 12.2. Changes in potential MGRS habitat inside the survey boundary are displayed between 1993, 1997, and 2003. Potential MGRS habitat was determined with pattern recognition and GIS for these three time periods. Potential refers to spectral properties of the forest canopy and not to microhabitat features that the squirrels might select for.

slopes at comparable elevations outside the survey boundary. The GIS-based model appeared to delineate the coniferous vegetation well, with oak thickets and other unsuitable vegetation being excluded. All of the sites had components necessary to be classified as mixed-conifer, but some also had isolated pine and oak scattered throughout. Thus, many of the southward- and westward-facing sites were transitional vegetation communities that made a clear-cut classification difficult.

Change Detection Analysis: 1993 to 2003

Potential MGRS habitat, as determined from the pattern recognition model, declined 8.0 percent between 1993 and 2003 (fig. 12.2). In 1993, there were 3,769 ha of potential MGRS habitat, which decreased by 3.2 percent in 1997, and by another 4.8 percent by 2003. The Clark Peak fire of 1996 was clearly responsible for the decline in potential MGRS habitat between 1993 and 1997. In contrast, the decline of potential MGRS habitat between 1997 and 2003 was largely due to insect damage in the spruce-fir forest.

Survey Boundary

Middens were detected at different elevations according to aspect, and their frequency of occurrence was inversely related to the lower-elevation bounds at which they were first detected. Midden occurrence, by elevation, differed within the four aspect classes depending on which side of the Pinaleño backbone (NW/SE axis) they were located. Middens that were on the NE side of the backbone were found much lower on northward- or eastward-facing slopes when compared with the SW side of the backbone. This pattern was true for middens found on westward-facing slopes as well, but the difference was not as pronounced as the northward/eastward slopes. In contrast, the southern slopes, regardless of their orientation to the backbone, contained middens at similar elevations.

Based upon the topographic analysis, I temporarily adjusted the survey boundary on the NE side of the Pinaleño backbone to the 2,353 m contour and identified all potential MGRS habitat within Zone 2 (fig. 12.3). This resulted in 1,596 ha of potential MGRS habitat that has had few or no surveys, areas that might contain middens.

Discussion

Topographic Analysis

Elevation, temperature, and aspect are important factors in the formation of biotic communities in the Southwest (Merriam and Steineger 1890; Brown 1994), and this is especially apparent in the Pinaleño Mountains.

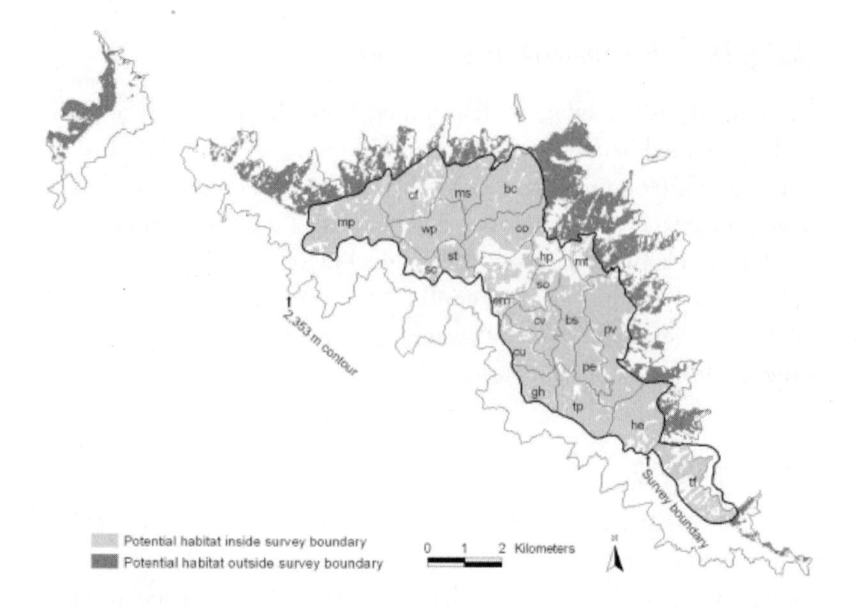

Figure 12.3. Potential MGRS habitat, determined from a pattern recognition model, within the survey boundary and within zone 2—the area between the 2,353 m contour (the lowest elevation that MGRS were observed) and the current survey boundary.

The fact that MGRS middens occurred at different elevations, according to aspect, is consistent with the life zone (biome) concept. Slopes that face eastward or northward are cooler and moister than southward- or westward-facing slopes and result in mixed-conifer forests occurring at lower elevations. The influence of slope on MGRS placement is less clear because many of the steeper northern and eastward slopes (>30°) on Mt. Graham have not been searched. These steeper slopes present difficulties and dangers to MGRS surveyors and will not likely be searched systematically or repeatedly.

Spectral Analysis and Accuracy Assessment

The MGRS pattern recognition model is a simple tool for identifying MGRS habitat based upon spectral and structural characteristics of the forest canopy and should be useful as long as TM imagery is available.

The close agreement between my MGRS habitat estimate and an earlier effort (USFS 1988) provided an independent check of the pattern recognition model.

It would benefit MGRS managers if additional GIS-based models were developed to monitor specific habitat types (e.g., mixed-conifer, ecotone, spruce-fir) or to rank the quality of MGRS habitat. The pattern recognition model presented in this chapter cannot be used to monitor changes in the three habitat types effectively because the spectral classes were collapsed into a single class. Of the six spectral classes found suitable, only class 1 contained a relatively pure habitat type (88.5 percent spruce-fir). It is probable that the higher-resolution satellite imagery now commonly available might improve model performance by reducing spectral confusion that resulted when two or more features (e.g., trees, boundaries) occurred within a single image cell. Higher-resolution imagery would also reduce omission errors that resulted when MGRS middens were <30 m from the edge of the forest, resulting in spectral confusion.

Another weakness of the MGRS pattern recognition model is that it has no statistical equation that can rank the probability or quality of MGRS habitat. A logistic regression approach to modeling (Hosmer and Lemeshow 1989) similar to that taken by Pereira and Itami (1991) might provide a more robust model for assessing MGRS habitat. I did not attempt logistic regression modeling because accurate records have not been kept for all areas that have been searched for middens. However, for logistic regression modeling, a subset of MGRS middens could be extracted from the AGFD midden database from areas where complete surveys have occurred. This subset would not include the most rugged areas of the Pinaleño Mountains, but it would include many of the survey units within the current survey boundary.

Change Detection

Potential MGRS habitat declined significantly between 1993 and 2003, raising concerns about the stability of their habitat. The decrease between 1993 and 1997 was attributable almost exclusively to the Clark Peak fire of 1996, while the decline between 1997 and 2003 was related to insect damage in the spruce-fir habitat zone. The full impact of the spruce-fir insect infestation was not evident in 2003 when the latest change-detection work was done. Since then, additional declines in MGRS habitat have occurred from insects, and a large fire occurred in 2004. It would benefit MGRS managers if the quantity of MGRS habitat was assessed annually or biannually. Remote sensing coupled with a GIS appears to be a promising tool that can assist managers in monitoring the amount and health of MGRS habitat over time.

Survey Boundary

There is evidence to support lowering the survey boundary on the NE ramparts of the Pinaleños to 2,353 m (the lowest elevation at which a midden was observed). Middens were found noticeably lower on the NE side of the Pinaleños' backbone (NE/SW axis), particularly on northward- and eastward-facing slopes. In contrast, there was little evidence to support substantially altering the survey boundary on the SW side of the Pinaleños' backbone, where MGRS habitat appeared marginal outside the survey boundary. Zone 2 was significant in our analysis because it corresponded to an elevation zone (>2,353 m) where middens have been located near Turkey Flat, but much of the current survey boundary skirts above 2,353 m. Thus, most of the potential habitat within Zone 2 has never been surveyed and remains undocumented. It stands to reason that if the survey boundary were lowered, more middens would be discovered on similar-facing slopes. Potential MGRS habitat identified by the pattern recognition model can be used to locate new survey areas outside the survey boundary: areas like West Peak, Ladybug Peak, and Mt. Graham.

Population Estimates

It is important that managers have access to the most accurate information when calculating the MGRS population. Based upon our analysis, I conclude that not all MGRS middens have been accounted for in the last decade, and until the survey boundary is modified and the entire area searched, this situation will continue. Fortunately, it appears that the population has been underestimated by some fraction, meaning that the current population estimates are conservative. Given the decline in

MGRS habitat over the last decade, it will become increasingly important to identify all potential MGRS habitat and to revise the population estimates as appropriate.

Acknowledgments

Primary funding for this project was provided by voluntary contributions to the Arizona Game and Fish Department's Heritage Fund (Lottery Dollars Working for Wildlife); Arizona's Nongame Wildlife Checkoff; and Project W-95–M, under the Federal Aid in Wildlife Restoration Act (Pittman-Robertson Act).

I thank Bill Van Pelt, Linda Allison, Sue Boe, Terry Johnson, and Genice Froehlich (U.S. Forest Service) for reviewing Nongame Technical Report 160, from which part of this document is derived; and Mike Pruss, Lisa Haynes, and Lin Piest for assisting me in the accuracy assessment. I also appreciate Dr. Paul Young's (University of Arizona) support and advice, and permission for AGFD personnel to stay at the University of Arizona's research camp on Mt. Graham. I thank Bill Krausman (U.S. Forest Service) for loaning me a 1997 TM image for part of the change detection. Lastly, I thank Reed Sanderson for providing the encouragement and assistance to complete this analysis and for pulling together the Mt. Graham Red Squirrel Symposium.

Literature Cited

Aronoff, S. 1989. *Geographic information systems: A management perspective.* WDL Publications, Ottawa, Ontario, Canada.

Avery, T. E., and G. L. Berlin. 1992. *Fundamentals of remote sensing and airphoto interpretation,* 5th ed. Macmillan Publishing Company, New York.

Brown, D. E. 1994. *Biotic communities of the southwestern United States and northwestern Mexico.* University of Utah Press, Salt Lake City.

ESRI. 1992. *Cell-based modeling with GRID.* Environmental Systems Research Institute, Redlands, California.

Hosmer, D. W., and S. Lemeshow. 1989. *Applied logistic regression,* 1st ed. John Wiley and Sons, New York.

Jensen, J. R. 1983. Biophysical remote sensing. *Annals of the Association of American Geographers* 73:111–132.

Merriam, C. H., and Steineger, L. 1890. *Results of a biological survey of the San Francisco mountain region and the desert of the Little Colorado, Arizona.* North American

Fauna Report 3. U.S. Department of Agriculture, Division of Ornithology and Mammalia, Washington, D.C.

Pereira, Jose M. C., and R. M. Itami. 1991. GIS-based habitat modeling using logistic multiple regression: A study of the Mt. Graham red squirrel. *Photogrammetric Engineering and Remote Sensing* 57(11):1475–1486.

Schrader, S., and R. Pouncey. 1997. *ERDAS field guide*, 4th ed. ERDAS, Inc., Atlanta, Georgia.

Story, M., and R. G. Congalton. 1986. Accuracy assessment: A user's perspective. *Photogrammetric Engineering and Remote Sensing* 52:397–399.

Tou, J. T., and R. C. Gonzalez. 1974. *Pattern recognition principles.* Addison-Wesley, Reading, Massachusetts.

U.S. Forest Service. 1988. *Mount Graham red squirrel—An expanded biological assessment.* Coronado National Forest, Tucson, Arizona.

CHAPTER 13

Site Selection for the Establishment of New Middens of Mt. Graham Red Squirrels

Sadie R. Bertelsen and John L. Koprowski

SELECTION OF A HOME RANGE or territory site is believed to be critical to lifetime reproductive success of most mammals (Wauters and Dhondt 1993; Wauters et al. 1994; Tannerfeldt and Angerbjorn 1996; Wauters et al. 2001). A poor-quality territory can have fitness consequences to individuals, and therefore, site selection should be based upon characteristics that will maximize fitness (Wauters and Dhondt 1993; Wauters et al. 1994, 2001). However, many factors related to habitat condition and spatial arrangement may constrain choices available to animals, including availability of habitat, lack of connectivity that permits access to sites, quality of sites from which to select, and previous experience with habitat types (Larsen and Boutin 1994; Bakker and Van Vuren 2004; Haughland and Larsen 2004a, 2004b). Social factors also present barriers. Residents are rarely displaced by an intruder (Grafen 1987; Cristol et al. 1990; Haley 1994), and thus, occupied sites may not be considered as available. In many mammals, natal dispersal is the rule, and young often are obligated to find their own territory some distance from their natal area (Greenwood 1980; Waser and Jones 1983).

Red squirrels are excellent models for the study of site selection. Individuals live alone in well-defined, traditional territories (Streubel 1968; Finley 1969; Steele 1998), and females may bequeath these territories to offspring (Price and Boutin 1993). Near the territory center, middens of cone scales form and are used for caching cones (Finley 1969). Larderhoards facilitate identification of activity centers and are likely required for long-term persistence of the resident (Gurnell 1987). In the case of endangered Mt. Graham red squirrels, knowledge of key characteristics associated with middens is likely critical to the management and conservation of this federally endangered species (U.S. Fish and Wildlife Service 1993). Middens of Mt. Graham red squirrels are in non-random locations within the forest; sites can be characterized as having greater tree densities, downed logs, and structural complexity relative to random locations in the forest (Smith and Mannan 1994). Given that many midden sites appear to be used for extended periods of time and are thus traditional (Finley 1969), present-day conditions at midden sites may represent past choices when forest and environmental conditions differed. An important question is whether or not middens constructed under recent forest conditions are in sites similar to traditional middens. Ultimately, if the Mt. Graham red squirrel population is to increase, new middens need to be formed; however, data on the habitat characteristics of new midden sites are not available. Herein, we use a long-term census database to address the question of whether traditionally occupied midden sites differ from newly established sites.

Methods

The study area (252 ha) is located in the high-elevation forests (2,680 m to 3,267 m) of the Pinaleño Mountains, Graham County, Arizona. Two habitat types (transition forest and spruce-fir forest) are found on the study area. Transition forest is found in the ecotone of spruce-fir forest and mixed-conifer forests at elevation ranges from 2,680 m to 3,048 m. Dominant tree species found in this forest type are Engelmann spruce, corkbark fir, and Douglas-fir, interspersed rarely with ponderosa pine, southwestern white pine, white fir, and quaking aspen. Spruce-fir forest is found in elevations above 3,048 m and consists mainly of Engelmann spruce and corkbark fir. Understory of both forest types is generally open and consists of seasonal grasses and other small herbaceous plants.

Middens were located by intensive searching of the study area. Middens found from 1987 to 1990 were classified as traditional, and middens found between 1990 and 1994 were classified as new. We measured habitat variables within a 10 m radius of the midden center. Aspect of slope, percent slope, canopy cover, volume of downed logs, and diameter and snag class of all trees within the plot were measured at each plot after Smith and Mannan (1994). We quantified diversity using Shannon-Weiner diversity index and evenness measures (Magurran 1988).

We conducted monthly censuses on the study area beginning in 1989 and continuing through 1996, after which censuses were performed quarterly (Young in this vol.). Census data were used to determine number and distribution of occupied middens. We determined occupancy based on feeding sign (e.g., cone scales, dried mushrooms, and conifer clippings) or cone caching, but attempts were made on subsequent midden visits to confirm occupancy by observing a red squirrel (Koprowski et al. 2005).

Our univariate approach used Mann-Whitney *U* tests to compare all characteristics between traditional and new middens. To determine if percentage of time a midden was occupied was related to a characteristic measured within the plot, we used Spearman's rank correlation analyses. For multivariate examination, we used principal components analysis, applying the correlation matrix to collapse variables into components that accounted for >15 percent of the variation. Scores on each principal component were compared between traditional and new sites using two-sample t-tests. Discriminant function analysis was used to distinguish between traditional and new middens. To avoid multiple collinearity, we removed variables where correlation coefficients were $r > 0.70$; only the most useful of the pair of highly correlated predictive variables was retained.

Results

Univariate Analyses

Traditional and new middens differ when examined with a univariate approach (table 13.1). Traditional middens ($n = 82$) were more abundant than new middens ($n = 22$) in spruce-fir forest, whereas traditional ($n = 30$) and new ($n = 25$) middens were similar in number within the transition forest.

Traditional midden sites in both transition and spruce-fir forests were found on steeper slopes than new middens. Aspect of the slope did not differ in cardinal direction between traditional and new middens in either forest type (transition: $\chi^2 = 14.68$, df = 35, $P = 1.00$; spruce-fir: $\chi^2 = 35.80$, df = 35, $P = 0.43$). In transition forest, basal area and log volume at traditional middens were greater than those at new middens (table 13.1).

Table 13.1. Site characteristics at traditional and new Mt. Graham red squirrel middens in the transition and spruce-fir forest*

Median and interquartile ranges are presented for each characteristic. Mann-Whitney *U* test statistics are shown.

Characteristic	Midden Type: Traditional	Midden Type: New	*U*	df	*P*
Transition Forest					
Basal Area (m^2)	2.39 (1.98–3.11)	1.92 (1.70–2.40)	522.0	54	0.012
Slope (%)	17.0 (8.75–22.75)	9.0 (6.0–16.0)	500.0	54	0.034
Canopy Cover (%)	85.50 (79.0–89.25)	82.67 (78.39–86.94)	438.0	54	0.290
Log Volume (m^3)	11.34 (8.97–16.37)	6.75 (2.51–12.67)	511.0	54	0.021
# Spruce	8.0 (2.0–18.25)	5.0 (2.50–12.0)	419.0	54	0.462
# Corkbark Fir	46.5 (15.75–58.0)	46.0 (37.0–58.5)	341.5	54	0.577
Total # Trees	64.0 (42.75–75.75)	58.0 (47.0–74.0)	390.0	54	0.798
Species Richness	3	4			
Diversity	0.687 (0.545–0.854)	0.678 (0.398–0.818)	654.0	54	0.442
Evenness	0.524 (0.450–0.642)	0.503 (0.328–0.674)	646.0	54	0.366
Spruce-Fir Forest					
Basal Area (m^2)	2.70 (2.33–3.045)	2.10 (1.84–2.49)	1412.0	103	<0.0001
Slope (%)	19.0 (13.75–24.0)	11.0 (6.75–14.0)	1425.5	103	<0.0001
Canopy Cover (%)	87.0 (84.0–91.25)	83.5 (81.64–87.47)	1233.5	103	0.0082
Log Volume (m^3)	9.05 (5.63–13.37)	3.26 (2.04–5.57)	1345.0	103	<0.0001
# Spruce	46.0 (30.75–62.25)	32.5 (21.25–43.75)	1228.5	103	0.0093
# Corkbark Fir	40.0 (7.75–54.0)	26.0 (7.75–33.0)	1108.0	103	0.1009
Total # Trees	85.0 (62.0–111.5)	63.0 (48.8–75.5)	1389.5	103	0.0001
Species Richness	2	2			
Diversity	0.677 (0.560–0.707)	0.653 (0.481–0.691)	952.5	98	0.406
Evenness	0.700 (0.618–0.969)	0.841 (0.518–0.988)	1138.5	98	0.451

*Pinaleño Mountains, Arizona.

Table 13.2. Principal components analysis of traditional and new Mt. Graham red squirrel middens*

Variable	Transition Forest		Spruce-Fir Forest	
	PC1	PC2	PC1	PC2
Diversity	0.340	0.360	−0.146	0.654
Basal Area	−0.261	0.276	−0.589	−0.074
Slope	−0.404	0.319	−0.153	−0.197
Aspect	−0.078	0.283	0.054	−0.136
Canopy Cover	−0.181	0.148	−0.466	−0.164
Log Volume	−0.371	0.141	−0.064	0.231
Spruce	−0.043	0.315	−0.468	−0.389
Corkbark Fir	−0.467	−0.382	−0.406	0.527
Douglas-fir	−0.133	0.376		
Aspen	0.204	−0.314		
Total Stems	−0.443	−0.292		
Eigenvalue	2.4011	2.0847	2.1043	1.5948
Proportion	0.218	0.190	0.263	0.199

* Pinaleño Mountains, Arizona.

All other characteristics were greater at traditional sites, including canopy cover, number of Engelmann spruce trees, number of corkbark fir trees, and stem density, diversity, and evenness, but not significantly so.

In spruce-fir forests, basal area, canopy cover, log volume, number of Engelmann spruce trees, and stem density were significantly greater at traditional sites than at new sites. Tree diversity and number of corkbark fir trees were greater at traditional sites than at new sites, but the differences were not significant.

Multivariate Analyses

A multivariate approach also elucidates important differences between traditional and new midden sites (table 13.2). Two principal components that accounted for >15 percent of the variation were identified for transition and spruce-fir forests. The first principal component in each forest

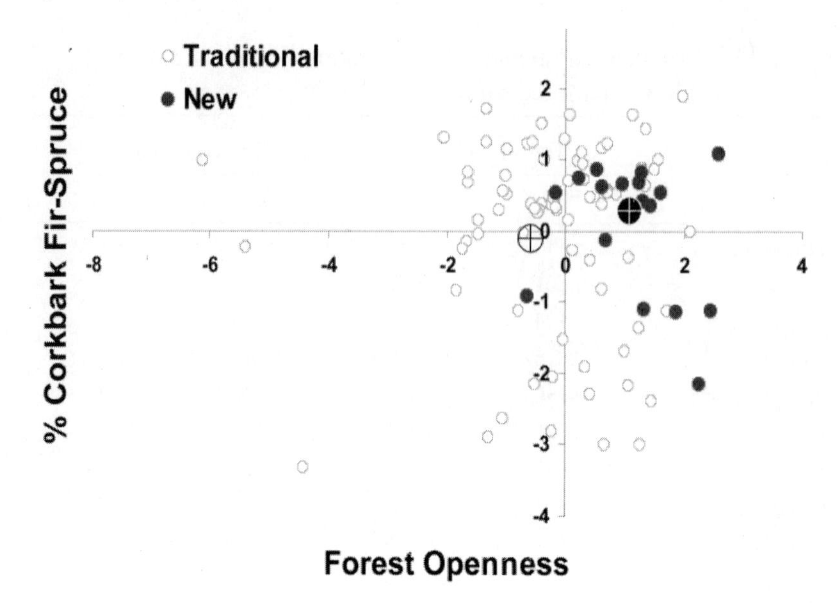

Figure 13.1. Comparison of principal component scores on PC1 and PC2 for traditional and new Mt. Graham red squirrel middens, Pinaleño Mountains, Arizona. High PC1 scores are associated with open forests with low numbers of spruce and corkbark fir stems, low basal area, low log volume, and low canopy cover. High PC2 scores refer to areas with high numbers of corkbark fir stems, log volume, and tree diversity with low numbers of spruce stems. Large circles with crosshairs refer to the bivariate mean for each midden group.

type consisted primarily of an axis dominated by density of trees. This component was not useful in distinguishing traditional and new middens in transition forests (traditional PC1 score: -0.22 ± 0.28 *SE*; new PC1 score: 0.27 ± 0.31 *SE*; $t = -1.17$, df = 50, $P = 0.25$); however, in spruce-fir forests, new middens were in open forests relative to traditional sites (traditional PC1 score: -0.25 ± 0.16 *SE*; new PC1 score: 1.14 ± 0.22 *SE*; $t = -5.12$, df = 36, $P < 0.001$).

In transition forest, the second principal component consisted primarily of high positive loadings for number of spruce and Douglas-fir stems as well as diversity and basal area with low occurrence of corkbark fir and aspen stems; new sites tended to be in more open sites with corkbark fir

Table 13.3. Discriminant functions to distinguish between traditional and newly established Mt. Graham red squirrel middens*

	Transition Forest		Spruce-Fir Forest	
	Traditional	New	Traditional	New
Constant	–104.64	–101.27	–193.37	–179.48
Diversity	43.48	42.95	37.32	37.42
Basal Area	–1.21	–2.08	0.35	–0.27
Slope	–0.29	–0.33	0.29	0.21
Aspect	0.0	0.04	0.03	0.03
Canopy Cover	1.95	1.94	4.18	4.06
Log Volume	0.79	0.71	1.10	0.93
Spruce	–0.47	–0.48	–0.36	–0.41
Corkbark Fir	0.24	0.30	–0.50	–0.55
Douglas-fir	0.25	0.30		
Aspen	0.48	0.52		
Total Stems	0.03	–0.02	0.19	0.24

* Pinaleño Mountains, Arizona

and aspen common (traditional PC2 score: 0.35 ± 0.26 *SE*; new PC2 score: –0.42 ± 0.28 *SE*; $t = 2.05$, df = 51, $P = 0.046$). In spruce-fir forest, the second principal component was highly influenced by diversity, corkbark fir, and log volume with low occurrence of spruce; however, PC2 was not useful in distinguishing traditional from new sites (traditional PC2 score: –0.01 ± 0.15 *SE*; new PC2 score: 0.044 ± 0.23 *SE*; $t = -0.19$, df = 31, $P = 0.85$) (fig. 13.1).

In transition forest, traditional and new middens could only be classified correctly 60.0 percent of the time (cross-validation corrected rate, 60 percent of 30 traditional sites, 60 percent of 25 new sites correctly classified) using discriminant function analysis (table 13.3). Spruce-fir traditional and new middens could be classified correctly 74.5 percent of the time (cross-validation corrected rate, 74 percent of 77 traditional sites, 76.5 percent of 17 new sites correctly classified) using discriminant function analysis (table 13.3).

Likelihood of Occupancy

The proportion of middens occupied for traditional (0.596 ± 0.054 *SE*) and new middens (0.626 ± 0.039 *SE*) did not differ in transition forest (Mann-Whitney $U = 889.0$, df = 55, $P = 0.879$). In spruce-fir forest, however, traditional sites (0.418 ± 0.039 *SE*) were marginally less likely to be occupied relative to new middens (0.556 + 0.072 *SE*, $U = 4122.5$, df = 103, $P = 0.084$). In general, all sites in the transition forests (0.610 ± 0.034 *SE*) are occupied more frequently than sites in spruce-fir forest (0.448 ± 0.035 *SE*, $U = 5091.5$, df = 156, $P = 0.009$). Traditional sites tended to be occupied more frequently in transition forest than spruce-fir forest ($U = 2142.5$, df = 111, $P = 0.016$); however, new sites were equally likely to be occupied in the forest types ($U = 659$, df = 47, $P = 0.865$). Correlates of likelihood of occupancy were few in each forest type, with only slope demonstrating a strong negative relationship ($r = -0.259$, df = 1, 54, $P = 0.056$; all other $P > 0.23$) in transition forest. In spruce-fir forest, only slope ($r = -0.286$, df = 1, 102, $P = 0.003$) and log number ($r = -0.204$, df = 1, 102, $P = 0.044$) were correlated with occupancy (all other $P > 0.21$). When examining only new middens, occupancy was similarly correlated with only a few variables. In transition forest, only number of aspen stems ($r = -0.471$, df = 1, 24, $P = 0.018$; all other $P > 0.16$) was related to occupancy rates. Similarly, in spruce-fir forest, only diversity ($r = -0.387$, df = 1, 21, $P = 0.075$; all other $P > 0.18$) approached a relationship with occupancy rate.

Discussion

Suitability of forest habitat for Mt. Graham red squirrels depends heavily upon a reliable cone crop and a microclimate to preserve cones within middens (U.S. Fish and Wildlife Service 1993). Large middens might be decades old and represent accumulations of cone scales by many generations of resident animals (Streubel 1968; Finley 1969). Well-established middens can harbor a large supply of food and may contain a valuable food source that would not be found in a site previously unoccupied by a squirrel (Finley 1969; Larsen and Boutin 1994). Traditional midden sites of Mt. Graham red squirrels appear to have identifiable structural components and perhaps unique microclimates (Smith and Mannan

1994). The establishment of a new midden is a considerable undertaking that involves finding a site with suitable structure and a microclimate conducive to cone storage as well as an abundance of cones that can be used to provision the midden, permitting over-winter survival. The costs of poor selection of a new site are likely to be extreme.

Many habitat features distinguished traditional and new red squirrel middens. All characteristics considered essential to squirrel survival (Froehlich 1990; Smith and Mannan 1994) had higher values at traditional sites, most significantly so. In transition forest, basal area and log volume were greater at traditional sites. This trend is also reflected in spruce-fir forest where canopy cover, number of spruce trees, and stem density were also greater in traditional relative to new sites. Basal area provides an indication of how much woody matter is at a site and thus likely structural complexity and productivity of tree seeds at that site (Greene et al. 2002). Downed logs are also very important as feeding sites, cache sites, and traveling medium (Vahle and Patton 1983; Patton and Vahle 1986; Bakker and Van Vuren 2004). Moist, decaying logs are ideal for mushroom growth (Smith and Mannan 1994), thus providing an important supplemental food source for the red squirrels. A high number of trees in close proximity to the midden will also provide accessible escape and travel routes as well as reduce foraging time (Rothwell 1979; Bakker and Van Vuren 2004).

Multivariate analyses for the lower mixed-conifer forest sites supported the univariate analyses that suggested slope and aspect were of some import, but middens were found in many varied microhabitats. In spruce-fir forest, however, new middens were almost exclusively in relatively open areas with low stem densities, low ground and canopy cover, and low diversity. These sites would appear to be sub-optimal relative to traditional sites (Smith and Mannan 1994), suggesting that animals were relegated to these sites at a time when population levels were high (Snow in this vol.). If one assumes that the location of traditional middens on the mountain represents the historical distribution of quality sites, then new middens would represent the marginal sites that remain. Alternatively, new sites could represent preferred locations as suggested by the slightly higher occupancy rates for new sites in spruce-fir forest. Such patterns in occupancy also could reflect simply the natural life cycle of middens where new middens are established, often occupied, and become less likely to be occupied with age. Unfortunately, little is known about the longevity and persistence of middens.

Red squirrels depend heavily on conifer species for forage (Finley 1969; U.S. Fish and Wildlife Service 1993; Steele and Koprowski 2001). In spruce-fir forests, tree density, canopy cover, and basal area certainly reflect a likely increase in availability of tree seeds. Red squirrels tend to settle in habitats that are similar to those of their natal area and spend little time exploring habitats that differ greatly from natal areas (Haughland and Larsen 2004a, 2004b). The tree species dominating the high-elevation forest of the Pinaleños are Engelmann spruce and cork-bark fir. These two species provide ideal habitat for a midden because their branches are often low and dense, creating shade and cover near the ground, and their cone crops are relatively dependable (Finley 1969; Burns and Honkala 1990). Proximity of cone resources to the area of highest activity provides more foraging opportunities, especially during autumn caching activities (Smith and Mannan 1994). The tendency for more spruce and fir at traditional sites than at new sites suggests that new sites are marginally sub-optimal and traditional sites represent the preferred parameters that permit long-term persistence. Why are traditional sites found on steeper slopes than new sites? Steeper slopes may be more exposed to sun and wind and thus are more dry, have a shorter fire interval, retain winter snows longer, and favor the growth of Douglas-fir and southwestern white pine (Jones 1974; Dieterich 1983; Burns and Honkala 1990) of significant sizes due to difficult access for timber extraction.

The habitat of this federally endangered species is often subject to pressures of human activity. While active management is needed to encourage a healthy forest, it is important that available information is used to avoid degrading present red squirrel habitat. Also, a large number of trees and downed logs, especially those that provide shade and a medium for storage and growth of resources, are essential and must be retained. The composition and structure of typical midden sites, as described here, need to be preserved. Conclusions drawn from evaluating choices made in midden site selection may be important in identifying forest structure and composition characteristics essential to the survival of red squirrels in the Pinaleño Mountains.

Acknowledgments

We thank V. Greer, K. Hutton Kimple, C. Coates, and P. Young for assistance in data collection. A. Smith and M. Grinder collected much of the early data on vegetation. The University of Arizona; U.S. Forest Service, Rocky Mountain Research Station; and Arizona Game and Fish Department provided funding for field research and data analyses. Field efforts were conducted under permits from the U.S. Forest Service, Arizona Game and Fish Department, U.S. Fish and Wildlife Service, and the University of Arizona's Institutional Animal Care and Use Committee.

Literature Cited

Bakker, V. J., and D. H. Van Vuren. 2004. Gap-crossing decisions by the red squirrel, a forest-dependent small mammal. *Conservation Biology* 18:689–697.

Burns, R. M., and B. H. Honkala. 1990. *Silvics of North America.* Vol 1. *Conifers*; Vol 2. *Hardwoods.* U. S. Forest Service, Agriculture Handbook 654, Washington, D.C.

Cristol, D. A., V. Nolan Jr., and E. D. Ketterson. 1990. Effect of prior residence on dominance status of dark-eyed juncos, *Junco hyemalis. Animal Behaviour* 40:580–586.

Dieterich, J. H. 1983. Fire history of southwestern mixed conifer: A case study. *Forest Ecology* 6:13–31.

Finley, R. B., Jr. 1969. Cone cache and middens of *Tamiasciurus* in the Rocky Mountain region. Pages 233–273 in J. K. Jones Jr., ed. *Contributions in mammalogy.* University of Kansas Museum of Natural History Miscellaneous Publication 51.

Froehlich, G. F. 1990. Habitat use and life history of the Mount Graham red squirrel. Master's thesis, University of Arizona, Tucson.

Grafen, A. 1987. The logic of divisively asymmetric contests: Respect for ownership and the desperado effect. *Animal Behaviour* 35:462–467.

Greene, D. F., C. Messier, H. Asselin, and M. J. Fortin. 2002. The effect of light availability and basal area on cone production in *Abies balsamea* and *Picea glauca. Canadian Journal of Botany* 80:370–377.

Greenwood, P. J. 1980. Mating systems, philopatry, and dispersal in birds and mammals. *Animal Behaviour* 28:1140–1162.

Gurnell, J. 1987. *Natural history of squirrels.* Facts on File, New York.

Haley, M. P. 1994. Resource-holding power asymmetries, the prior residence effects, and reproductive payoffs in male northern elephant seal fights. *Behavioral Ecology and Sociobiology* 34:427–434.

Haughland, D. L., and K. W. Larsen. 2004a. Exploration correlates with settlement: Red squirrel dispersal in contrasting habitats. *Journal of Animal Ecology* 73:1024–1034.

———. 2004b. Ecology of North American red squirrels across contrasting habitats: Relating natal dispersal to habitat. *Journal of Mammalogy* 85:225–236.

Jones, J. R. 1974. *Silviculture of southwestern mixed conifers and aspen: The status of our knowledge.* Research Paper RM-122:1–44. U.S. Dept. of Agriculture, Forest Service, Rocky Mountain Forest and Range Experiment Station, Fort Collins, Colorado.

Koprowski, J. L., M. I. Alanen, and A. M. Lynch. 2005. Nowhere to run and nowhere to hide: Response of endemic Mt. Graham red squirrels to catastrophic forest damage. *Biological Conservation* 126:491–498.

Larsen, K. W., and S. Boutin. 1994. Movements, survival, and settlement by red squirrel (*Tamiasciurus hudsonicus*) offspring. *Ecology* 75:214–223.

Magurran, A. 1988. *Ecological diversity and its measurement.* Princeton University Press, Princeton, New Jersey.

Patton, D. R., and J. R. Vahle. 1986. Cache and nest characteristics of the red squirrel in an Arizona mixed-conifer forest. *Western Journal of Applied Forestry* 1:48–51.

Price, K., and S. Boutin. 1993. Territorial bequeathal by red squirrel mothers. *Behavioral Ecology* 4:144–150.

Rothwell, R. 1979. Nest sites of red squirrels (*Tamiasciurus hudsonicus*) in the Laramie Range of southeastern Wyoming. *Journal of Mammalogy* 60:404–405.

Smith, A. A., and R. W. Mannan. 1994. Distinguishing characteristics of Mount Graham red squirrel midden sites. *Journal of Wildlife Management* 58:437–445.

Steele, M. A. 1998. *Tamiasciurus hudsonicus. Mammalian Species* 586:1–9.

Steele, M. A., and J. L. Koprowski. 2001. *North American tree squirrels.* Smithsonian Institution Press, Washington, D.C.

Streubel, D. P. 1968. Food storing and related behavior of red squirrels (*Tamiasciurus hudsonicus*) in interior Alaska. Master's thesis, University of Alaska, Fairbanks.

Tannerfeldt, M., and A. Angerbjorn. 1996. Life history strategies in a fluctuating environment: Establishment and reproductive success in the artic fox. *Ecography* 19:209–220.

U.S. Fish and Wildlife Service. 1993. *Mount Graham Red Squirrel* (Tamiasciurus hudsonicus grahamensis) *Recovery Plan*. Albuquerque, New Mexico.

Vahle, J. R., and D. R. Patton. 1983. Red squirrel cover requirements in Arizona mixed conifer forests. *Journal of Forestry* 81:14–15, 22.

Waser, P. M., and W. T. Jones. 1983. Natal philopatry among solitary mammals. *Quarterly Review of Biology* 58:355–390.

Wauters, L. A., and A. A. Dhondt. 1993. Lifetime reproductive success and its correlates in female Eurasian red squirrels. *Oikos* 72:402–410.

Wauters, L. A., J. Gurnell, D. Preatoni, and G. Tosi. 2001. Effects of spatial variation in food availability on spacing behaviour and demography of Eurasian red squirrels. *Ecography* 24:525–538.

Wauters, L. A., E. Matthysen, and A. A. Dhondt. 1994. Survival and lifetime reproductive success in dispersing and resident red squirrels. *Behavioral Ecology and Sociobiology* 34:197–201.

CHAPTER 14

Habitat Characteristics of the Midden Sites of Mt. Graham Red Squirrels

Do Sex Differences Exist?

MARIT I. ALANEN, JOHN L. KOPROWSKI, MARTHA I. GRINDER, VICKI L. GREER, CAROL A. COATES, AND KELLY A. HUTTON KIMPLE

FITNESS-LIMITING RESOURCES often differ between the sexes in mammals and can have a profound impact on the ecology of the sexes (Clutton-Brock 1989). Sex differences in the social and mating systems of ground-dwelling squirrels have been well documented for many species (Armitage 1981; Michener 1983); however, these systems are not well described for tree squirrels (*Sciurus* and *Tamiasciurus*) (Koprowski 1998). In non-territorial tree squirrels that scatterhoard food resources (*Sciurus* spp.), males range farther than females despite the lack of sexual dimorphism in body size (Thompson 1977; Wauters and Dhondt 1992; Koprowski 1998). Little is known about sex differences in territorial, larderhoarding tree squirrels of the genus *Tamiasciurus* (red squirrels [*T. hudsonicus*] and Douglas squirrels [*T. douglasii*]). Unlike scatterhoarding tree squirrels, *T. hudsonicus* displays modest sexual dimorphism in body size (adult male:female = 1.05:1.10), although this does not appear to affect survival rates (Boutin and Larsen 1993). Boutin and Larsen (1993) suggest that other factors present in field conditions may affect survival more than body size. Differences in large-scale habitat characteristics between midden sites could affect microclimate, nesting sites, and food availability within a midden area and therefore potentially influence an individual's survival and reproductive success.

The existence of sex differences might suggest the potential for sex-specific conservation strategies within this endangered small mammal. We assessed large-scale habitat characteristics between male and female

Mt. Graham red squirrel midden sites to determine if sex differences in middens could be distinguished.

Methods

Study Sites

Habitat characteristics of 159 red squirrel middens in the Pinaleño Mountains were measured in 1989–1990 (113 middens) and 1994 (46 middens). The large-scale characteristics and methods for measurement are defined in Smith and Mannan (1994) and were followed in 1994 when the 46 additional middens were assessed. Data from the 1989–1990 measurements were used in combination with occupancy data identifying male and female squirrels from 1991–1993. Data from the 1994 measurement were used in combination with occupancy data from 1994–1996. With the exception of catastrophic events, these large-scale characteristics are slow to change within a midden area; therefore, we felt the measurements gathered in one year could apply to several subsequent years.

Midden Selection and Occupancy

All middens were visited at least twice during June, July, August, and September, and occupancy was determined. Typically 2–10 hours were expended at each occupied midden in an attempt to note the sex of the resident squirrel. During 1991–1996, a midden was classified as "male" or "female" if the occupant was identified as the same sex for at least two of these months and there was no occurrence of the opposite sex occupying the midden during these months. Additionally, the midden had to be occupied for all four summer months, whether or not the sex was determined (we assumed the resident squirrel remained the same). Summer months were used because the squirrels are reproductively active and differences in large-scale habitat characteristics between "male" and "female" middens should be most obvious during these months. During these years, 56.9 percent (±7.3 percent) of the available middens were unoccupied for at least two months during the summer, suggesting that squirrels had the opportunity to select from a number of available midden sites.

Table 14.1. Number of "male" and "female" Mt. Graham red squirrel middens qualifying for analyses for each year

Italicized values indicate years that were not analyzed using univariate statistics due to the small sample size.

	Year					
Forest Type	1991	1992	1993	1994	1995	1996
Spruce-Fir Forest						
Number of males	20	13	7	6	8	7
Number of females	11	6	*0*	3	2	5
Transitional Forest						
Number of males	6	6	5	*0*	3	2
Number of females	7	7	3	5	2	*1*

Data Collection

Circular 10 m radius plots (0.03 ha) were centered on the midden, defined as the tree or snag with the greatest degree of activity surrounding it (e.g., cached cones and cone scales). Slope and aspect (in degrees) were measured at the midden center. Percent canopy closure was also measured at the midden center and at 5 and 10 m from the midden center in each of the four cardinal directions using a spherical densiometer (Strickler 1959); the measurements were averaged to determine the percent canopy closure for the entire midden. In each plot, species and diameter at breast height (dbh) were recorded for all trees and snags >3 cm dbh, and basal area was calculated. The diameter and length of all downed logs >2 m long and >20 cm in diameter at the small end were measured to calculate log volume. Simpson's and Shannon's diversity indices (Magurran 1988) were calculated for each midden using live tree data.

Statistical Analysis

Data were transformed where possible to meet the assumption of normality (Shapiro-Wilk W test, $P > 0.05$). Where normality could not be achieved, we used non-parametric techniques. In the spruce-fir and transition habitats, we used Student's univariate t-test to determine differences between the habitat characteristics of male and female middens. Additionally, we used two multivariate tests: (1) principal components analysis (PCA) to determine patterns of variation in habitat characteristics of middens, and (2) stepwise logistic regression to determine the factors identifying "male" and "female" middens. For the multivariate tests, we pooled all middens that qualified for inclusion in the univariate tests and discarded middens where the sex of the resident squirrel varied between years. We combined the spruce-fir and transition habitats and included a nominal variable, "area," to differentiate between the two habitat types.

Results

The number of "male" and "female" middens qualifying for inclusion in the analyses varied from year to year (table 14.1). In the spruce-fir habitat, the years 1993 and 1995 were excluded, and in the transition habitat, the years 1994–1996 were excluded due to insufficient sample size.

Univariate Tests

The only variable that differed between the middens of male and female squirrels was slope in the spruce-fir habitat; females were found on significantly steeper slopes than males in 1991 ($t = -2.71$, df = 29, $P = 0.01$) and 1994 ($t = -2.57$, df = 7, $P = 0.04$) (tables 14.2 and 14.3). Additionally, this trend was consistent across all years and habitat types, whether or not it was significant (fig. 14.1). No other variables differed between male and female middens.

Multivariate Tests

Principal components analysis of the habitat variables did not show any distinct pattern of difference between male and female middens (table 14.4; fig. 14.2). The first principal component (PC1) appears to be a woody density factor, which separates middens with lower numbers of live trees, snags, and logs from those with relatively higher numbers.

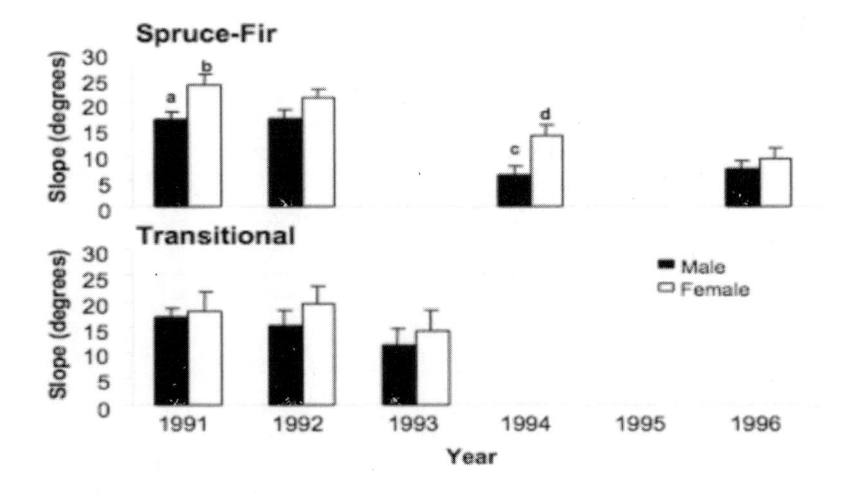

Figure 14.1. Differences in slope (mean±SE) between male and female Mt. Graham red squirrel middens. Different letters indicate significantly different values.

This axis accounted for approximately 35 percent of the variability within the habitat data. We interpreted the second principal component (PC2) as a woody volume factor, which separates middens with lower log volume and basal area of trees and snags from those with relatively higher amounts. This axis accounted for approximately 18 percent of the variability within the habitat data (table 14.4). Both males and females are fairly evenly distributed across both axes, indicating they are both found in middens with relatively low and high amounts of woody density and volume (PC1: t = 1.107, df = 60, P = 0.27; PC2: t = 1.191, df = 60, P = 0.24) (fig. 14.2).

Stepwise logistic regression indicated that slope (P <0.01) and average canopy cover (P = 0.02) significantly affected whether a midden was "male" or "female." The chances of a midden being "male" increased with a decrease in slope and an increase in the transformed variable of canopy cover. Using a jackknife resubstitution method and assuming the odds of a midden being "male" or "female" are equivalent, or a 50:50 ratio, the model correctly classified the data 62.9 percent of the time, with an R^2 value of 0.16.

Table 14.2. Mean ± SE for "male" and "female" Mt. Graham red squirrel middens for large-scale habitat characteristics in the spruce-fir forest

Italicized values are significantly different between "male" and "female" middens. The years 1993 and 1995 are excluded due to insufficient sample size.

	Year			
Characteristic	1991	1992	1994	1996
Male				
Slope	*17.0 ±1.4*	17.2 ± 1.7	*6.2 ± 1.7*	7.3 ± 1.6
Aspect	204.8 ± 27.7	213.8 ± 34.3	194.5 ± 42.8	215.4 ± 41.1
Avg Canopy Cover	86.8 ± 1.6	89.3 ± 1.3	83.3 ± 1.3	86.0 ± 1.5
Basal Area (Live)	180.4 ± 27.6	183.6 ± 34.5	52.0 ± 11.1	49.0 ± 11.7
# of Live Trees	79.9 ± 6.6	90.6 ± 8.8	55.7 ± 5.6	49.9 ± 4.4
Basal Area (Snag)	55.4 ± 10.4	61.3 ± 13.3	7.8 ± 3.3	12.8 ± 5.8
# of Snags	21.5 ± 2.8	25.4 ± 2.8	8.5 ± 3.5	9.7 ± 2.9
Volume of Logs	113.7 ± 20.1	93.8 ± 22.9	45.1 ± 8.5	39.1 ± 13.0
# of Logs	16.9 ± 1.6	13.7 ± 1.7	6.0 ± 1.0	5.3 ± 1.0
Shannon's DI	0.54 ± 0.05	0.58 ± 0.03	0.45 ± 0.10	0.55 ± 0.05
Simpson's DI	1.70 ± 0.08	1.70 ± 0.07	1.51 ± 0.16	1.60 ± 0.12
Female				
Slope	*23.6 ± 2.1*	21.2 ± 1.7	*13.7 ± 2.2*	9.2 ± 2.3
Aspect	170.9 ± 38.4	183.0 ± 56.9	225.0 ± 25.2	153.8 ± 44.5
Avg Canopy Cover	88.2 ± 1.2	87.6 ± 2.0	83.6 ± 3.4	81.3 ± 2.3
Basal Area (Live)	206.7 ± 42.9	197.7 ± 68.7	22.1 ± 6.3	38.5 ± 8.5
# of Live Trees	78.1 ± 9.0	76.5 ± 16.3	43.0 ± 1.5	42.8 ± 4.7
Basal Area (Snag)	62.8 ± 13.4	59.1 ± 18.5	4.6 ± 0.9	11.5 ± 3.9
# of Snags	17.4 ± 2.8	18.7 ± 5.2	7.0 ± 2.6	8.6 ± 2.1
Volume of Logs	137.1 ± 28.5	114.1 ± 36.4	25.9 ± 3.3	27.3 ± 4.7
# of Logs	18.5 ± 1.9	18.7 ± 3.1	3.3 ± 1.7	5.8 ± 0.7
Shannon's DI	0.50 ± 0.07	0.50 ± 0.11	0.52 ± 0.10	0.55 ± 0.14
Simpson's DI	1.60 ± 0.12	1.63 ± 0.18	1.58 ± 0.22	1.81 ± 0.20

Table 14.3. Mean ± SE for "male" and "female" Mt. Graham red squirrel middens for large-scale habitat characteristics in the transitional forest

The years 1994 through 1996 are excluded due to insufficient sample size.

Characteristic	Year		
	1991	1992	1993
Male			
Slope	17.0 ± 1.7	15.3 ± 2.9	11.6 ± 3.1
Aspect	293.3 ± 20.7	263.3 ± 34.7	227.0 ± 41.3
Avg Canopy Cover	87.9 ± 2.1	88.6 ± 1.6	83.6 ± 3.3
Basal Area (Live)	497.1 ± 86.3	579.7 ± 160.9	528.4 ± 173.0
# of Live Trees	51.8 ± 4.0	54.2 ± 4.0	65.6 ± 8.0
Basal Area (Snag)	135.4 ± 39.0	178.4 ± 68.6	304.5 ± 110.0
# of Snags	8.5 ± 1.8	13.3 ± 4.2	19.4 ± 7.5
Volume of Logs	357.4 ± 49.4	331.1 ± 78.1	517.0 ± 291.3
# of Logs	30.2 ± 5.4	32.2 ± 4.2	24.6 ± 3.8
Shannon's DI	0.63 ± 0.09	0.49 ± 0.09	0.68 ± 0.12
Simpson's DI	1.65 ± 0.16	1.45 ± 0.13	1.63 ± 0.16
Female			
Slope	18.1 ± 3.6	19.6 ± 3.4	14.3 ± 3.9
Aspect	205.0 ± 57.3	253.6 ± 45.0	268.3 ± 44.8
Avg Canopy Cover	80.1 ± 4.0	81.2 ± 3.9	82.1 ± 3.6
Basal Area (Live)	513.1 ± 120.7	642.8 ± 156.5	493.2 ± 106.8
# of Live Trees	56.0 ± 7.4	63.9 ± 9.5	78.0 ± 20.0
Basal Area (Snag)	239.1 ± 83.2	169.0 ± 34.9	90.8 ± 41.9
# of Snags	15.4 ± 5.9	8.7 ± 1.7	9.3 ± 1.8
Volume of Logs	527.2 ± 201.1	486.5 ± 110.2	300.1 ± 78.8
# of Logs	27.4 ± 1.4	30.3 ± 3.0	29.3 ± 10.3
Shannon's DI	0.74 ± 0.08	0.67 ± 0.07	0.40 ± 0.11
Simpson's DI	1.72 ± 0.12	1.62 ± 0.10	1.24 ± 0.07

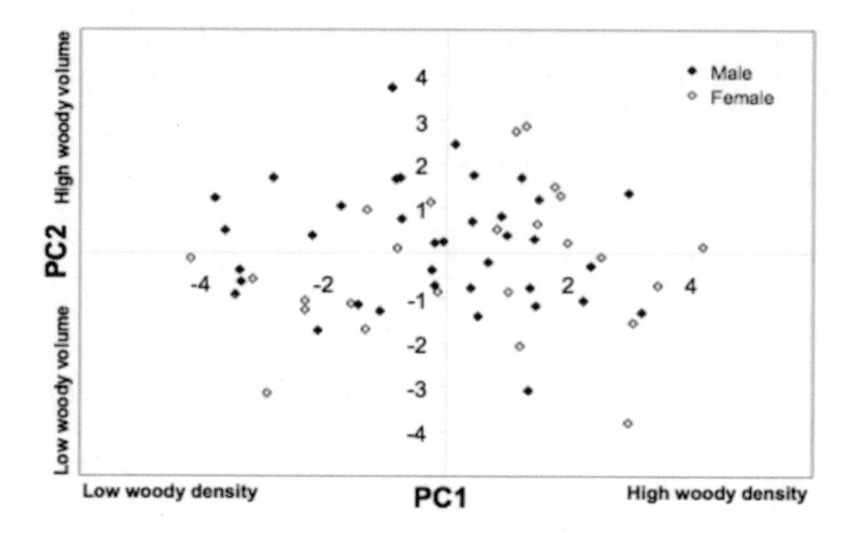

Figure 14.2. Principal component 1 (PC1), a woody density axis, and principal component 2 (PC2), a woody volume axis, for large-scale habitat characteristics between the middens of male and female Mt. Graham red squirrels.

Discussion

Few large-scale variables appear to distinguish between male and female *T. hudsonicus* midden sites. Large-scale habitat variables were significantly different between randomly selected sites and Mt. Graham red squirrel midden sites irrespective of sex; consequently, there appears to be some degree of selection by red squirrels when establishing a midden (Smith and Mannan 1994). This suggests that while components of middens may be different from random sites, selection of middens based upon large-scale habitat characteristics of these sites by males and females may be insignificant.

In spite of the fact that 56.9 percent of the middens were unoccupied during half of the summer months from 1991–1996, red squirrels did not appear to differentially select midden sites based upon sex. Male and female red squirrel offspring compete equally for all available sites (Larsen and Boutin 1998), indicating large-scale habitat characteristics may not be important when a squirrel initially selects a midden site. We

Table 14.4. Results of the principal components analysis on the correlation matrix of the large-scale habitat variables in Mt. Graham red squirrel middens

Italicized values indicate significant correlation between variables and principal components.

	Principal Components		
	1	2	3
Eigenvalue	4.23	2.14	1.87
% Variance Explained	35.23	17.83	15.55
Cumulative % Variance Explained	35.23	53.06	68.61
Correlations of Original Variables with PCs			
Area (Spruce-Fir or Transitional Forest)	−0.30	0.27	0.31
Slope	0.22	0.26	−0.06
sin (Aspect)	0.09	−0.02	−0.17
logit (Average Canopy Cover)	0.11	*0.43*	0.11
ln (Live Basal Area)	*0.46*	0.02	−0.07
ln (# Live Trees)	0.12	*0.50*	0.11
ln (Snag Basal Area)	*0.42*	0.07	0.07
ln (# Snags)	0.18	*0.41*	0.40
ln (Log Volume)	*0.44*	−0.11	−0.10
# Logs	0.39	−0.07	−0.25
cos (Shannon's Diversity Index)	−0.21	0.39	*−0.48*
Simpson's Diversity Index	0.12	−0.30	*0.61*

did, however, find that slope was significantly different between male and female middens for two years, with females being found on steeper slopes than males. It is unlikely that slope in itself is biologically meaningful, but it could indicate that other microclimatic factors are important. This lack of large-scale differences in habitat characteristics between the middens of male and female Mt. Graham red squirrels might be because red squirrels are territorial and collect food into a central larderhoard (Smith 1968; Koford 1982; Larsen and Boutin 1994), requiring both males and females to use similar areas. Several studies indicate that scatterhoarding tree squirrels display sexual dimorphism in home range size (Thompson

1977; Wauters and Dhondt 1992; Koprowski 1998), but this may not be possible for larderhoarding tree squirrels. Unlike non-territorial tree squirrels, red squirrels have a limited distance they can travel and still protect their midden (Smith 1968, 1981), placing males and females under similar biological restrictions when selecting a midden site and defending their cache. Differences in large-scale habitat characteristics for male and female Mt. Graham red squirrels may not be possible under the constraints of their territorial social system.

Our study did not investigate space use by Mt. Graham red squirrels; however, four events within the past decade have altered the landscape used by the squirrel and could potentially affect midden site selection. Multiple insect outbreaks killed most of the Engelmann spruce and corkbark fir; two fires burned approximately 2,718 ha and 12,029 ha of the Pinaleño Mountains in 1996 and 2004, respectively, including both transitional and spruce-fir habitat; and the construction of three astrophysical telescopes occurred on 3.5 ha in the spruce-fir habitat (Istock 1995). Consequently, the habitat of the Mt. Graham red squirrel has been fragmented to some degree in a relatively short period of time. In one study on the Siberian flying squirrel (*Pteromys volans*), habitat fragmentation differentially influenced space use by male and female squirrels in southern Finland (Selonen et al. 2001). Male flying squirrels were found to inhabit a greater number of preferred habitat patches in more fragmented than continuous landscapes, whereas females seemed to select patches large enough for their home range and were not sensitive to size variation in those patches. While our study did not focus on space use by red squirrels, it is possible that sex differences in space use could exist, influenced by microclimatic conditions or behavioral strategies, and further study is recommended.

Acknowledgments

For data collection, we thank A. Smith, B. Mannan, the Mt. Graham Red Squirrel Monitoring Program, and the numerous field assistants who worked on the mountain throughout the years. We also thank Paul Young for his guidance; Jonathan Grinder for his support; the U.S. Forest Service, U.S. Fish and Wildlife Service, and Arizona Game and Fish Department for their assistance; and the University of Arizona for logistical support and continued funding of this project.

Literature Cited

Armitage, K. B. 1981. Sociality as a life-history tactic of ground-squirrels. *Oecologia* 48:36–49.

Boutin, S., and K. W. Larsen. 1993. Does food availability affect growth and survival of males and females differently in a promiscuous small mammal, *Tamiasciurus hudsonicus? Journal of Animal Ecology* 62:364–370.

Clutton-Brock, T. H. 1989. Mammalian mating systems. *Proceedings of the Royal Society of London B* 126:339–372.

Istock, C. A. 1995. Telescopes, red squirrels, Congress, courtrooms, and conservation. Pages 19–35 in C. A. Istock and R. S. Hoffman, eds. *Storm over a mountain island: Conservation biology and the Mt. Graham affair.* University of Arizona Press, Tucson.

Koford, R. R. 1982. Mating system of a territorial tree squirrel (*Tamiasciurus douglasii*) in California. *Journal of Mammalogy* 63:274–283.

Koprowski, J. L. 1998. Conflict between the sexes: A review of social and mating systems of the tree squirrels. Pages 33–41 in M. A. Steele, J. F. Merritt, and D. A. Zegers, eds. *Ecology and evolutionary biology of tree squirrels.* Special Publications, Virginia Museum of Natural History, Martinsville.

Larsen, K. W., and S. Boutin. 1994. Movements, survival, and settlement of red squirrel (*Tamiasciurus hudsonicus*) offspring. *Ecology* 75:214–223.

———. 1998. Sex-unbiased philopatry in the North American tree squirrel: (*Tamiasciurus hudsonicus*). Pages 21–32 in M. A. Steele, J. F. Merritt, and D. A. Zegers, eds. *Ecology and evolutionary biology of tree squirrels.* Special Publication no. 6. Virginia Museum of Natural History, Martinsville.

Magurran, A. E. 1988. *Ecological diversity and its measurement.* Princeton University Press, Princeton, New Jersey.

Michener, G. R. 1983. Kin identification, matriarchies, and the evolution of sociality in ground-dwelling sciurids. Pages 528–572 in J. F. Eisenberg and D. G. Kleiman, eds. *Recent advances in the study of mammalian behavior.* Special Publications, American Society of Mammologists, Lawrence, Kansas.

Selonen, V., I. K. Hanski, and P. C. Stevens. 2001. Space use of the Siberian flying squirrel *Pteromys volans* in fragmented forest landscapes. *Ecography* 24:588–600.

Smith, A. A., and R. W. Mannan. 1994. Distinguishing characteristics of Mount Graham red squirrel midden sites. *Journal of Wildlife Management* 58:437–445.

Strickler, G. S. 1959. Use of the densiometer to estimate density of forest canopy on permanent sample plots. U.S. Forest Service, Pacific Northwest Forest and Range Experiment Station, Research Note, No. 180.

Thompson, D. C. 1977. The social system of the grey squirrel. *Behaviour* 64:305–328.

Wauters, L., and A. A. Dhondt. 1992. Spacing behavior of red squirrels, *Sciurus vulgaris*: Variation between habitats and the sexes. *Animal Behaviour* 43:297–311.

Part V

Mt. Graham Red Squirrel Behavior and Ecology

. . . as the squirrels manifest no concern whether the woods will bear chestnuts this year or not.
—Henry David Thoreau

TRENDS IN ABUNDANCE for red squirrels on Mt. Graham suggest that the carefree life of Thoreau's squirrels may not characterize the population of this endangered species. Despite the availability of long-term data sets on the abundance of Mt. Graham red squirrels, little was known of their ecology. Often, assumptions were made that the ecology and behavior would be identical to other subspecies found across the red squirrel's exceedingly wide distribution. The dearth of knowledge was the result of a number of conditions ranging from the rarity of the squirrel, the difficulty of capturing and observing the species, the rugged terrain, and the extreme seasonality of climate. Recent advances in capture techniques and monitoring technologies have permitted rapid increases in our knowledge of the detailed life history of endangered Mt. Graham red squirrels.

In Part V, we review the state of knowledge of the behavior and ecology of Mt. Graham red squirrels in an attempt to gain insight into the plight of the species. Vicki Greer and John Koprowski describe how Mt. Graham red squirrels allocate their time to various tasks throughout the year and demonstrate the considerable investment placed into energy acquisition. Diane Angell examines food-hoarding tactics and describes a dynamic process of storage including larder- and scatterhoarding. The significant variation in cone and fungal energy availability and perishability are shown to have likely import by Bill Miller and Bill Yoder and by Craig Frank and Stephanie Cox. Finally, the natural history of Mt. Graham red squirrels is reviewed by Munroe and others, and compared to other populations of red squirrels. Their results suggest that Mt. Graham red squirrels demonstrate lower reproductive rates and may have relatively large home ranges relative to more northerly populations. Such insights into the Mt. Graham red squirrel life history continue to provide a solid framework on which to build effective conservation strategies.

CHAPTER 15

Time Budget of the Mt. Graham Red Squirrel

VICKI L. GREER AND JOHN L. KOPROWSKI

A SUCCESSFUL ENERGETIC STRATEGY for any animal necessitates allocation of finite resources towards maintenance, growth, and reproduction in an ever-changing environment (Gadgil and Bossert 1970). Time budgets can provide insight into strategies an animal employs and into how strategies change under different ecological conditions (Duncan 1985) and may be combined with measures of metabolism to determine energetic costs of different activities (Wolf and Hainsworth 1971; Williams et al. 1997). Changes in food availability may drive changes in time budget as animals adjust time spent feeding and resting (Bronikowski and Altmann 1996; Poulsen et al. 2001). Risk of predation may result in more time spent in vigilance and less time provisioning young (Sasvari and Hegyi 2000). Differential strategies to maximize fitness may be noted in time budgets of animals in marginal habitats (Caudron et al. 2001; Huang et al. 2003). Changes in weather can affect time spent in feeding and energy conservation (Wishart and Sealy 1980; Nash 1998). Other factors, such as parasite load, can cause shifts in patterns of parental care (Tripet et al. 2002). Social structure in a population may also be inferred from how animals apportion time (Duncan 1980; Cote et al. 1997).

Investigating time budgets of males and females can elucidate strategies used by each sex throughout reproductive cycles. Males spend more time traveling in search of mates and in vigilance or territory defense during the breeding season, whereas females change activity to deal with demands of production and rearing of young through increased foraging and feeding and/or decreased movement (Boyd et al. 1991; Buech 1995). Parents shift time from feeding to protect and provision young (Seddon and Nudds 1994; Aviles 2003). Age differences in time budgets are often found; juveniles spend more time feeding and resting during parental care (Boy and Duncan 1979; Alonso and Alonso 1993).

Time budgets can help management and conservation decisions (Plumpton and Andersen 1997; Petrie and Petrie 1998) and assessments of behavioral changes when habitat degrades or is manipulated (Gostomski and Evers 1998; Passamani 1998). For species about which little is known, construction of a time budget is often the first step towards understanding their ecology (Verts and Gehman 1991; Todorovsky 1997).

Investigations of sciurids have employed time budgets. Male Richardson's ground squirrels (*Spermophilus richardsonii*) reduced feeding, increased monitoring of estrous females, and interacted more with males, whereas estrous females evaded males and increased travel (Michener and McLean 1996). Interspecific competition between Eurasian red squirrels (*Sciurus vulgaris*) and introduced eastern gray squirrels (*S. carolinensis*) has been examined by comparison of time budgets for animals in areas with and without overlap (Wauters and Gurnell 1999; Wauters et al. 2001). Female Mexican fox squirrels (*S. nayaritensis chiricahuae*) varied little in activities throughout the year and focused on food acquisition, whereas males increased feeding before the breeding season and increased locomotion during the breeding season, likely in search of mates (Koprowski and Corse 2005).

For species of precarious conservation status, time budgets may inform management decisions. Big Cypress fox squirrels (*S. niger avicennia*) in urban areas have more stable foraging time budgets than in native habitats, due in part to a good annual food supply. Managers can use activity budgets to plan future developments (Jodice and Humphrey 1992).

From 1989–1991, over 3,100 hours of observations were made to assess the behavior of endangered Mt. Graham red squirrels relative to their distance from construction activity at an astrophysical site. Construction of the observatory had no discernable effect on behavior of red squirrels (Young et al. 2000). Herein, we re-examine a subset of these data to assess activity of squirrels by sex, reproductive condition, and season.

Methods

The study area encompassed 337.9 ha of the three highest peaks in the Pinaleño Range in southeast Arizona and included two forest habitat types: mixed-conifer forest (MC) and spruce-fir (SF) forest. Middens were characterized by denser foliage and canopy, greater basal area and

Table 15.1. Behaviors and vocalizations of Mt. Graham red squirrels, Graham County, Arizona

Behaviors	
Feeding	eating any food item (cone, mushroom, bud, insect, etc.)
Foraging	food gathering or searching, in trees or on ground
Breeding	mating chases with one or several squirrels, copulations
In Nest	squirrel known to be in a bolus or cavity nest
Out of Sight	squirrel out of sight of observer, still in midden area
Caching	placing food items in snag, log, or cone scale pile
Movement	general movement in midden area that can't be classified
Territory Defense	inter- and intra-specific, active chasing and/or territorial calls
Vigilant	alert posture, response to humans, predators, construction activity
Miscellaneous	groom, bask, play, drink, defecate, urinate, nest maintenance, unusual
Vocalizations	
Chatter	territorial defense, announce presence, signature call for individual recognition
Bark	alarm or agitation call, indicates presence of predator (# barking bouts tallied)
Squeak	indicates agitation, often precedes barks

tree density, and more large logs and snags than surrounding forest. In addition, trees in SF were older than those in MC (A. Smith and Mannan 1994).

Behaviors were observed for 2 hours in 15 sec intervals between 0700 to 1600 h in a 10 m radius of occupied middens from May to October 1991. Activities were categorized into 10 behavior types, and three vocalization types were common enough to be tallied: chatter, bark, and squeak (table 15.1). Behavioral data on adult animals were examined by reproductive status (females without litters, females with litters, and males) and season. Seasons were May and June (breeding season), July and August (litter emergence), and September to October (post-juvenile dispersal and food gathering).

Most animals were not marked for individual identification due to permit limitations; however, middens were under frequent observation and turnover of residents was generally detected. To ensure independence of observations at a midden, only observations made ≥1 week apart were included. We compared behaviors, time spent at midden, and vocalizations of reproductive groups in each habitat (MC and SF) irrespective of season. We also assessed variation in the above parameters within and across seasons for each reproductive group. For within- and across-seasons comparisons, only data from SF were included, as sample sizes in MC were too small.

Kruskal-Wallis tests were used to assess initial intergroup differences. Due to the endangered status and rarity of squirrels, significance for Kruskal-Wallis tests was implied when $P \leq 0.05$ and potential biological significance was noted when $P < 0.10$; tests were followed by additional post hoc multiple comparisons (Conover-Inman method—StatsDirect software; Conover 1999). Bonferroni adjustments to probabilities yielded significance when $P \leq 0.017$ and potential biological significance when $P < 0.05$.

Results

Seasons Combined—Time Budget

Mt. Graham red squirrels expend unequal effort on behaviors (fig. 15.1). Food acquisition (feeding, foraging, and caching) and resting in nest accounted for large proportions of time spent in middens in MC (♀ without litters, 48 percent; ♀ with litters, 25 percent; ♂, 30 percent). Differences in time spent caching occurred (Kruskal-Wallis: $H = 5.37$, df = 2, $P = 0.068$), as females without litters tended to spend less time caching than males (Conover-Inman multiple comparison: $C = 2.01$, df = 47, $P = 0.044$).

In SF, a similar pattern was noted, with combined food-related and resting behaviors accounting for a large proportion of time budgets (♀ without litters, 38 percent; ♀ with litters, 39 percent; ♂, 36 percent). Time spent feeding differed among reproductive groups ($H = 8.76$, df = 2, $P =$ 0.013); females with litters spent more time feeding than males ($C = 1.97$, df = 181, $P = 0.005$). Time spent caching food also differed among groups ($H = 9.32$, df = 2, $P = 0.01$), with males caching more than females ($C =$

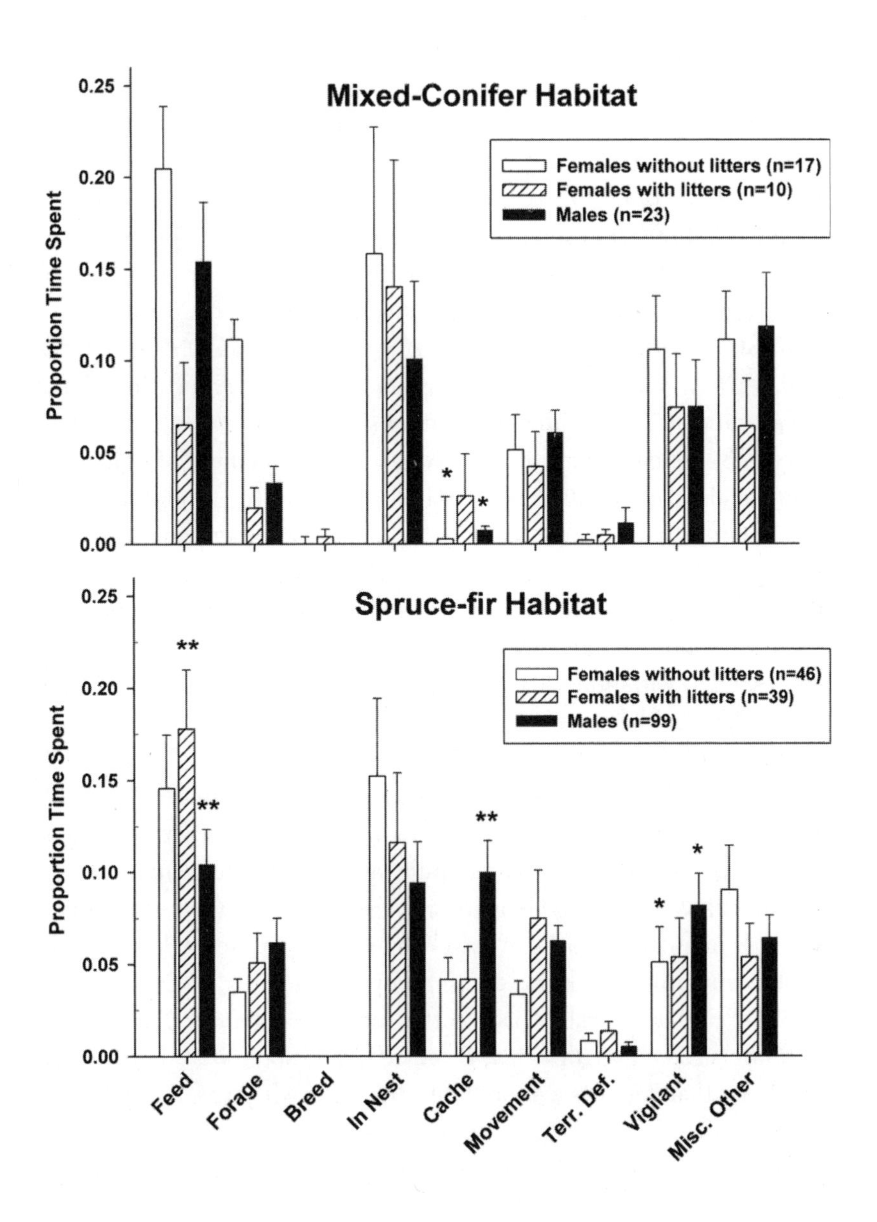

Figure 15.1. Mean time spent by Mt. Graham red squirrels in behaviors, as a proportion of total time spent in midden for each reproductive group and habitat in 1991, seasons grouped. N = number of 2 h observations. Significance is shown as $^{**}P \leq 0.017$, $^{*}P = 0.018 - 0.05$, Conover-Inman post hoc multiple comparisons with Bonferroni adjustment.

1.97, df = 181; ♀ with litters, $P = 0.011$; ♀ without litters, $P = 0.031$). Time devoted to vigilance also differed ($H = 5.74$, df = 2, $P = 0.057$), with males tending towards more vigilance than females without litters (C = 1.97, df = 181, $P = 0.02$).

Time out of sight accounted for the largest individual proportion of time spent in the midden for all reproductive groups in both habitats (MC: ♀ without litters, 0.23 ± 0.07; ♀ with litters, 0.54 ± 0.1; ♂, 0.41 ± 0.06; SF: ♀ without litters, 0.42 ± 0.05; ♀ with litters, 0.39 ± 0.05; ♂, 0.40 ± 0.03). Time out of sight was similar among groups in SF, but groups differed in the time spent out of sight in MC ($H = 6.56$, df = 2, $P = 0.038$). Females with litters in MC were recorded out of sight more than females without litters (C = 2.01, df = 47, $P = 0.01$). Out-of-sight data were retained so as not to confound the analysis of other behaviors.

Time spent at midden. Squirrels were in their middens >50 percent of the 2 hr observation periods: (mean proportion ± *SE*) MC: ♀ without litters, 0.62 ±0.09; ♀ with litters, 0.67 ±0.08, ♂, 0.54 ±0.05; SF: ♀ without litters, 0.61 ±0.05; ♀ with litters, 0.51 ±0.05, ♂, 0.52 ±0.03). In MC, groups varied in number of visits to the midden (fig. 15.2; $H = 10.27$, df = 2, $P = 0.006$). Females with litters (C = 2.01, df = 47, $P = 0.008$) and males (C = 2.01, df = 47, $P = 0.003$) visited middens more than females without litters. Number of visits to middens was similar among SF groups (fig. 15.2).

Total time spent at middens and average duration of visits were similar among reproductive groups in each habitat. Average interval between visits was similar in MC; however, SF groups differed (fig. 15.2: $H = 10.61$, df = 2, $P = 0.005$). Males had shorter inter-visit intervals than females with litters (C = 1.98, df = 121, $P = 0.004$), and somewhat shorter inter-visit times than females without litters (C = 1.98, df = 121, $P = 0.020$).

Vocalizations. Vocalizations varied between habitats dependent on the emitter of the call (fig. 15.3) In SF, number of vocalizations varied among groups (H= 9.21, df = 2, $P = 0.01$). Females with litters vocalized more than males (C = 1.97, df = 181, $P = 0.003$) and somewhat more than females without litters (C = 1.97, df = 181, $P = 0.037$). In MC, number of vocalizations was similar among groups.

Vocalization types also differed. In MC, chatters varied among groups ($H = 5.78$, df = 2, $P = 0.056$); females with litters tended to chatter more

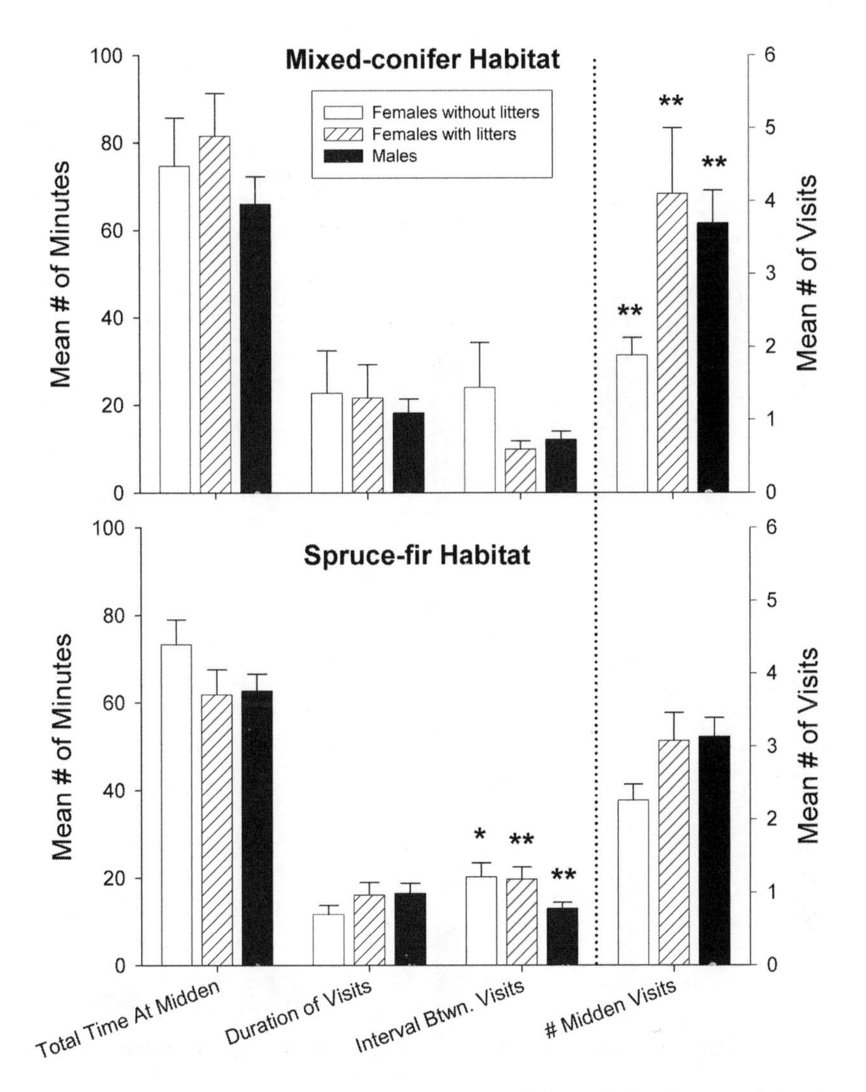

Figure 15.2. Measures of time spent at middens (left of vertical line) and visits to middens (right of vertical line) by Mt. Graham red squirrels, for each reproductive group and habitat in 1991, seasons grouped. Significance is shown as $^{**}P \leq 0.017$, $^{*}P = 0.018 - 0.05$, Conover-Inman post hoc multiple comparisons with Bonferroni adjustment.

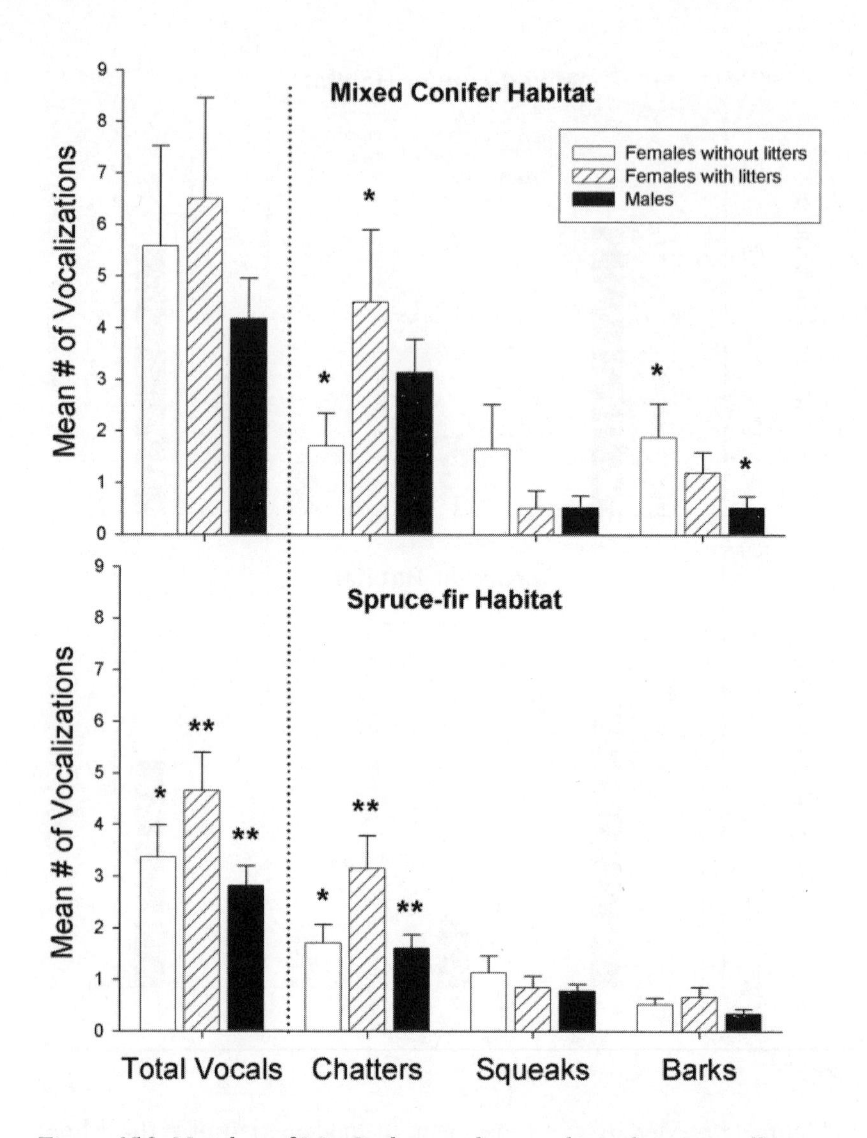

Figure 15.3. Number of Mt. Graham red squirrel vocalizations: all types combined (left of vertical line) and individual types (right of vertical line) for each reproductive group and habitat in 1991, seasons grouped. Significance is shown as $^{**}P \leq 0.017$, $^{*}P = 0.018 - 0.05$, Conover-Inman post hoc multiple comparisons with Bonferroni adjustment.

than females without litters (C = 2.01, df = 47, $P = 0.028$). Differences in chatters between groups existed in SF ($H = 8.35$, df = 2, $P = 0.015$); again, females with litters chattered more than other groups (C = 1.97, df = 181, ♂, $P = 0.007$; ♀ without litters, $P = 0.034$). Number of barks differed among groups in MC ($H = 6.28$, df = 2, $P = 0.043$), but not in SF. Females without litters tended to bark more than males (C = 2.01, df = 47, $P = 0.042$).

Within Seasons—Time Budget

In May–June, feeding time differed among groups (table 15.2: $H = 6.15$, df = 2, $P = 0.046$). Males fed less than females (C = 2.02, df = 43, $P = 0.022$). Foraging also varied among groups ($H = 5.66$, df = 2, $P = 0.059$). Females without litters tended to forage more than females with litters (C = 2.02, df = 43, $P = 0.043$). Vigilance differed among groups ($H = 5.50$, df = 2, $P = 0.064$); females with litters tended to be more vigilant than females without litters (C = 2.02, df = 43, $P = 0.026$).

In July–Aug, groups differed in time spent feeding (table 15.2: $H = 8.34$, df = 2, $P = 0.015$); males spent less time feeding than females, especially females with litters (C = 2.00, df = 69, $P = 0.004$). Time spent in movement differed among groups ($H = 6.48$, df = 2, $P = 0.039$), as males tended to move more than females without litters (C = 2.00, df = 69, $P = 0.02$). In Sep–Oct, time budgets of reproductive groups did not differ.

Time spent at midden. In May–Jun, only total time spent at midden varied among groups (table 15.3: $H = 8.95$, df = 2, $P = 0.011$), and no measures differed in Jul–Aug. Males spent less time at middens than females without litters (C = 2.02, df = 43, $P = 0.004$), and marginally less at middens than females with litters (C = 2.02, df = 43, $P = 0.043$). In Sep–Oct, however, groups differed in visits to middens ($H = 6.74$, df = 2, $P = 0.035$). Females with litters visited middens more than females without litters (C = 2.00, df = 63, $P = 0.013$) and tended to visit more than males (C = 2.00, df = 63, $P = 0.048$). Inter-visit interval varied among groups ($H = 6.64$, df = 2, $P = 0.036$); males tended towards shorter intervals than both groups of females (C = 2.02, df = 44; ♀ with litters, $P = 0.031$; ♀ without litters, $P = 0.039$).

Table 15.2. Time spent by Mt. Graham red squirrels within spruce-fir in behaviors (mean minutes ± SE), by reproductive group, *within* seasons

N = number of 2 h observations.
Data with significant differences are shown and noted in bold as **$P < 0.017$, *$P = 0.018 – 0.05$, Conover-Inman post hoc multiple comparisons with Bonferroni adjustment.

Season	Behavior	♀ no litter	♀ w/ litter	♂
		(*n* = 19)	(*n* = 6)	(*n* = 21)
May–Jun	Feed	***14.79** ± 4.10	15.00 ± 7.36	***4.29** ± 1.42
	Forage	***2.25** ± 0.57	***0.29** ± 0.19	1.43 ± 0.79
	Breed	0.00 ± 0.00	0.00 ± 0.00	0.00 ± 0.00
	In Nest	21.55 ± 8.80	22.67 ± 15.33	3.91 ± 2.41
	Out of Sight	36.80 ± 9.18	36.17 ± 16.31	25.06 ± 6.76
	Cache	1.26 ± 0.87	0.17 ± 0.17	4.93 ± 1.75
	Movement	2.21 ± 0.57	2.08 ± 1.12	2.51 ± 0.63
	Territory Defense	0.73 ± 0.58	0.25 ± 0.25	0.17 ± 0.13
	Vigilant	***1.71** ± 0.37	***6.00** ± 1.94	4.00 ± 1.23
	Misc Other Behavior	1.63 ± 0.73	1.79 ± 1.69	1.94 ± 0.92
		(*n* = 11)	(*n* = 18)	(*n* = 43)
Jul–Aug	Feed	8.21 ± 3.99	****12.11** ± 2.96	****3.51** ± 0.76
	Forage	1.41 ± 0.61	2.92 ± 0.97	2.37 ± 0.47
	Breed	0.00 ± 0.00	0.00 ± 0.00	0.00 ± 0.00
	In Nest	5.64 ± 2.72	10.17 ± 6.19	13.35 ± 4.32
	Out of Sight	23.68 ± 6.95	25.05 ± 6.86	31.71 ± 4.58
	Cache	5.84 ± 3.41	6.47 ± 3.94	8.52 ± 2.07
	Movement	***1.41** ± 0.67	1.83 ± 0.49	***2.66** ± 0.36
	Territory Defense	1.59 ± 0.80	1.21 ± 0.54	0.43 ± 0.11
	Vigilant	3.09 ± 1.55	2.63 ± 0.63	3.75 ± 0.70
	Misc Other Behavior	***7.23** ± 3.26	***1.60** ± 0.71	3.97 ± 1.10
		(*n* = 16)	(*n* = 15)	(*n* = 35)
Sep–Oct	Feed	3.55 ± 1.22	6.68 ± 1.91	6.05 ± 1.44
	Forage	2.53 ± 0.97	1.90 ± 0.80	5.68 ± 1.80
	Breed	0.00 ± 0.00	0.00 ± 0.00	0.00 ± 0.00
	In Nest	7.06 ± 4.87	3.93 ± 2.81	4.86 ± 1.94
	Out of Sight	40.30 ± 9.69	25.75 ± 5.80	30.59 ± 5.15
	Cache	3.50 ± 1.76	1.35 ± 0.73	4.61 ± 1.63
	Movement	1.73 ± 0.41	3.12 ± 0.58	2.61 ± 0.41
	Territory Defense	0.11 ± 0.08	0.72 ± 0.46	0.26 ± 0.14
	Vigilant	2.20 ± 1.00	0.85 ± 0.27	2.64 ± 0.48
	Misc Other Behavior	11.39 ± 5.85	5.90 ± 3.55	4.90 ± 1.46

Table 15.3. Time spent by Mt. Graham red squirrels within spruce-fir at middens (mean minutes ± SE), and vocalizations by reproductive group, *within* seasons.

N = number of 2 h observations.
Data with significant differences are shown and noted in bold as **$P < 0.017$, *$P = 0.018 – 0.05$, Conover-Inman post hoc multiple comparisons with Bonferroni adjustment.

Season	Measure	♀ no litter	♀ w/ litter	♂
		(n = 19)	(n = 6)	(n = 21)
May–Jun	# Visits to Midden	1.84 ± 0.22	1.17 ± 0.17	2.81 ± 0.73
	Visit Duration	12.25 ± 4.55	19.75 ± 19.25	12.36 ± 2.24
	Interval Between	18.61 ± 3.76	35.50 ± —	18.79 ± 6.11
	Total Time at Midden	****82.93** ± 8.01	***84.41** ± 18.67	****48.23** ± 8.09
	Total # Vocals	4.42 ± 1.35	4.00 ± 1.81	1.86 ± 0.48
	# Chatters	1.90 ± 0.73	2.50 ± 1.36	0.62 ± 0.27
	# Squeaks	1.80 ± 0.71	0.50 ± 0.34	0.71 ± 0.23
	# Barks	0.74 ± 0.20	1.00 ± 0.68	0.52 ± 0.30
		(n = 11)	(n = 18)	(n = 43)
Jul–Aug	# Visits to Midden	3.00 ± 0.59	3.11 ± 0.63	3.63 ± 0.41
	Visit Duration	14.94 ± 3.82	17.45 ± 4.82	10.87 ± 2.13
	Interval Between	18.63 ± 7.38	22.70 ± 5.36	12.48 ± 1.56
	Total Time at Midden	58.09 ± 10.67	63.97 ± 8.76	70.27 ± 5.23
	Total # Vocals	2.55 ± 0.76	****4.67** ± 0.71	****2.91** ± 0.69
	# Chatters	1.36 ± 0.53	2.33 ± 0.59	1.30 ± 0.37
	# Squeaks	0.73 ± 0.47	1.28 ± 0.38	1.21 ± 0.27
	# Barks	0.46 ± 0.21	***1.06** ± 0.32	***0.33** ± 0.13
		(n = 16)	(n = 15)	(n = 35)
Sep–Oct	# Visits to Midden	****2.25** ± 0.40	****3.80** ± 0.57	***2.71** ± 0.32
	Visit Duration	8.31 ± 1.67	14.15 ± 3.44	24.11 ± 4.92
	Interval Between	23.68 ± 6.39	15.66 ± 1.92	***10.99** ± 1.43
	Total Time at Midden	72.38 ± 10.65	50.17 ± 6.65	62.19 ± 6.82
	Total # Vocals	2.69 ± 0.64	4.93 ± 1.59	3.29 ± 0.62
	# Chatters	1.75 ± 0.47	4.40 ± 1.36	2.57 ± 0.56
	# Squeaks	0.63 ± 0.30	0.47 ± 0.29	0.29 ± 0.11
	# Barks	0.31 ± 0.25	0.07 ± 0.07	0.26 ± 0.10

Vocalizations. Vocalizations did not differ between groups in May–Jun or Sep–Oct. However, in Jul–Aug, number of vocalizations differed among groups (table 15.3: $H = 8.09$, df = 2, $P = 0.018$). Females with litters vocalized more than males ($C = 2.00$, df = 69, $P = 0.005$), and marginally more than females without litters ($C = 2.00$, df = 69, $P = 0.053$). Number of barks also varied among groups ($H = 8.28$, df = 2, $P = 0.016$), as females with litters tended to bark more than males ($C = 2.00$, df = 69, $P = 0.02$).

Across seasons—time budget. Females without litters differed seasonally in time spent feeding (table 15.4: $H = 4.89$, df = 2, $P = 0.087$) with more in May–Jun than in Sep–Oct ($C = 2.02$, df = 43, $P = 0.032$). Caching differed across seasons ($H = 5.96$, df = 2, $P = 0.051$), as females without litters cached slightly less in May–Jun than in Sep–Oct ($C = 2.02$, df = 43, $P = 0.046$).

For females with litters, feeding behavior was similar across seasons with only slight differences in foraging (table 15.4: $H = 5.09$, df = 2, $P = 0.078$), with less foraging in May–Jun than in Jul–Aug ($C = 2.03$, df = 36, $P = 0.036$). Vigilance by females with litters also varied by seasons ($H = 9.51$, df = 2, $P = 0.009$) and was elevated in May–Jun and reduced in Sep–Oct ($C = 2.03$, df = 36, $P = 0.003$).

Foraging (table 15.4: $H = 9.5$, df = 2, $P = 0.009$), but not feeding, by males differed across seasons, gradually increasing through the year (Sep–Oct vs. May–Jun: $C = 1.99$, df = 96, $P = 0.003$).

Time spent at midden. Time spent at middens varied by sex (table 15.4). Females without litters spent similar amounts of time at middens across seasons. Females with litters varied seasonally in visits to middens ($H = 9.97$, df = 2, $P = 0.007$), visiting more in May–Jun ($C = 2.03$, df = 36; Jul–Aug, $P = 0.021$; Sep–Oct, $P = 0.001$).

Males' visits to middens differed by season ($H = 4.91$, df = 2, $P = 0.086$), and they tended to visit less in May–Jun than Jul–Aug ($C = 1.99$, df = 96, $P = 0.04$). Time spent at middens by males also varied across seasons ($H = 4.75$, df = 2, $P = 0.093$), with more visits in Jul–Aug than in May–Jun ($C = 1.99$, df = 96, $P = 0.03$).

Table 15.4. Time spent by Mt. Graham red squirrels within spruce-fir in behaviors (mean minutes ± SE), time spent at the midden, and vocalizations by reproductive group, *across* seasons

N = number of 2 h observations. Only data with significant differences are shown and noted in bold as **$P < 0.017$, *$P = 0.018$ - 0.05, Conover-Inman post hoc multiple comparisons with Bonferroni adjustment.

Repro	Behavior	May–Jun	Jul–Aug	Sep–Oct
Time Spent in Behaviors				
		(n = 19)	(n = 11)	(n = 16)
♀ no litter	Feed	***14.79** ± 4.10	8.21 ± 3.99	***3.55** ± 1.22
	Cache	***1.26** ± 0.87	5.84 ± 3.41	***3.50** ± 1.76
	Misc Other Behavior	***1.63** ± 0.73	***7.23** ± 3.26	11.39 ± 5.85
		(n = 6)	(n = 18)	(n = 15)
♀ w/ litter	Forage	***0.29** ± 0.19	***2.92** ± 0.97	1.90 ± 0.80
	Vigilant	****6.00** ± 1.94	***2.63** ± 0.63	****0.85** ± 0.27
		(n = 21)	(n = 43)	(n = 35)
♂	Forage	****1.43** ± 0.79	2.37 ± 0.47	****5.68** ± 1.80
Time Spent at Midden				
		(n = 6)	(n = 18)	(n = 15)
♀ w/ litter	# Visits to Midden	****1.17** ± 0.17	***3.11** ± 0.63	****3.80** ± 0.57
		(n = 21)	(n = 43)	(n = 35)
♂	# Visits to Midden	***2.81** ± 0.73	***3.63** ± 0.41	2.71 ± 0.32
	Tot. Time - Midden	***48.23** ± 8.09	***70.27** ± 5.23	62.19 ± 6.82
Vocalizations				
		(n = 19)	(n = 11)	(n = 16)
♀ no litter	# Barks	***0.74** ± 0.20	0.46 ± 0.21	***0.31** ± 0.25
		(n = 6)	(n = 18)	(n = 15)
♀ w/ litter	# Barks	1.00 ± 0.68	****1.06** ± 0.32	****0.07** ± 0.07
		(n = 21)	(n = 43)	(n = 35)
♂	# Chatters	****0.62** ± 0.27	1.30 ± 0.37	****2.57** ± 0.56
	# Squeaks	0.71 ± 0.23	****1.21** ± 0.27	****0.29** ± 0.11

Vocalizations. Sex and reproductive status influenced vocalizations (table 15.4). Females without litters vocalized similarly throughout the year, except for barking (H = 5.28, df = 2, P = 0.071); barking was slightly more common in May–Jun than in Sep–Oct (C = 2.02, df = 43, P = 0.055). Females with litters increased barking throughout seasons (H = 8.3, df = 2, P = 0.016); more barking was recorded in Jul–Aug than in Sep–Oct (C = 2.03, df = 36, P = 0.016).

Males also gradually increased vocalizations from May–Jun through Sep–Oct. Chatter calls differed across seasons (H = 10.23, df = 2, P = 0.006), being infrequent in May–Jun and the most in Sep–Oct (C = 1.99, df = 96, P = 0.003). Squeaks also differed (H = 9.17, df = 2, P = 0.01), with the least in Sep–Oct compared to Jul–Aug (C = 1.99, df = 96, P = 0.008).

Discussion

Mt. Graham red squirrels spend a large proportion of their time in food-related and resting behaviors. Red squirrels from other populations that are not of precarious conservation status similarly apportion time (C. Smith 1968; Ferron et al. 1986; Benhamou 1996). Mt. Graham red squirrels spent about twice as much time in food acquisition and rest than other observed behaviors combined. In addition, much time recorded as out of sight was likely at rest, basking or in a nest.

Females generally spent more time feeding than males. When feeding is considered across seasons, females spent most time feeding in spring, as several females were lactating in May and some litters had emerged by late June. Increased nutritional demands and the quest to retain physical condition may explain this allocation of effort. Reproductive success is often dependant on food availability in other squirrel populations (Becker 1993; Humphries and Boutin 1996; Larsen 1997). Females without litters spent nearly equal time feeding as females with litters in the May–Jun period, but fed less in the latter two seasons as energy demands likely increased for gravid and nursing females.

Males fed least often in early seasons, gradually increasing throughout the year. Males spent considerably less time in nests than females in May–Jun and were only seen in their middens approximately half the time that females were seen. May–Jun is near the end of the breeding season (six breeding chases seen in May, none in June), when males devote more time to breeding activities (C. Smith 1968; Rusch and Reeder 1978; Fancy 1981).

Mt. Graham red squirrels, like other red squirrels in coniferous forests (Finley 1969; C. Smith and Reichman 1984), larderhoard and defend the main food cache. Active defense consists of chasing and/or vocalizing to deter intruders (Kemp and Keith 1970; Gurnell 1984; Lair 1990). Chatter calls are most often associated with territory defense (Embry 1970; C. Smith 1978; Lair 1990) and were most common in Mt. Graham red squirrels. Females with litters chattered more than either females without litters or males. This more intense territorial defense by breeding females has also been noted in other red squirrels (Price et al. 1990), as females with litters may "value" territories the most to safely rear young and can bequeath territories to young (Boutin and Schweiger 1988; Price 1989; Price and Boutin 1993).

High levels of caching observed in Jul–Aug on Mt. Graham are surprising, rather than in Sep–Oct, when caching behavior is generally most prevalent (M. Smith 1968; Streubel 1968; Finley 1969). However, the 1991 cone crop was poor, and squirrels may have gathered cones as soon as this scarce resource became available. Other squirrels exhibit this sort of differential response to variable food resources (Deutch 1978; Ferron et al. 1986; Wauters et al. 1992).

Seasonality of energy availability typical of most non-hibernating temperate tree squirrels (Steele and Koprowski 2001) demands accumulation of energy stores during times of abundance. Time budget analyses confirm the considerable time that Mt. Graham red squirrels allocate to resource acquisition and conservation. While inter-habitat comparisons were beyond the scope of the current research, some patterns did emerge: time spent feeding and caching by reproductive groups was dissimilar between the habitats. Because differences in structure and composition exist between the two habitats (Stromberg and Patten 1991; A. Smith and Mannan 1994), how might these differences influence behavior? Yearly differences in cone abundance, relative population densities, and anthropogenic changes in forest structure also may affect how squirrels allocate time. Investigation of these influences can provide further insights into the strategies that Mt. Graham red squirrels employ to survive in a spatially and temporally varying environment.

Acknowledgments

We thank P. Young and R. Davis for design of observation protocols. K. Kost, W. Kreisel Jr., M. Maghini, A. Mohammed, M. Morgan, S. Nelson, D. Oliver, C. Russworm, M. Stegman, S. Winter, and P. Zoeller assisted in data collection. Funding and in-kind support for this research was provided by the University of Arizona's Office of the Vice President for Research, Steward Observatory, U.S. Forest Service, and Arizona Game and Fish Department.

Literature Cited

Alonso, J. A., and J. C. Alonso. 1993. Age-related differences in time budgets and parental care in wintering common cranes. *Auk* 110(1):78–88.

Aviles, J. M. 2003. Time budget and habitat use of the common crane wintering in dehesas of southwestern Spain. *Canadian Journal of Zoology* 81(7):1233–1238.

Becker, C. D. 1993. Environmental cues of estrus in the North American red squirrel (*Tamiasciurus hudsonicus* Bangs). *Canadian Journal of Zoology* 71:1326–1333.

Benhamou, S. 1996. Space use and foraging movements in the American red squirrel (*Tamiasciurus hudsonicus*). *Behavioural Processes* 37:89–102.

Boutin, S., and S. Schweiger. 1988. Manipulation of intruder pressure in red squirrels (*Tamiasciurus hudsonicus*): Effects on territory size and acquisition. *Canadian Journal of Zoology* 66:2270–2274.

Boy, V., and P. Duncan. 1979. Time-budgets of Camargue horses. I. Developmental changes in the time-budgets of foals. *Behaviour* 71:187–202, Parts 3–4.

Boyd, I. L., N. J. Lunn, and T. Barton. 1991. Time budgets and foraging characteristics of lactating Antarctic fur seals. *Journal of Animal Ecology* 60(2):577–592.

Bronikowski, A. M., and J. Altmann. 1996. Foraging in a variable environment: Weather patterns and the behavioral ecology of baboons. *Behavioral Ecology and Sociobiology* 39(1):11–25.

Buech, R. R. 1995. Sex differences in behavior of beavers living in near-boreal lake habitat. *Canadian Journal of Zoology* 73(11):2133–2143.

Caudron, A. K., C. R. Joiris, and J. C. Ruwet. 2001. Comparative activity budget among grey seal (*Halichoerus grypus*) breeding colonies—the importance of marginal populations. *Mammalia* 65(3):373–382.

Conover, W. J. 1999. *Practical nonparametric statistics*, 3rd ed. John Wiley and Sons, New York.

Cote, S. D., J. A Schaefer, and F. Messier. 1997. Time budgets and synchrony of activities in muskoxen: The influence of sex, age, and season. *Canadian Journal of Zoology* 75(10):1628–1635.

Deutch, R. S. 1978. Seasonal activity budget of the red squirrel (*Tamiasciurus hudsonicus*) in southern Ohio. Master's thesis, University of Dayton, Dayton, Ohio.

Duncan, P. 1980. Time-budgets of Camargue horses. II. Time-budgets of adult horses and weaned sub-adults. *Behaviour* 72:26–49, Parts 1–2.

——. 1985. Time-budgets of Camargue horses. III. Environmental influences. *Behaviour* 92:188–208.

Embry, P. C. 1970. Vocal communication of the red squirrel, *Tamiasciurus hudsonicus*. Master's thesis, University of Montana, Missoula.

Fancy, S. G. 1981. Daily movements of red squirrels, *Tamiasciurus hudsonicus*. *The Canadian Field-Naturalist* 95(3):348–350.

Ferron, J., J. P. Ouellet, and Y. Lemay. 1986. Spring and summer time budgets and feeding behaviour of the red squirrel (*Tamiasciurus hudsonicus*). *Canadian Journal of Zoology* 64:385–391.

Finley, R. B., Jr. 1969. Cone caches and middens of *Tamiasciurus* in the Rocky Mountain region. Pages 233–273 in J. K. Jones Jr., ed. *Contributions in mammalogy*. University of Kansas Museum of Natural History Miscellaneous Publication 51.

Gadgil, M., and W. H. Bossert. 1970. Life historical consequences of natural selection. *American Naturalist* 104:1–24.

Gostomski, T. J., and D. C. Evers. 1998. Time-activity budget for common loons (*Gavia immer*) nesting on Lake Superior. *Canadian Field Naturalist* 112(2):191–197.

Gurnell, J. 1984. Home range, territoriality, caching behaviour and food supply of the red squirrel (*Tamiasciurus hudsonicus fremonti*) in a subalpine lodgepole pine forest. *Animal Behavior* 32:1119–1131.

Huang, C. M., F. W. Wei, M. Li, Y. B. Li, and R. Y. Sun. 2003. Sleeping cave selection, activity pattern and time budget of white-headed langurs. *International Journal of Primatology* 24(4):813–824.

Humphries, M. M., and S. Boutin. 1996. Reproductive demands and mass gains: A paradox in female red squirrels (*Tamiasciurus hudsonicus*). *Journal of Animal Ecology* 65:332–338.

Jodice, P. G. R., and S. R. Humphrey. 1992. Activity and diet of an urban population of Big Cypress fox squirrels. *Journal of Wildlife Management* 56(4):685–692.

Kemp, G. A., and L. B. Keith. 1970. Dynamics and regulation of red squirrel (*Tamiasciurus hudsonicus*) populations. *Ecology* 51(5):763–779.

Koprowski, J. L., and M. C. Corse. 2005. Sex differences in time budgets, activity periods, and behavior of Mexican fox squirrels. *Journal of Mammalogy* 86:947–952.

Lair, H. 1990. The calls of the red squirrel: A contextual analysis of function. *Behaviour* 115:254–282.

Larsen, K. W. 1997. Effects of hoard manipulations on life history and reproductive success of female red squirrels (*Tamiasciurus hudsonicus*). *Journal of Mammalogy* 78:192–203.

Michener, G. R., and I. G. McLean. 1996. Reproductive behaviour and operational sex ratio in Richardson's ground squirrels. *Animal Behaviour* 52:743–758.

Nash, L. T. 1998. Vertical clingers and sleepers: Seasonal influences on the activities and substrate use of *Lepilemur leucopus* at Beza Mahafaly Special Reserve, Madagascar. *Folia Primatologica* 69:204–217, Supplement 1.

Passamani, M. 1998. Activity budget of Geoffroy's marmoset (*Callithrix geoffroyi*) in an Atlantic forest in southeastern Brazil. *American Journal of Primatology* 46(4):333–340.

Petrie, S. A., and V. Petrie. 1998. Activity budget of white-faced whistling-ducks during winter and spring in northern Kwazulu-Natal, South Africa. *Journal of Wildlife Management* 62(3):1119–1126.

Plumpton, D. L., and D. E. Andersen. 1997. Habitat use and time budgeting by wintering ferruginous hawks. *The Condor* 99(4):888–893.

Poulsen, J. R., C. J. Clark, and T. B. Smith. 2001. Seasonal variation in the feeding ecology of the grey-cheeked mangabey (*Lophocebus albigena*) in Cameroon. *American Journal of Primatology* 54(2):91–105.

Price, K. 1989. Territorial defence and bequeathal by red squirrel (*Tamiasciurus hudsonicus*) mothers in the northern boreal forest. Master's thesis, Simon Fraser University, Burnaby, British Columbia, Canada.

Price, K., and S. Boutin. 1993. Territorial bequeathal by red squirrel mothers. *Behavioral Ecology* 4(2):144–150.

Price, K., S. Boutin, and R. Ydenberg. 1990. Intensity of territorial defense in red squirrels: An experimental test of the asymmetric war of attrition. *Behavioural Ecology and Sociobiology* 27:217–222.

Rusch, D. A., and W. G. Reeder. 1978. Population ecology of Alberta red squirrels. *Ecology* 59:400–420.

Sasvari, L., and Z. Hegyi. 2000. Avian predators influence the daily time budget of lapwings (*Vanellus vanellus*). *Folia Zoologica* 49(3):211–219.

Seddon, L. M., and T. D. Nudds. 1994. The costs of raising nidifugous offspring—brood rearing by giant Canada geese (*Branta canadensis maxima*). *Canadian Journal of Zoology* 72(3):533–540.

Smith, A. A., and W. R. Mannan. 1994. Distinguishing characteristics of Mt. Graham red squirrel midden sites. *Journal of Wildlife Management* 58(3):437–445.

Smith, C. C. 1968. The adaptive value of social organization in the genus of three [*sic*] squirrels *Tamiasciurus*. *Ecological Monographs* 38:31–63.

———. 1978. Structure and function of the vocalizations of tree squirrels (*Tamiasciurus*). *Journal of Mammalogy* 59:793–808.

Smith, C. C., and O. J. Reichman. 1984. The evolution of food caching by birds and mammals. *Annual Review of Ecology and Systematics* 15:329–351.

Smith, M. C. 1968. Red squirrel responses to spruce cone failure in interior Alaska. *Journal of Wildlife Management* 32:305–317.

StatsDirect, Ltd. 2005. StatsDirect statistical software. www.statsdirect.com. Altrincham, Cheshire, Great Britain.

Steele, M. A., and J. L. Koprowski. 2001. *North American tree squirrels*. Smithsonian Institution Press, Washington, D.C.

Streubel, D. P. 1968. Food storing and related behavior of red squirrel (*Tamiasciurus hudsonicus*) in interior Alaska. Master's thesis, University of Alaska, Fairbanks.

Stromberg, J. C., and D. T. Patten. 1991. Dynamics of the spruce-fir forests on the Pinaleño Mountains, Graham County, Arizona. *Southwestern Naturalist* 36(1):37–48.

Todorovsky, D. 1997. On the working time budget of the university teacher. *Scientometrics* 40(1):13–21.

Tripet, F., M. Glaser, and H. Richner. 2002. Behavioural responses to ectoparasites: Time-budget adjustments and what matters to Blue Tits (*Parus caeruleus*) infested by fleas. *Ibis* 144(3):461–469.

Verts, B. J., and S. D. Gehman. 1991. Activity and behavior of free-living *Sylvilagus nuttallii*. *Northwest Science* 65(5):231–237.

Wauters, L. A., and J. Gurnell. 1999. The mechanism of replacement of red squirrels by grey squirrels: A test of the interference competition hypothesis. *Ethology* 105(12):1053–1071.

Wauters, L. A., J. Gurnell, A. Martinoli, and G. Tosi. 2001. Does interspecific competition with introduced grey squirrels affect foraging and food choice of Eurasian red squirrels? *Animal Behaviour* 61:1079–1091.

Wauters, L. A., C. Swinnen, and A. A. Dhondt. 1992. Activity budget and foraging behaviour of red squirrels (*Sciurus vulgaris*) in coniferous and deciduous habitats. *Journal of Zoology, London* 227:71–86.

Williams, J. B., M. D. Anderson, and P. R. K. Richardson. 1997. Seasonal differences in field metabolism, water requirements, and foraging behavior of free-living aardwolves. *Ecology* 78(8):2588–2602.

Wishart, R. A., and S. G. Sealy. 1980. Late summer time budget and feeding- behavior of marbled godwits (*Limosa fedoa*) in southern Manitoba. *Canadian Journal of Zoology* 58(7):1277–1282.

Wolf, L. L., and F. R. Hainsworth. 1971. Time and energy budgets of territorial hummingbirds. *Ecology* 52:980–988.

Young, P. J., V. L. Greer, J. E. Lowry, E. Bibles, N. Ferguson, and E. Point. 2000. *The Mt. Graham red squirrel monitoring program: 1989–1998*. Steward Observatory, University of Arizona, Tucson.

CHAPTER 16

Removal Rates and Fate of Two Cone Species Collected by Red Squirrels

DIANE K. ANGELL

QUANTITATIVE RESEARCH on the foraging ecology of the Mt. Graham red squirrel is of particular interest because of its status as an endangered species. Red squirrel foraging has been well studied (e.g., C. Smith 1968; Kemp and Keith 1970; Gurnell 1984; Wheatley et al. 2002), and the Mt. Graham red squirrel has been the focus of much research during the past decade (Pereira and Itami 1991; A. Smith and Mannan 1994; Young et al. 2002; Schauffert et al. 2002). Despite this research, little is known about the foraging and caching behavior of this population.

During the early 1990s, an effort was made to eliminate the trapping of red squirrels on Mt. Graham to avoid the stresses of capture and marking (Young 1995). Previous experience with other populations of red squirrels has shown that it is possible to infer foraging and caching behavior by tracking the cones red squirrels rely on as their main food resource (Gurnell 1984; Angell 1993). This study monitored foraging by tracking conifer cones, as cones were gathered during the fall harvesting period and then cached or stored for winter. Cones were presented at evenly dispersed points distributed throughout red squirrel territories and were individually numbered with metal tags so that they could be relocated with a metal detector. Red squirrels specialize on cones, and observations confirm that no other species collected these cones. Foraging behavior was therefore monitored unobtrusively by inferring foraging choices from the removal, movement, and caching of individually marked cones.

Three aspects of the foraging behavior of the Mt. Graham red squirrel were addressed using this cone-marking and -tracking technique. Knowledge of cone preferences can be inferred by quantifying the likelihood of cone removal. Cone preference information would be useful as future management options for this species are evaluated. Although there are a number of different cone species available on Mt. Graham, here I focus on two of the most common species, Douglas-fir and Engelmann spruce. Cone preference was determined by comparing the numbers of each cone species removed from large gridded areas where cones were laid out at regular intervals on the forest floor during the fall caching period.

An understanding of the caching patterns adopted by this population can be inferred by monitoring the fate of cones tracked to their caching location. Quantification of the caching behavior of this population is important because current estimates of red squirrel population sizes focus on counting active middens (Young 1995). Red squirrels are generally considered midden cachers or larderhoarders, carrying cones to one or several large clusters or middens within their territories (C. Smith 1968; Gurnell 1984; Vander Wall 1990). Recently, a number of populations have been identified where red squirrels scatterhoard, or disperse, their stored cones in individual, less noticeable, caches away from the midden (Hurly and Robertson 1987, 1990; Dempsey and Keppie 1993; Hurly and Lourie 1997). The accuracy of the current census technique depends on the assumption that red squirrels on Mt. Graham primarily cache in middens, or larderhoard. Extensive scatterhoarding may result in middens being classified as inactive, thus potentially skewing population estimates. Caching pattern was quantified by determining the proportion of recovered cones cached in middens.

Finally, information concerning the kinds of habitats within which red squirrels are foraging most actively can be evaluated by comparing cone removal in different locations. There is some debate concerning the kinds of habitats red squirrels prefer. Despite initial focus on high-elevation habitats, surveys of active middens at different elevations have provided evidence that red squirrels may not be found at high densities in the highest-elevation spruce-fir forests (Young 1995). In many years, high densities of active middens are found in the middle elevation or ecotonal areas as well as the lower-elevation mixed-conifer habitats. Here, I evaluate red squirrel cone removal activities at three different elevations (high, middle, and low), chosen to broadly represent the spruce-fir, ecotonal, and mixed-conifer habitats respectively.

Methods

Study Sites

The low-elevation site (approximately 2,800 m) contained mostly Douglas-fir with some white fir. The middle-elevation site (approximately 3,000 m) contained primarily Engelmann spruce with some corkbark fir, and the high-elevation site (approximately 3,200 m) contained mainly Engelmann spruce. Sites were chosen in terrain that was relatively flat (so cones did not get washed away by rain), and in consultation with U.S. Forest Service personnel. The size and shape of study sites were determined by the terrain and the location of middens. The low-elevation study site was located south of the road to Riggs Flat Recreation Area. The middle-elevation site was northwest of the University of Arizona's access road to the Mt. Graham International Observatory. The high-elevation site was just east of Hawk Peak.

Grids on each area were sized to fit the available area and to avoid steep slopes; obstacles such as boulders, roads, and trails; and open areas with few trees on a 20 × 20 m spacing to provide as many line intersections, or "presentation points," as possible. Consequently, the number of presentation points was different on each area; however, all areas were equally saturated with cones.

During 1990, the low-elevation site contained 81 presentation points, the middle-elevation site contained 42 presentation points, and the high-elevation site contained 52 presentation points. The high site had a low density of active middens in September 1990, when the grid was created; therefore, the high site is actually made up of two separate grids because I was not able to include more than one midden within a reasonable grid size. In 1991, there were many fewer cones available, and access to the high-elevation site was difficult as a result of construction. In that second year, therefore, I restricted my study to the middle- and low-elevation sites and reduced the cone-presentation area in size at the middle (to 35 presentation points) and low sites (to 40 presentation points).

Cone Presentation Procedure

In 1990, cones were presented approximately weekly from September 9 to October 19. In 1991, cones were set out only twice, on September 19 and September 28, due to the small numbers of cones available. Cones were not set out until red squirrels at sites had begun cutting and caching cones. Because the kinds of cones available varied among my three study sites, I did reciprocal cone transplants, thereby presenting all cone species to squirrels at all sites. Cones were collected from a variety of sources. Some were picked directly from small trees, or trees that were cut during preparation of telescope sites. Others were collected from the forest floor both within and outside of the study area after they had been cut from trees by squirrels. Few cones were collected in any one area to ensure that the reduction in cone numbers in any one location was very small. In 1991, I was able to collect some Douglas-fir cones produced in that year; however, no Engelmann spruce or any other cone species were available (Young 1995). Therefore, in 1991, I used Engelmann spruce cones that had been collected from cut trees in 1990 and stored in freezers. These cones were readily collected by squirrels (see Results).

Cones were presented and checked on the same dates at each study site to minimize the effects of weather on cone removal rates. Cones were typically checked every other day in 1990 and more frequently in 1991. Cones were checked more frequently in the second year because they were removed much more quickly. Overall, I presented a total of 1,344 marked cones in 1990 and 282 cones in 1991.

Cone Marking and Recovery Procedure

Cones were marked and set out on the ground (presented) in groups at the corners of each 20 × 20 m quadrant in the grid (presentation points). Cones were individually marked with metal tags so they could be more efficiently located from underground storage locations using a metal detector (Sork 1984; Steele et al. 2001). Metal tags consisted of blunted carpet tacks approximately 5 mm long. Each tack was painted, individually numbered, and inserted into the apical end of cones, an area that is normally discarded by squirrels after the seeds have been removed. During the fall, winter, and spring, I recovered cones by searching the forest floor and middens with a metal detector (Fisher M-Scope, model 1225-X, Fisher Research Laboratory, Inc., Los Banos, CA). Marked recovered cones were left in place if they had been cached and were removed if they had been stripped of seeds or eaten. Cones that were recovered stripped of seeds or "eaten" during the fall caching period were excluded from recovery data.

Throughout the season, presentation points and middens were systematically searched for marked cones on a weekly basis using the metal detector by moving outward in progressively wider circles. At the end of the caching season, each study site was systematically surveyed. Each of these final site surveys included roughly 64 person-hours of searching. Many cones were recovered in multiple locations. For example, cones found cached in one location would be found cached in a different location later in the season or found eaten in yet a different location the following year. Clearly, red squirrels relocate and re-cache some cones. When cones were located more than once, the final location of the cone was used to determine carrying distance. Cones recovered within 3 m of a large aggregation of cached cones and cone debris were categorized as having been cached in middens. I contrast midden-cached cones with cones cached individually a short distance from where they were presented (scatterhoarded). In red squirrels, scatterhoarded cones remain dispersed throughout an individual's territory.

Although no other species were observed eating or moving conifer cones in any of these study sites, it is possible that small mammals other than red squirrels removed some cones. Because such a large number of cones were presented, this study assumes that red squirrels collected the vast majority of cones.

Results

Cone Preferences

After a marked Douglas-fir cone and Engelmann spruce cone were presented or set out on grids, they were checked at two-day intervals. In some cases, one or the other of these cones had been removed, while the other remained at the presentation point. When only one of the two cones had been removed, it was typically the Douglas-fir cone that was removed first (85 percent, 219/258). The proportion of cones removed within two days of being presented varied significantly according to cone type, with Douglas-fir cones more likely to have been removed than Engelmann spruce ($\chi^2 = 17.70$, $P < 0.0001$) (fig. 16.1). This analysis focused on cones presented during the first year, since in the second year very few cones remained. Douglas-fir cones were also more likely to be removed by the

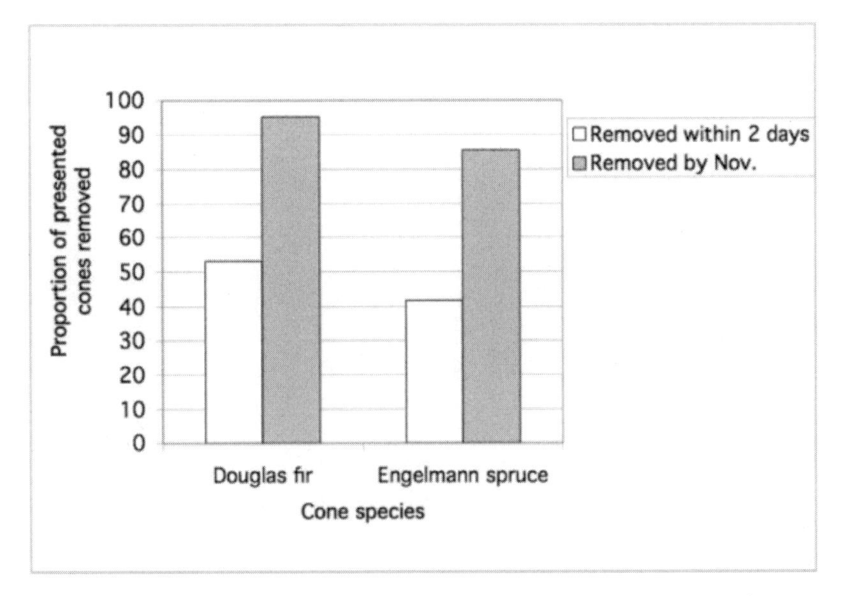

Figure 16.1. Cone removal on Mt. Graham during 1990. Open bars represent removal in the first two days; shaded bars represent removal by end of November. Douglas-fir cones were significantly more likely to be removed from study sites during the fall caching period than Engelmann spruce cones. A total of 672 individually marked cones of each species were presented.

final cone check towards the end of the caching period in November than Engelmann spruce cones ($\chi^2 = 25.18$, $P < 0.0001$) (fig. 16.1). This analysis focused on only the first year of data and excluded the middle-elevation site since all cones were removed at that site.

If Douglas-fir cones are collected more readily over Engelmann spruce cones, then perhaps Douglas-fir cones may be collected from locations more distant from the midden than Engelmann spruce cones or might be more likely to be cached at middens. There was no clear relationship between cone species and carrying distance or caching location. At each site, however, the cone carried the farthest (i.e., maximum carrying distance) was a Douglas-fir cone. At the low-elevation site, a single Douglas-fir cone was carried 79.0 m, at the middle-elevation site the distance was 63.0 m, whereas at the high site the distance was 45.5 m.

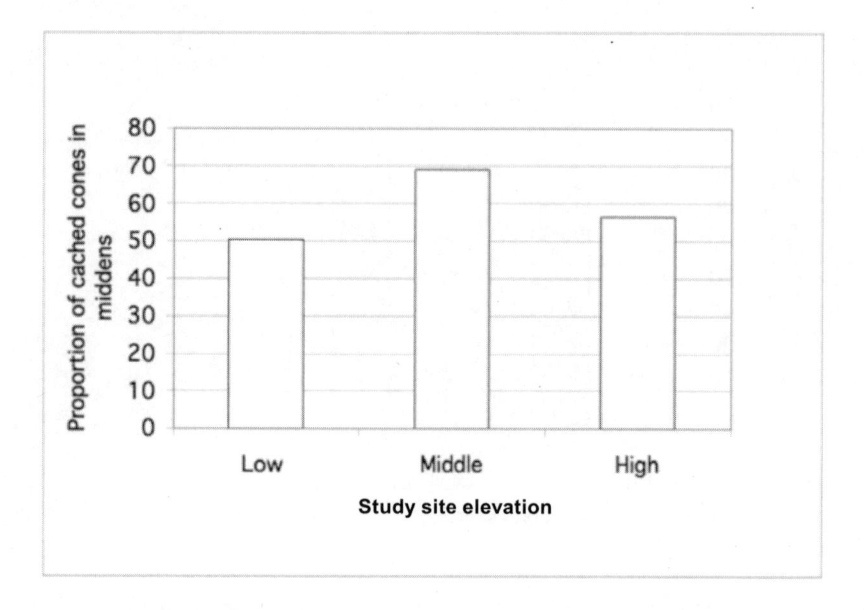

Figure 16.2. Caching location of recovered cones at three different elevations (approximately 2,800 m, 3,000 m, and 3,200 m) on Mt. Graham during 1990. Red squirrels at all elevations cached abundantly at middens. Midden caching was significantly more common at the middle-elevation site (total cached cones recovered = 97) than at either the low-elevation site (total cached cones recovered = 165) or the high-elevation site (total cached cones recovered = 85).

Midden Caching

Most cached cones were recovered from red squirrel middens at all three study sites; however, the proportion of cached cones found at middens differed significantly among the three study sites ($\chi^2 = 10.13$, $P < 0.01$). Recovered cones were designated as cached in middens if they were within 3 m of the center of a large aggregation of cones or cone debris. Cones were most likely to have been cached at middens at the middle-elevation site and least likely to have been cached at middens at the low site. The high-elevation site had an intermediate value (fig. 16.2).

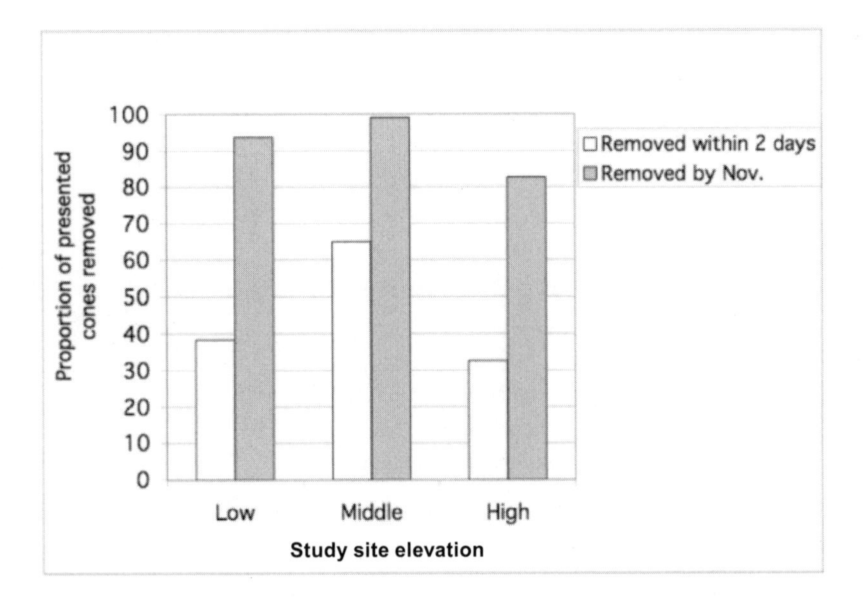

Figure 16.3. Cone removal at three different elevations (approximately 2,800 m, 3000 m, and 3,200 m) on Mt. Graham during 1990. Open bars represent removal in the first two days; shaded bars represent removal by end of November. Cones were significantly more likely to be removed at the middle-elevation site (where a total of 420 cones were presented) than at either the low-elevation site (where a total of 640 cones were presented) or the high-elevation site (where a total of 284 cones were presented).

Variation among Sites

The proportion of cones removed within two days of being presented was significantly different between sites ($\chi^2 = 88.77$, $P < 0.0001$), as was the proportion of cones removed after a final cone check in November 1990 ($\chi^2 = 78.9$, $P < 0.0001$) (fig. 16.3). The greatest proportions of cones were removed from the middle-elevation site, whether they were removed within two days or by the end of the final cone check in November. The proportion removed was smaller at the low-elevation site and high-elevation site. This analysis was not carried out in 1991 due to the small numbers of cones available and the monitoring of only two sites.

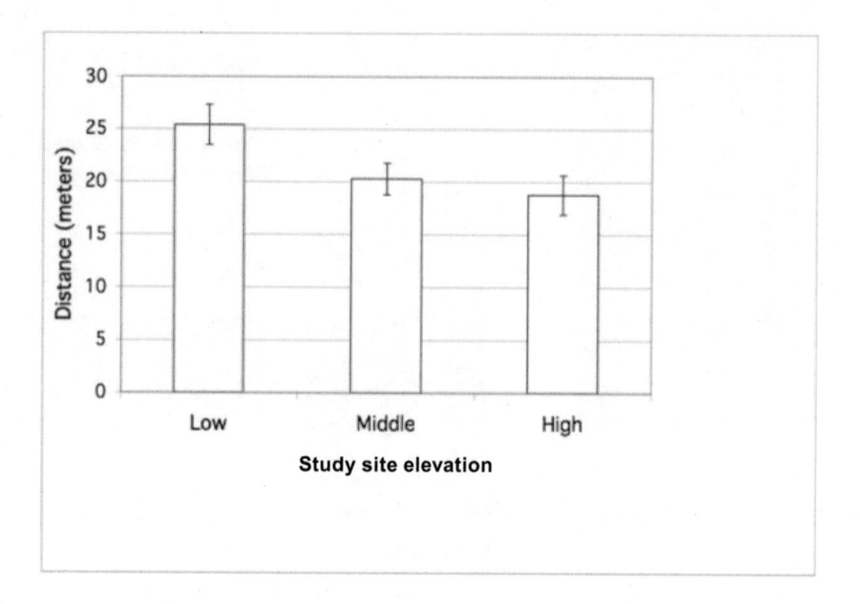

Figure 16.4. Distance recovered cones were moved from presentation locations to middens at three different elevations (approximately 2,800 m, 3,000 m, and 3,200 m) on Mt. Graham in 1990 (N ± SE). At the low-elevation site ($n = 83$) cones were moved significantly greater distances than at either the middle-elevation ($n = 67$) site or high-elevation site ($n = 48$).

Cone-movement distances were measured as the distance between the presentation point and the location where the cone was recovered. All recovered cached cones were separated into two categories. Cones found in middens (i.e., cached within 3 m of large aggregations of other cones or cone debris) were separated from cones found cached individually near where they were presented. Cones cached at middens differed significantly in how far they were moved at each study site (fig. 16.4). A two-way ANOVA including both the effects of cone species and study site showed that these site differences were significant ($F_{2,170}$, $P < 0.05$), while cone differences were not significant ($F_{1,170}$, $P = 0.61$). Cones presented at the low-elevation site were moved the greatest distances, with cones at the middle-elevation site moved intermediate distances and cones at the high-elevation site moved the shortest distances.

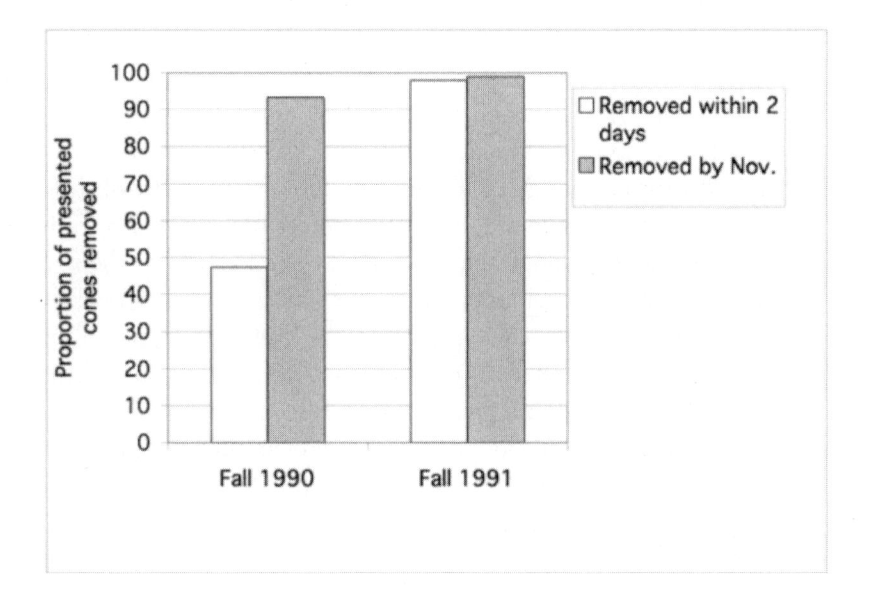

Figure 16.5. Cone removal during two years on Mt. Graham. Open bars represent removal in the first two days; shaded bars represent removal by end of November. During 1990 when cone crops were heavy, cone removal was significantly slower (total cones presented = 1,344) than 1991 when cone crops were reduced in size (total cones presented = 282).

Variation between Years

Years varied dramatically in cone production (Young 1995), with cone production being dramatically reduced in 1991 after a very large crop in 1990. There were significant differences in cone removal between the two years, with the proportion of cones removed from presentation sites lower in the first year when cones were abundant than in the second year when cone crops were lower ($\chi^2 = 11.92$, $P = 0.0006$) (fig. 16.5).

The first and second year differed significantly in the likelihood of recovered cones being cached in middens ($\chi^2 = 45.60$, $P < 0.001$). Overall, in the first year, 57.1 percent (198/347) of cached cones were found at middens, whereas in the second year, 95.5 percent (84/88) were found at middens. Carrying distance did not vary between the two years.

Discussion

Individually marking cones and tracking their movement yielded several important kinds of information on the foraging behavior of the Mt. Graham red squirrel. First, Mt. Graham red squirrels were more likely to remove Douglas-fir cones than Engelmann spruce cones from large grids at all three elevations. Douglas-fir cones were more likely to be the first cone of a Douglas-fir–Engelmann spruce pair removed, were more likely to be removed within two days, and were more likely to be removed by November, towards the end of the caching season. Such removal patterns suggest Douglas-fir cones are preferred over Engelmann spruce cones. Miller and Yoder (in this vol.) found that crude protein content was similar in these two cone species, while gross energy (cal/g) was actually slightly greater in Engelmann spruce seed. Cone preference, therefore, may have more to do with handling time, since Engelmann spruce cones have more, but smaller, seeds than Douglas-fir. As a result, handling time per gram of seed may be higher for Engelmann spruce cones. Cone preference information is a valuable addition to the debate concerning squirrel habitat and elevation preferences (Young 1995). Many factors affect the places red squirrels establish territories and middens (e.g., A. Smith and Mannan 1994), but the species of cones available must also be important. Douglas-fir, the cone species preferentially collected here, is primarily a member of the low-elevation ecotonal and mixed-conifer habitats. Thus, red squirrels may occur in low-elevation locations because their preferred cones, Douglas-fir, are available in those locations.

This study also quantified the proportion of cones cached in middens. Caching pattern can be surprisingly variable in red squirrels (Hurly and Robertson 1987, 1990; Angell 1993; Dempsey and Keppie 1993; Hurly and Lourie 1997), and red squirrel population estimates on Mt. Graham focus on quantifying the number of active middens (Young 1995). Because of this reliance on counting active middens, it is important to monitor the frequency of midden caching (or larderhoarding). Half or more of all cones at all sites were cached in middens, indicating that red squirrels in all areas on Mt. Graham use middens extensively. Counting middens, therefore, is a relatively reliable way of obtaining population estimates when trapping and permanent marking of individuals is not possible. Furthermore, in years when cones were scarce (in 1991 as compared to 1990), cones were more likely to be cached in middens. Therefore, when a year of low cone crops follows a year of heavy cone crops, counts of active middens should remain reliable.

The three study sites varied dramatically in cone removal activities, with the middle-elevation site being the most active. This rapid cone removal rate or this high foraging activity must be a function of one or a combination of possibilities. Removal rates may be a function of (1) the speed with which squirrels collect cones, (2) the number of squirrels in an area, and (3) the total number of cones available in the area. For simplicity, we may assume that red squirrels collect cones at some average speed so that differences between sites at different elevations are not a result of differences in speed. Furthermore, all sites during the first year of the study had bumper crops of cones; therefore, the number of existing cones in the area is also likely similar. As long as cone crops are similar at different elevations, the greater cone removal activity at the middle-elevation site may well be a result of differences in squirrel density. Clearly, without trapping and marking individuals, it is impossible to distinguish between these alternatives. Nevertheless, a rapid cone removal rate at least suggests high squirrel density at this middle-elevation site, supporting assertions that low-elevation habitats support large numbers of red squirrels on Mt. Graham (Young 1995).

Cone removal rates also varied across years. The two years of this study varied in cone availability and red squirrel population size. Cone production was much lower during the fall of 1991, while population estimates based on midden counts increased. Both these factors should and did result in very rapid cone removal rates. In fact, most experimental cones were removed within a matter of hours rather than days.

In addition to evaluating red squirrel cone preferences, caching patterns, and cone removal activities at different altitudes, this technique provided other kinds of information. The middle-elevation site was the location with the most recovered cones cached in middens. The cause of this variation is not clear, but if densities of squirrels are greater at the middle-elevation site, then competition from neighboring individuals may increase midden caching or larderhoarding if cones in middens are safer from theft than cones cached elsewhere. It is certainly also possible that the middle-elevation site has microhabitats more conducive to caching large quantities of cones in middens than other sites (Hurly and Robertson 1990).

Sites differed in how far cones were moved between presentation points and middens. Cones at the low-elevation site were moved much greater distances before ending up in middens. Red squirrels at the low-elevation site, therefore, must collect cones from a much wider area than at other elevations. Such a pattern has several possible causes. Low-elevation, mixed-conifer forests do seem open, so trees may be at lower densities than at other elevations. When trees are at low densities, squirrels may be forced to collect cones from a wider area to collect a number sufficient for winter survival.

This study clearly shows that important management-related information can be collected in an unobtrusive manner. Trapping in this population was specifically banned as a result of several trap deaths. For populations of small size, this may not be an uncommon problem. By monitoring conifer cones as they were gathered during the fall harvesting period and cached or stored for winter, this study contributes to our understanding of the foraging ecology of this species. Such information is critical as we continue to evaluate the long-term survival of this small isolated population

Acknowledgments

I would like to thank the University of Arizona and the U.S. Forest Service for access to study sites, in particular G. Froelich for assistance in selecting study sites. Thanks also to R. Swearer for assistance during the field portion of this project, as well as S. Gaines and J. Waage for advice with many aspects of this research. Funding for this study was provided by the University of Arizona through the Mt. Graham Red Squirrel Study Committee, which is comprised of representatives from the Arizona Game and Fish Department, the University of Arizona, the U.S. Forest Service, and the U.S. Fish and Wildlife Service.

Literature Cited

Angell, D. K. 1993. Red squirrels and conifer cones: The ecology of caching and foraging decisions. PhD diss., Brown University, Providence, Rhode Island.

Dempsey, J. A., and D. M. Keppie. 1993. Foraging patterns of eastern red squirrels. *Journal of Mammalogy* 74:1007–1013.

Gurnell, J. 1984. Home range, territoriality, caching behaviour and food supply of the red squirrel (*Tamiasciurus hudsonicus fremonti*) in a subalpine lodgepole pine forest. *Animal Behaviour* 32:1119–1131.

Hurly, T. A., and S. A. Lourie. 1997. Scatterhoarding and larderhoarding by red squirrels: Site, dispersion, and allocation of hoards. *Journal of Mammalogy* 78(2):529–537.

Hurly, T. A., and R. J. Robertson. 1987. Scatterhoarding by territorial red squirrels: A test of the optimal density model. *Canadian Journal of Zoology* 65:1247–1252.

——. 1990. Variation in the food hoarding behavior of red squirrels. *Behavioral Ecology and Sociobiology* 26:91–97.

Kemp, G. A., and L. B. Keith. 1970. Dynamics and regulation of red squirrel (*Tamiasciurus hudsonicus*) populations. *Ecology* 51:763–779.

Pereira, J. M. C., and R. M. Itami. 1991. GIS-based habitat modeling using logistic multiple-regression: A study of the Mt. Graham red squirrel. *Photogrammetric Engineering and Remote Sensing* 57(11):1475–1486.

Schauffert, C. A., J. L. Koprowski, V. L. Greer, M. I. Alanen, K. A. Hutton, and P. J. Young. 2002. Interactions between predators and Mt. Graham red squirrels (*Tamiasciurus hudsonicus grahamensis*). *Southwestern Naturalist* 47(3):498–501.

Smith, A. A., and R. W. Mannan. 1994. Distinguishing characteristics of Mount Graham red squirrel midden sites. *Journal of Wildlife Management* 58(3):437–445.

Smith, C. C. 1968. The adaptive nature of social organization in the genus of three [*sic*] squirrels *Tamiasciurus*. *Ecological Monographs* 38:31–63.

Sork, V. L. 1984. Examination of seed dispersal and survival in red oak, *Quercus rubra* (Fagaceae), using metal-tagged acorns. *Ecology* 65:1020–1022.

Steele, M. A., G. Turner, P. D. Smallwood, J. O. Wolff, and J. Radillo.2001. Cache management by small mammals: Experimental evidence for the significance of acorn-embryo excision. *Journal of Mammalogy* 82(1):35–42.

Vander Wall, S. B. 1990. *Food hoarding in animals*. University of Chicago Press, Chicago.

Wheatley, L. A., K. W. Larsen, and S. Boutin. 2002. Does density reflect habitat quality for North American red squirrels during a spruce-cone failure? *Journal of Mammalogy* 83:716–727.

Young, P. J. 1995. Monitoring the Mt. Graham red squirrel. Pages 226–246 in C. A. Istock and R. S. Hoffman, eds. *Storm over a mountain island: Conservation biology and the Mt. Graham affair.* University of Arizona Press, Tucson.

Young, P. J., V. L. Greer, and S. K. Six. 2002. Characteristics of bolus nests of red squirrels in the Pinaleño and White Mountains of Arizona. *Southwestern Naturalist* 47(2):267–275.

## CHAPTER 17

Nutrient Content of Mt. Graham Red Squirrel Feedstuffs

WILLIAM H. MILLER AND WILLIAM E. YODER

Management of any wild species is based on the assumptions that provided with adequate cover, water, and space (1) the physical well-being of the species is a function of the quality and quantity of the diet, and (2) a species maintained on a high nutritional plane is more productive and less subject to losses due to natural causes (Moen 1973; Nelson and Leege 1982). In the case of the Mt. Graham red squirrel, we have limited knowledge about either the composition or the nutrient quality of their diet.

It is widely accepted that red squirrel diets consist primarily of the seeds of coniferous trees (Lindsay 1967, 1968; Vahle and Patton 1983; Brown 1984; Sullivan and Moses 1986). Ferron et al. (1986) observed that red squirrels spent 81 percent of their feeding activity on conifer trees, as opposed to 13.1 percent on deciduous trees, 3.6 percent on berries, and 1.8 percent on mushrooms. These data may be misleading, inasmuch as it requires longer time to obtain feedstuffs from cones than from berries and mushrooms. However, several studies have established the value of conifer cones as the primary winter storage feedstuff in middens (Finley 1969; Vahle and Patton 1983; Gurnell 1984; Ferron et al. 1986; Patton and Vahle 1986). Finley (1969) estimated that the average adult squirrel required between 8 to 24 bushels of cones to meet its energy requirements during the winter months.

Data on the nutrient content of typical red squirrel feedstuffs are not available. It was the objective of this study to catalog the crude protein (CP), gross energy (GE), crude fiber (CF), total ash, calcium, and phosphorus content of the major feedstuffs utilized by the Mt. Graham red squirrel during spring, summer, and fall collection periods.

The study area on Mt. Graham ranges in elevation from 2,600 m to

3,270 m. The true conifers are found in the upper elevations, transitioning to the mixed-conifer at lower elevations.

Methods

We subdivided the Mt. Graham study area into four collection areas for sampling purposes: Webb Peak, Hawk Peak, Heliograph Peak, and Columbine. Within each sampling area, we collected cone samples from the four major coniferous species found within the Mt. Graham study area (Engelmann spruce, corkbark fir, Douglas-fir, and white fir). We repeated these collections during three separate seasons: spring (late April–early May), summer (late July–early August), and fall (late September–early October) of 1990. The total sample size was 48 (sampling area × tree species × season). When available, samples of mushrooms species known to be consumed by Mt. Graham red squirrels were also collected in these areas.

Sample Processing

Upon return to the laboratory, cones were oven dried at 50°C for 96 hours to induce cone opening and seed release. Seeds with full endosperm were combined by species by season by collection site, and further dried at 50°C for 48 hours. Dried seed samples were then processed in a Wiley mill to a uniform size of 1 mm and stored in plastic containers until analyzed. All foodstuffs were analyzed by collection site, species, and season for crude protein (micro-kjeldahl Kjeltech auto nitrogen analyzer model 1040), gross energy (Parr adiabatic bomb calorimeter model 1080), crude fiber (Goering and Van Soest 1970), total ash, Ca (atomic absorption spectrophotometry), and P (colorimetric analysis). All results are reported by species and season on a dry weight basis.

Statistical Analysis

The statistical design of this study was a randomized block two-factor factorial design, with collection areas as blocks, and season and conifer species as the factors. The hypothesis tested was that there were no differences in nutrient quality in conifer seeds by species or season. Differences in nutrient content between species and season were tested using analysis of variance at $P = 0.10$. Means with significant differences were separated using a least squares difference test (Zar 1999). Values reported are the mean of at least four replications per species and season ($n = 48$).

Results

Conifer Species

There were significant differences in seed crude protein content between species and season (table 17.1). Seasonally, white fir had the highest CP in the spring (14.10 percent CP), but declined steadily over time, with Douglas-fir intermediate. In the summer, Engelmann spruce had the highest CP content (15.44 percent), with corkbark and Douglas-fir intermediate. In the fall, both Douglas-fir and Engelmann spruce had the highest CP, with white fir and corkbark fir protein levels as much as 60 percent less than the leading two species.

Gross energy content of the conifer seeds sampled was also highly variable with season (table 17.1). In the spring, corkbark fir had the highest GE, closely followed by Douglas-fir and white fir. Engelmann spruce had the lowest GE during this season. During the summer, both white fir and Engelmann spruce increased their GE content, while the GE content of corkbark and Douglas-fir decreased. By fall there was little difference in the GE content of white fir, corkbark fir, and Engelmann spruce, with Douglas-fir having significantly lower GE.

Crude fiber analysis is a measure of the amount of the plant cell wall present in plant materials. In conifer seeds, the seed coat is the primary crude fiber and is a measure of the structural carbohydrate content of the seed. For primary consumers with simple digestive systems such as the red squirrel, structural carbohydrates are poorly digestible, and the amount of crude fiber is an indirect inverse indicator of the nutritional value of a foodstuff. During the spring, we found Engelmann spruce to have the lowest crude fiber content, with white fir and corkbark fir intermediate, while Douglas-fir had the highest fiber content at 60.32 percent (table 17.1). Crude fiber content of white fir and corkbark fir fell dramatically in summer to 25.35 percent and 31.8 percent, respectively, and then increased to highs of 56.79 percent and 53.05 percent in fall, respectively

Table 17.1. Mean nutrient content of conifer seeds utilized by Mt. Graham red squirrels collected on Mt. Graham during three seasons in 1990 ($n = 48$)

[a,b] Means in rows with different letters are significantly different at $P < 0.1$.

Nutrient / Season	Food Source			
	White Fir	Corkbark Fir	Engelmann Spruce	Douglas-fir
Crude Protein (%)				
Spring	14.10[a]	5.83	5.57	10.42[b]
Summer	7.34	10.48[b]	15.44[a]	9.51[b]
Fall	6.49	5.64	17.52[a]	19.58[a]
Gross Energy (cal/g)				
Spring	5839[a]	6116[a]	5297	5932[a]
Summer	6020[a]	5642[b]	5838[ab]	5260
Fall	5795	5680	5743	5230[a]
Crude Fiber (%)				
Spring	44.67[a]	51.74[ab]	38.97	60.32[a]
Summer	25.35	31.80[b]	47.85[b]	61.76[a]
Fall	56.79[a]	53.05[a]	40.04	32.29
Cell Soluble (%)				
Spring	55.33[a]	48.26	61.03[a]	39.68
Summer	74.65[a]	68.20[a]	52.15	38.24
Fall	43.21	46.95	59.96[a]	67.71[a]
Total Ash (%)				
Spring	1.2	1.1	1.3	0.9
Summer	1.2	1.2	1.1	1.4
Fall	1.4	1.6	1.7	1.6
Calcium (%)				
Spring	0.45	0.28	0.16	0.39
Summer	0.58	0.48	0.51	0.53
Fall	0.49	0.38	0.49	0.47
Phosphorus (%)				
Spring	0.62	0.21	0.14	0.35
Summer	0.29	0.22	0.18	0.19
Fall	0.34	0.27	0.70	0.82

(table 17.1). Crude fiber content of Engelmann spruce increased slightly during the summer, then declined in the fall. While Douglas-fir had the highest CF content in the spring and summer (60.32 and 61.76 percent, respectively), it dropped to the lowest level of 32.29 percent in the fall.

The soluble portion of the cell consists of those cellular nutrients that are highly digestible, such as sugars, starch, proteins, and fats. As such, this value is a good direct indicator of the overall digestibility of the feedstuff. Overall, Engelmann spruce showed the least variation in the cell soluble content (table 17.1). During the summer, both white fir and corkbark fir had the higher cell soluble content at 74.65 percent and 68.20 percent, declining in fall to 43.21 percent and 46.95 percent, respectively (table 17.1). By fall, Engelmann spruce and Douglas-fir both increased in their cell soluble content, reaching 59.96 percent and 67.71 percent, respectively.

Total ash is a measure of the inorganic contents of the cell and includes macro and micro mineral nutrients as well as non-nutrients such as silica. There was relatively little variation in the total ash content over time in any of the conifer seeds analyzed. Of the macro minerals, the two most often limiting in feeds are calcium and phosphorus. The amounts of each of these nutrients found in the analyzed feeds are reported in table 17.1. For the most part, the levels of Ca and P over time show minor variation, with Ca ranging from 0.28 to 0.58 percent and P ranging from 0.21 to 0.82 percent. There was one exception: Engelmann spruce had very low levels of both macro minerals in spring (0.16 percent and 0.14 percent), which increased to a more normal level of 0.49 percent and 0.70 percent (Ca and P respectively) in the fall. Of greater nutritional significance is the Ca:P ratio. For optimal nutrient utilization of these two macro minerals, a Ca:P ratio between 2:1 and 1:2 is desired. With the exception of the summer for Engelmann spruce and Douglas-fir, this ratio exists at all times for all feeds analyzed.

Mushroom Species

Several authors have reported the seasonal use of mushrooms by red squirrels (Klugh 1927; Smith 1968; Cork and Kenagy 1989). There were three major species of mushrooms sampled on Mt. Graham: *Condina* spp., *Flyaminas* spp., and *Russula emetica* [U.S. Fish and Wildlife Service

Table 17.2. Mean nutrient content of select mushrooms utilized by Mt. Graham red squirrels collected on Mt. Graham during 1990 ($n = 12$)

Species	Crude Protein (%)	Gross Energy (cal/g)	Cell Soluble (%)	Total Ash (%)	Ca (%)	P (%)
Condina spp.	14.1	4252	46.80	T	0.64	0.57
Flyaminas spp.	19.9	4230	39.37	8.5	0.43	0.76
Russula emetica	18.6	4246	56.11	1.2	0.5	0.47

personnel collected this material and supplied these names]. The CP of the mushroom species analyzed ranged from a high of 19.9 percent CP in *Flyaminas* spp. to a low of 14.1 percent CP in *Condina* spp., with *Russula emetica* at 18.6 percent CP (table 17.2). The mushrooms were lower in GE content than the conifer seeds, all averaging about 4,242 cal/g.

Crude fiber content of mid-summer mushrooms was relatively high, ranging from 60.63 percent to 43.89 percent. Cell soluble content of mushrooms ranged from a low of 39.37 percent in *Flyaminas* spp., to a high of 56.11 percent in *Russula emetica*. The highest ash level was found in *Flyaminas* spp. in summer at 8.5 percent (table 17.2). All other mushrooms sampled had <2 percent total ash. While the calcium levels were similar to those of the conifer seeds, phosphorus content was as much as two times higher. The higher phosphorus levels yielded a Ca:P ratio closer to 1:1.

Discussion

Conifer Species

The scope of our study was to focus on those nutrients generally considered to have the greatest impact on individual performance. While a number of minerals and vitamins play important roles in the physiological and metabolic performance of individual animal species, generally they are not limiting factors in the performance of primary consumer species such as the Mt. Graham red squirrel (Moen 1973; Robbins 1983).

We, therefore, chose to focus on the availability of protein, energy, and two macro minerals (calcium and phosphorus).

Digestible protein requirements are classified into two groups: protein for maintenance, and protein for production. Maintenance protein requirements are designed to offset nitrogenous losses due to metabolic processes such as metabolic fecal loss, endogenous urinary loss, and hair replacement. The quantity of protein required to meet these needs is a function of the size of the animal and the type of diet. Brown (1984) and Gurnell (1984) suggest that the average red squirrel will weigh approximately 213 g. If we assume this to hold for the Mt. Graham red squirrel, the daily digestible protein requirement should be 0.33 g. This maintenance requirement is applicable for males, non-gestating or non-lactating females, and juveniles.

Evaluation of the ability of the seeds of the four conifer species examined in our study to meet the individual protein maintenance requirement is presented in table 17.3. These analyses are based on the fall nutrient content and would be appropriate for most individuals. Our evaluation suggests that to meet the daily protein requirement an individual's daily seed intake would be in the range of 2.1 to 5.2 g of conifer seeds, or from 1 to 6.8 cones per day depending on the species.

Like protein, daily metabolizable energy (ME) requirements of an animal can be divided into two categories: maintenance and production. The daily maintenance requirement, or standard metabolic rate (SMR), has been defined as:

$$SMR = 70\ W^{0.75}$$

Where W is defined as the weight of the individual in kg, and the output is in Kcal of ME. The ME required for production includes gestation, lactation, activity, and body growth. Unlike the digestible protein estimation, the ME estimator needs to be more than just maintenance. If we assume that the energy for activity and other physiological processes, such as gestation and lactation, amount to 2.7 times the SMR, then the average daily ME requirement of the red squirrel will be 59.26 Kcal.

Comparing the nutrient content of the four conifer seeds with the daily estimated requirement of the red squirrel, we would estimate that the red squirrel would have to consume between 13 and 14.4 g of conifer

Table 17.3. Evaluation of potential capability of the seeds of four conifer species to meet the daily metabolizable energy (ME) and digestible protein (DP) needs of Mt. Graham red squirrels in 1990

Nutrient/ Conifer spp.	Daily DP Requirement (g)	DP Content[d] (g DP/g seed)	Daily Intake Required (g)	Mean Seed Weight (g/seed)	Seeds Yield (seeds/cone)	Cone Requirement (cones/day)
Digestible Protein						
White Fir[a]	0.33	0.065	5.22	0.0286	185	1.0
Corkbark Fir[b]	0.33	0.074	4.60	0.0118	150	2.6
Engelmann Spruce[c]	0.33	0.162	2.09	0.0057	139	2.6
Douglas-fir[c]	0.33	0.135	2.51	0.0148	25	6.8
	Daily ME Requirement (Kcal)	Caloric Content[e] (Kcal ME/g)	Daily Intake Required (g)	Mean Seed Weight (g/seed)	Seeds Yield (seeds/cone)	Cone Requirement (cones/day)
Metabolizable Energy						
White Fir[a]	59.26	4.562	12.99	0.0286	185	2.5
Corkbark Fir[b]	59.26	4.471	13.25	0.0118	150	7.5
Engelmann Spruce[c]	59.26	4.521	13.11	0.0057	139	16.4
Douglas-fir[c]	59.26	4.117	14.39	0.0148	25	38.8

[a]Seed weight and seed yield from Fowells and Schubert, 1956.
[b]Seed weight and yield from http://fcmr.forestry.ca/SeedCentre/species-available-e.htm
[c]Seed weight and yield from personal communication, Safford Ranger District, Coronado National Forest, U.S. Forest Service, 1990.
[d]Protein digestion coefficient used was 0.96 (Cork and Kenagy 1989).
[e]Energy digestion coefficient used was 0.96 (Cork and Kenagy 1989), and metabolic energy conversion coefficient was 0.82.

Table 17.4. Evaluation of potential capability of mushroom species to meet the metabolizable energy (ME) and digestible protein (DP) needs of Mt. Graham red squirrel in 1990

Nutrient/ Mushroom spp.	Daily Requirement (g)	Digestibility Coefficient (%)	DP Content (g DP/g)	Daily Intake (g)
Digestible Protein				
Condina spp.	0.33	47.8	0.020	5.0
Flyaminas spp.	0.33	39.4	0.017	4.3
Russula emetica	0.33	55.2	0.023	3.3
	Daily Requirement (Kcal)	Digestibility Coefficient (%)	ME Content (Kcal/g)	Daily Intake (g)
Metabolizable Energy				
Condina spp.	59.26	47.8	1.667	35.55
Flyaminas spp.	59.26	39.4	1.367	43.36
Russula emetica	59.26	55.2	1.871	31.68

seed daily to meet its energy needs. Based on the estimates of seed weight and seed yield per cone, this would mean that the Mt. Graham red squirrel would need to harvest between 2.5 and 38.8 cones daily, depending on the conifer cone characteristics of conifer species being consumed, specifically the cone size and number and quality of seeds contained in each cone. This estimate would agree with the observations of Finley (1969), Smith (1970), and Gurnell (1984).

Mushroom Species

Unlike the seed coat of conifer seeds, the fiber component of mushrooms is distributed throughout the fungal body. The cell wall is composed of structural carbohydrates such as hemicellulose, cellulose, and lignin. In species with a simple digestive tract such as the red squirrel, these structural carbohydrates are relatively indigestible, thus limiting nutrient availability. For simple-stomached animals, the nutrient digestibility is roughly equivalent to the percent cell soluble. An evaluation of the nutrient quality of mushrooms is presented in table 17.4.

The mushroom species evaluated in our study tended to have low amounts of cell solubles, coupled with relatively low amounts of CP and GE. This combination resulted in low nutrient availability, and conversely, large amounts are needed to meet the nutrient demands. A red squirrel would have to consume between 3.3 and 5 g of mushrooms to meet its daily CP needs. This is two to three times the amount of conifer seed needed. Further, the red squirrel would have to consume 31.6 to 43.4 g daily to meet its ME requirement, again two to three times the amount of conifer seed required to meet the same energy requirement. If red squirrels were to try to meet their nutrient need solely from mushrooms, they would have to increase their intake from a more normal 7 percent of their body weight to as much as 20 percent of their body weight daily.

Conclusions

Little data are available on the nutrient requirements of red squirrels, let alone the requirements of the Mt. Graham red squirrel. However, what data are available would suggest that the average red squirrel requires between 2.5 and 38.8 cones/day, or roughly 13.5 g of conifer seeds/day, to meet its ME needs. Based on the assumptions set forth in the discussion, the nutrient value of most feeds must be in the range of 10 to 16 percent CP, and 5,230 to 5,795 cal/g GE, to meet the daily protein and energy requirements of the red squirrel.

It would further suggest that the nutrient content of conifer seeds varies significantly on a seasonal basis. In spring, the species with the highest protein content are white fir and Douglas-fir, with Engelmann spruce and corkbark fir more desirable in the summer. In the fall, and most likely winter, Engelmann spruce and Douglas-fir provide the most protein. Relative to ME, corkbark fir is the superior source in the spring, with white fir the major source in summer, and both white fir and Engelmann spruce the major source of energy in the fall.

While some authors have reported mushrooms as being a highly preferred food source in red squirrels, our results suggest that due to their limited nutritional quality, they are most likely only a food source of opportunity that would provide some diversity to their diet. Additionally, the coniferous seeds and the mushrooms consumed by Mt. Graham red squirrels have an adequate balance between Ca and P to meet their daily requirements.

We would warn the reader to use the evaluation of the adequacy of these feeds to meet the red squirrel's nutrient requirements with caution. The values reported were derived based on some assumptions of date of parturition, number of offspring, and amount of daily activity to calculate the daily requirements of the individual animal, and they need to be verified before being widely applied to management of the species.

The scope of this study was limited to a few select nutrients. There is a serious need to expand our knowledge of the macro and micro mineral and vitamin requirements of the red squirrel as well as the contents for the red squirrel foodstuffs before widespread management decisions can be made on the adequacy of the diet to meet the individual animal's needs for these specific nutrients.

Literature Cited

Brown, D. E. 1984. *Arizona's tree squirrels*. Arizona Game and Fish Department, Phoenix.

Cork, S. J., and G. J. Kenagy. 1989. Nutritional values of hypogeous fungus for a forest dwelling ground squirrel. *Ecology* 70:577–586.

Ferron, J., J. P. Ouellete, and Y. Lemay. 1986. Spring and summer time budgets and feeding behavior of the red squirrel (*Tamiasciurus hudsonicus*). *Canadian Journal of Zoology* 64:385–391.

Finley, R. B., Jr. 1969. Cone caches and middens of *Tamiasciurus* in the Rocky Mountain region. Pages 233–273 in J. K. Jones Jr., ed. *Contributions in mammalogy*. University of Kansas Museum of Natural History Miscellaneous Publication 51.

Fowells, H. A., and G. H. Schubert. 1956. *Seed crops of forest trees in the pine region of California*. USDA Technical Bulletin Number 1150.

Goering, N. K., and P. J. Van Soest. 1970. *Forage analysis (apparatus, reagents, procedures and some applications)*. USDA Agriculture Handbook Number 379.

Gurnell, J. 1984. Home range, territoriality, caching behaviour, and food supply of the red squirrel (*Tamiasciurus hudsonicus fremonti*) in a subalpine lodgepole pine forest. *Animal Behavior* 32:1119–1131.

Klugh, A. B. 1927. Ecology of the red squirrel. *Journal of Mammalogy* 8:1–32.

Lindsay, S. L. 1967. Geographic size variation in *Tamiasciurus douglasii*: Significance in relation to conifer morphology. *Journal of Mammalogy* 67:317–325.

——. 1968. Geographic size and non-size variation in Rocky Mountain *Tamiasciurus hudsonicus*: Significance in relation to Allen's rule and vicariant biogeography. *Journal of Mammalogy* 68:39–48.

Moen, A. N. 1973. *Wildlife ecology*. W. H. Freeman and Co., San Francisco.

Nelson, J. R., and T. A. Leege. 1982. Nutritional requirements and food habits. Pages 323–368 in J. T. Thomas and D. E. Toweill, eds. *Elk of North America: Ecology and management*. Stackpole Book, Harrisburg, Pennsylvania.

Patton, D. R., and J. R. Vahle. 1986. Cache and nest characteristics of the red squirrel in an Arizona mixed-conifer forest. *Western Journal of Applied Forestry* 1:48–51.

Robbins, C. T. 1983. Wildlife feeding and nutrition. Academic Press, New York.

Smith, C. C. 1968. The adaptive nature of social organization in the genus of three [*sic*] squirrels *Tamiasciurus*. *Ecological Monographs* 88:31–63.

——. 1970. The coevolution of pine squirrels (*Tamiasciurus*) and conifers. *Ecological Monographs* 40:349–371.

Sullivan, T. P., and R. A. Moses. 1986. Red squirrel populations in natural and managed stands of lodgepole pine. *Journal of Wildlife Management* 50:595–601.

Vahle, J. R., and D. R. Patton. 1983. Red squirrel cover requirements in Arizona mixed conifer forest. *Journal of Forestry* 81:14–15.

Zar, J. H. 1999. *Biostatistical analysis*, 4th ed. Prentice-Hall, Upper Saddle River, New Jersey.

CHAPTER 18

The Adaptive Significance of Seed Hoarding by the Mt. Graham Red Squirrel

CRAIG L. FRANK AND STEPHANIE R. COX

MAMMALIAN SPECIES in the orders Insectivora, Primates, Carnivora, Rodentia, and Lagomorpha store food for use during periods of scarcity (Sherry 1985; Vander Wall 1990). Food-hoarding mammals use one of two storage strategies: larderhoarding or scatterhoarding. Larderhoarding is the practice of storing collected food in one area occupying a small portion of the home range. Scatterhoarding is the formation of many small food caches throughout most of the home range (C. Smith and Reichman 1984). Most mammals that store food for long periods of time (months) hoard seeds. Factors that promoted the coevolution of food hoarding and granivory are unknown. It is assumed that seeds are the food most frequently hoarded because their dormancy permits them to retain their nutritional composition during storage (C. Smith and Reichman 1984). Post (1992) demonstrated that energy, lipid, and protein contents of two fruit species (roughleaf dogwood [*Cornus drummondii*] and green ash [*Fraxinus pennsylvanica*]) stored in underground burrows by eastern woodrats (*Neotoma floridana*) decreased during 120 days, whereas nutritional composition of Eastern redbud (*Cercis canadensis*) seeds stored under the same conditions remained constant.

Nutritional composition of some seed species may actually decrease during storage through germination, insect damage, and/or microbial colonization (C. Smith and Reichman 1984; Sherry 1985; Steele et al. 1996). Red squirrel diets, for example, consist almost entirely of conifer seeds. They usually strip conifer cones and ingest seeds at the same location within their territories, and the resulting debris accumulates in large piles called middens. Feeding experiments conducted with captive red squirrels revealed that they could be maintained on a diet consisting of fresh (recently collected from trees) black spruce (*Picea mariana*) and white spruce (*Picea glauca*) seeds, but they quickly lost weight and died when fed black spruce and white spruce seeds that had been stored in their middens for one winter (Rusch and Reeder 1978).

We thus propose that (1) naturally stored seeds decrease in nutritional composition as frequently as they retain their nutritional composition during the caching period, and (2) the magnitude of nutritional loss during a given storage period varies with seed species. Many species of food-hoarding animals store items of more than one type. Red squirrels, for example, also store the fruiting bodies (sporocarps) of mycorrhizal fungi for winter use. Sporocarps are stored by placing them inside cavities of either downed logs or standing trees (Dice 1921; Cram 1924; C. Smith 1968). This method of storage desiccates the sporocarps and conserves their nutritional content (Frank in this vol.). Sporocarps comprise 45 to 95 percent of red squirrel winter diets, with conifer seeds accounting for the remainder (Vernes et al. 2004). Nutritional value of conifer seeds relative to that of fungal sporocarps is not known, however. Hardwood seeds/nuts ingested by fox squirrels (*Sciurus niger*) have substantially greater metabolizable energy contents than the sporocarp species consumed by them (C. Smith 1995). We thus also predict that (3) the nutritional content of conifer seeds is generally greater than that of fungal sporocarps, and (4) stored conifer seeds have an overall greater nutritional content than stored fungal sporocarps during winter use. These hypotheses were tested in field experiments on the nutritional compositions of conifer seeds hoarded by Mt. Graham red squirrels. Experiments involved the seeds of Douglas-fir, white fir, white pine, corkbark fir, and Engelmann spruce.

Methods

Seed Species and Field Experiments

Mt. Graham red squirrels store the cones (seeds) of corkbark fir, white fir, Engelmann spruce, Douglas-fir, and white pine in their middens during late summer for winter/spring use (Froehlich and Smith 1990; A. Smith and Mannan 1994; Angell in this vol). This study involved six middens located in a spruce-fir forest and seven middens in a transition-zone forest on the slopes of Mt. Graham. The first field experiment detailed the effects of a three-month storage period on seed nutritional composition, whereas the second examined the influence of an eight-month storage period. These storage periods were examined because red squirrels feed on seeds stored during the late summer throughout the following winter and spring (C. Smith 1968). The first experiment involved corkbark fir, Engelmann spruce, southwestern white pine, and Douglas-fir cones; no white fir cones were produced that year. All cones selected were green and closed and were knocked from a tree within 24 h of collection. Cones from corkbark fir and Engelmann spruce were buried inside three middens in the spruce-fir forest, whereas corkbark fir, Engelmann spruce, white pine, and Douglas-fir cones were placed inside three middens in the transition-zone forest. All cones were buried on the same day of collection during the last week of September at a depth of 4 cm in groups of 2–17 cones each to simulate the manner in which free-ranging red squirrels cache cones in their middens. A total of 10 Engelmann spruce and 2 corkbark fir cones were placed inside each spruce-fir midden, whereas 10 Engelmann spruce, 2 corkbark fir, 1 white pine, and 34 Douglas-fir cones were buried in each transition-zone midden. Each group of buried cones was marked with two flags. Mt. Graham red squirrels were observed storing cones during this period. Some of the cones collected were immediately stored at –20°C for later biochemical analyses; they became the control groups for this experiment. All experimental cones were removed from middens during the last week of December and stored at –20°C.

The second field experiment involved cones from corkbark fir, Engelmann spruce, white pine, Douglas-fir, and white fir collected during the last week of September during the second year. A number of experimental cones were removed from the middens during the first experiment, presumably by red squirrels. Cones for the second experiment were thus immediately sealed inside galvanized wire mesh (3.2 mm × 3.2 mm) envelopes that were 8.0 cm wide by 10.0 cm long. Only cones of the same species were placed inside a particular envelope. These envelopes permitted exposure of cones to the natural environment while preventing their consumption. They were then buried in four middens located in transition-zone forest and three middens found in the spruce-fir forest following the procedures described previously. A total of 10 Engelmann spruce and 7 corkbark fir cones were placed inside each spruce-fir midden, whereas 10 Engelmann spruce, 7 corkbark fir, 20 Douglas-fir, 2 white fir, and 5 white pine cones were inserted into each transition zone midden. Samples of each cone species collected during this period were frozen at –20°C for later biochemical analysis, and they represent the control for this experiment. All experimental cones were retrieved from the middens and stored at –20°C in the first week of the following June.

Laboratory Analyses of Seeds

Ash, carbon, hydrogen, crude protein, crude lipid, and energy contents for each seed species and treatment (control or stored) were determined by first drying cones at 60°C for 24 h in a laboratory oven. All seeds were removed from each cone. Wings were removed from each seed prior to chemical analysis, and the seed coat was also removed from each white pine seed. Wings and white pine seed coats were removed prior to nutritional analysis because observations of free-ranging Mt. Graham red squirrels revealed that they usually remove these parts prior to ingesting seeds (P. Young, personal communication). Dead white insect larvae measuring 2 mm in length were found inside many cones. Many of the seeds found in the cones containing larvae had holes in them and parts of their kernels were missing (see Results). Thus, only intact and whole seeds that did not appear to have been attacked by larvae were chosen for chemical analysis. All intact seeds collected from the same tree during the same period were pooled to provide a single control sample for each experiment; each control sample is thus from a different tree. All intact seeds collected from the same cone species stored in the same midden were likewise pooled to produce a single stored sample for each experiment; each stored sample thus represents a different midden.

Ash contents were measured by placing 1 g samples at 500°C for 30 minutes in a muffle furnace (Brower et al. 1997). The only chemical bonds in food that can be used by animals as an energy source are carbon-carbon and carbon-hydrogen bonds (Robbins 1993). Carbon and hydrogen contents were determined to further estimate the amount of digestible energy in each seed type. Amounts of carbon, hydrogen, and nitrogen in 1.5–2.5 mg samples of each species/treatment category were assayed using a PerkinElmer 2400 CHNS/O analyzer. Nitrogen contents were used to calculate crude protein contents following the methods of Robbins (1993). Energy contents were determined by combusting 0.7–1.0 g samples of each species/treatment group using a Parr model 1341 calorimeter using the methods outlined by Paine (1971). Crude lipid contents were determined by petroleum ether extraction in a Soxhlet apparatus using the techniques of Dobush et al. (1985).

Conifer seed oil contents can be as high as 62 percent (Duke and Atchley 1986). Seed oils are triacylglycerols, which are esters of a single glycerol molecule and three fatty acid molecules. Seed oils usually contain a mixture of different triacylglycerols, each with a different fatty acid composition. Linoleic acid is usually the only polyunsaturated fatty acid found in seed oils (Andrews and Ohlrogge 1990) and is required in the diets of mammals for proper health and survival (Robbins 1993). Linoleic acid contents were measured by gas-liquid chromatography. All lipids were first extracted from diet samples using a chloroform-methanol procedure (Folch et al. 1957). All fatty acids were isolated from the extracted lipids by transesterification with 1.0N methanolic HCl, producing fatty acid methyl esters (Christie 1989). Fatty acid methyl esters were identified and quantified using a Hewlett-Packard model 5890 gas chromatograph fitted with a J&W Scientific, Inc. model DB-23 capillary column (30 m). Seeds selected for fatty acid analyses were not oven dried in order to preserve their lipids. An Onset Computer Corporation model TBI32 temperature logger was buried inside each midden at the beginning (September) of the second experiment. Each logger was buried at a depth of 4 cm and programmed to record temperature at this depth once every 6 h. Loggers were retrieved at the end of this experiment (following June) to determine thermal conditions to which the buried cones were exposed. Nutritional compositions of control and stored seeds were compared using a two-way ANOVA (general linear models), examining for species, treatment (control or stored), and species/treatment interaction effects. If a significant species/treatment interaction was found, then compositions of the control and stored groups for each species were compared with Student's t-tests. All statistical methods were performed using SYSTAT version 11 software.

Seed/Sporocarp Comparisons

Nutritional compositions of the fresh (control) and stored seeds collected from the seed-hoarding experiments were compared to those of fungal sporocarps from corresponding sporocarp-hoarding experiments that were conducted at the same study sites during the same period. Sporocarp data are summarized in Frank (in this vol.). The sporocarp study involved seven tree cavities found in a transition-zone forest on the slopes of Mt. Graham. The first sporocarp experiment detailed the effects of a three-month (93 d) storage period on sporocarp nutritional composition, whereas the second examined the influence of an eight-month (229 d) storage period. The first experiment involved 1–3 kg each of *Amanita* spp., *Auricularia* spp., *Lactarius* spp., *Leccinum* spp., *Lycoperdon* spp., and *Russula* spp. sporocarps collected from the spruce-fir forests during September. All of these species are in the fungal class Basidiomycetes, and all except *Auricularia* spp. and *Lycoperdon* spp. produce sporocarps commonly recognized as "mushrooms" (Alexopoulus and Mims 1979). All 6 species are ingested and stored by Mt. Graham red squirrels (Young 1996). Sporocarps of only 6 species were collected because they were the only 6 of 15 species listed as ingested and stored by Mt. Graham red squirrels by Young (1996) present at the study site during this particular year in large quantities, and only *Auricularia* spp., *Lycoperdon* spp., and *Russula* spp. were selected for the food-hoarding experiment because sufficient quantities (3 kg each) of other sporocarps were not available. Half of the *Auricularia* spp., *Lycoperdon* spp., and *Russula* spp. sporocarps collected were immediately frozen at –20°C for later analysis. The remaining *Auricularia* spp., *Lycoperdon* spp., and *Russula* spp. were used to determine the effects of hoarding by Mt. Graham red squirrels on sporocarp nutritional composition.

Four 500 g samples of each species were immediately sealed inside galvanized wire mesh (3.2 mm × 3.2 mm) envelopes (one sample/envelope) that were 8.0 cm wide by 10.0 cm long. Only sporocarps of the same species were placed inside a particular envelope. These envelopes permitted exposure of the sporocarps to the natural environment while preventing their consumption by mammals. One envelope of each species was then placed at a height of 1–2 m above ground inside the cavities of four standing quaking aspen trees with hollow main trunks to simulate hoarding by red squirrels. Mt. Graham red squirrels frequently store sporocarps in this type of tree cavity (personal observation). The second field experiment involved 1–3 kg each of *Amanita* spp., *Boletus* spp., *Lactarius* spp., *Leccinum* spp., *Lycoperdon* spp., and *Russula* spp. sporocarps collected from the spruce-fir forests of the Pinaleño Mountains during August; all six species are ingested and stored by Mt. Graham red squirrels (Young 1996).

Half of the *Lactarius* spp., *Leccinum* spp., and *Russula* spp. collected were immediately frozen at –20°C for later analysis. The remaining *Lactarius* spp., *Leccinum* spp., and *Russula* spp. were used to determine the effects of an eight-month hoarding period by Mt. Graham red squirrels on sporocarp nutritional composition. Field experiments involving these remaining sporocarps were conducted using the methods described previously.

Control seed and sporocarp data were statistically compared by first averaging all the seed or sporocarp means for a particular nutritional parameter of each seed or sporocarp species to produce grand means and standard errors for both seeds and sporocarps. The grand means were then statistically compared using a Student's t-test. Seed and sporocarp grand means were calculated and statistically compared for the three-month and eight-month storage experiments in the same manner.

Results

Three-Month Seed-Storage Experiment

Most of the cones buried during the first experiment were removed from the middens before the end of this period, presumably by red squirrels. Only the seeds of cones found in experimental caches that had not been disturbed were analyzed and included in the study to avoid any biases. Seeds from the first experiment were thus analyzed only for ash, hydrogen, carbon, and crude protein contents due to the limited amount of material available (table 18.1). Storage for three months significantly increased the ash contents of all seed species examined (F = 986.060, df = 1,15, P <0.001), and there was a significant species/treatment interaction (F = 22.445, df = 3,15, P <0.001). Storage did not significantly influence carbon contents (F = 3.794, df = 1,49, P = 0.06), however. Hydrogen contents of all seeds significantly decreased during storage (F = 17.057, df = 3,49, P <0.001), with no significant species/treatment interaction (F = 1.610, df = 3,49, P = 0.199). Storage also significantly influenced crude protein content (F = 10.408, df = 1, 49, P = 0.002), with a significant species/treatment interaction (F = 10.124, df = 3, 49, P <0.001). Protein contents of both Engelmann spruce and white pine seeds significantly decreased after three months of storage, whereas those of corkbark fir and Douglas-fir seeds were not affected.

Table 18.1. Mean (± SE) seed compositions in the first (three-month) field experiment

Nutrient (% dry mass)	*A. lasiocarpa*		*P. engelmannii*		*P. strobiformis*		*P. menziesii*	
	Control	Stored	Control	Stored	Control	Stored	Control	Stored
Ash	2.9 ± 0.1	10.5 ± 0.6*	3.9 ± 0.1	11.6 ± 0.8*	2.3 ± 0.1	16.1 ± 0.5*	4.1 ± 0.3	14.7 ± 0.8*
	(5)	(5)	(5)	(5)	(5)	(5)	(5)	(5)
Carbon	57.4 ± 1.8	57.5 ± 0.6	59.1 ± 0.2	56.6 ± 0.5	52.6 ± 0.3	50.4 ± 0.1	56.9 ± 0.3	55.5 ± 0.6
	(10)	(5)	(10)	(5)	(5)	(5)	(9)	(3)
Hydrogen	7.2 ± 0.2	7.0 ± 0.1*	7.9 ± 0.1	7.1 ± 0.1*	6.8 ± 0.1	6.2 ± 0.1*	7.5 ± 0.1	7.3 ± 0.1*
	(10)	(5)	(10)	(5)	(5)	(5)	(5)	(5)
Crude Protein	9.8 ± 0.5	10.9 ± 0.1	21.6 ± 0.6	18.2 ± 0.8*	10.7 ± 1.2	3.4 ± 0.5*	18.6 ± 0.7	20.1 ± 2.8
	(10)	(5)	(10)	(5)	(5)	(5)	(9)	(3)

*Significantly different from the corresponding control mean at the $P < 0.05$ level; number in parentheses represents the number of middens sampled (n) for that mean.

Eight-Month Seed-Storage Experiment

Many of the seeds examined had holes, and parts of their kernels were missing. White insect larvae measuring 2 mm in length were found in the scales of cones that contained seeds with holes. The holes in these seeds are thus apparently due to larval feeding. The entire kernels of white pine seeds stored for eight months were replaced by fungal hyphae, apparently consumed by these organisms. All white pine seeds from the second experiment consequently were not analyzed for nutritional composition. Intact corkbark fir kernels had a mean (± SE) mass of 0.84 ± 0.06 mg/seed (n = 10), which was significantly greater than 0.20 ± 0.06 mg/seed for the kernels (n = 5) that had been partially consumed by insect larvae (t = 6.73, P = 0.0001). Whole Engelmann spruce kernels had a mean (± SE) mass of 3.93 ± 0.11 mg/seed (n = 10), whereas mass decreased to 0.43 ± 0.07 mg/seed (n = 5) after larval infestation (t = 21.06, P = 0.0001). Intact kernels of white pine had a mean (± SE) mass of 48.22 ± 2.11 mg/seed (n = 10), but this significantly decreased (t = –5.33, P = 0.0001) to 30.66 ± 2.38 mg/seed (n = 6) after larval feeding. The mean (± SE) kernel mass of intact Douglas-fir was 7.85 ± 0.52 mg/seed (n = 10), but it was reduced significantly (t = 14.09, P = 0.0001) to 0.45 ± 0.07 mg/seed by larval infestation (n = 4).

Only intact seed kernels that had no exit holes or larvae were used in all nutritional analyses (table 18.2). Mean ash contents of all seed species significantly increased during storage (F = 34.957, df = 1,34, P <0.001), with a species/treatment interaction (F = 38.811, df = 3,34, P <0.001). Storage for eight months significantly influenced carbon content (F = 61.993, df = 1,19, P <0.001), with a species/treatment interaction (F = 3.132, df = 3, 19, P = 0.05). Carbon contents of both corkbark fir and white fir significantly decreased after storage, but those of Engelmann spruce and Douglas-fir were not affected. Storage for eight months affected hydrogen contents (F = 34.705, df = 1,19, P <0.001), with the same species/treatment interaction (F = 3.658, df = 3,19, P = 0.031). Hydrogen contents of both corkbark fir and white fir significantly decreased after storage, but those of Engelmann spruce and Douglas-fir did not change. Crude protein contents were changed significantly by storage (F = 4.202, df = 1,19, P = 0.05), with a species/treatment interaction (F = 7.409, df = 3,19, P = 0.002). Protein contents of corkbark fir, white fir, and Douglas-

Table 18.2. Mean (± SE) seed compositions in the second (eight-month) field experiment

Nutrient	*A. lasiocarpa*		*P. engelmannii*		*P. menziesii*		*A. concolor*	
(% dry g)	Control	Stored	Control	Stored	Control	Stored	Control	Stored
Ash	2.2 ± 0.1	2.5 ± 0.1*	2.4 ± 0.7	4.9 ± 0.2*	4.0 ± 0.8	8.8 ± 0.7*	4.3 ± 0.2	2.2 ± 0.1*
	(5)	(14)	(3)	(3)	(4)	(4)	(5)	(5)
Carbon	63.1 ± 0.3	58.2 ± 0.3*	58.7 ± 0.5	56.6 ± 0.6	60.0 ± 0.2	58.8 ±0.9	60.6 ± 0.3	56.4 ± 0.7*
	(4)	(12)	(4)	(8)	(4)	(4)	(9)	(3)
Hydrogen	9.2 ± 0.1	8.1 ± 0.1*	8.0 ± 0.1	7.6 ± 0.2	8.8 ± 0.1	8.6 ± 0.1	8.2 ± 0.2	7.3 ± 0.1*
	(4)	(12)	(4)	(8)	(4)	(4)	(5)	(5)
Crude Protein	21.6 ± 0.8	13.0 ± 1.3*	20.2 ± 1.5	25.8 ± 2.1	36.3 ± 1.2	32.9 ± 1.5*	10.8 ± 0.6	8.5 ± 0.4*
	(4)	(12)	(4)	(8)	(4)	(4)	(4)	(4)
Crude Lipid	8.9 ± 0.3	10.4 ± 1.1	4.9 ± 0.3	2.9 ± 0.3*	19.3 ± 1.1	7.2 ± 0.2*	15.5 ± 0.6	5.7 ± 0.6*
	(4)	(4)	(4)	(4)	(4)	(4)	(4)	(4)
% Infested Seeds/	17.3 ± 11.3	51.5 ± 5.4	38.8 ± 4.1	36.3 ± 4.9	15.4 ± 2.8	31.9 ± 3.9	39.5 ± 17.1	8.8 ± 4.0
Cone	(7)	(29)	(23)	(22)	(34)	(16)	(3)	(4)
Energy	27.7 ± 0.8	26.0 ± 1.5	25.2 ± 0.6	25.9 ± 0.4	23.1 ± 0.3	23.8 ± 0.6	27.7 ± 0.3	24.2 ± 0.3
(kJ/ash–free g)	(4)	(4)	(3)	(3)	(3)	(3)	(3)	(3)
Linoleic Acid	34.3 ± 2.1	36.0 ± 1.2	22.1 ± 0.5	13.0 ± 0.6*	75.6 ± 1.6	24.2 ± 1.0*	15.2 ± 2.4	5.5 ± 0.3*
(mg/g)	(4)	(4)	(4)	(4)	(4)	(4)	(4)	(4)

*Significantly different from the corresponding control mean at the $P < 0.05$ level; number in parentheses represents the number of middens sampled (n) for that mean.

fir significantly decreased after storage, but those of Engelmann spruce remained constant. Crude lipid contents were also significantly affected by storage ($F = 104.749$, df = 1,18, $P < 0.001$), with a species/treatment interaction ($F = 41.09$, df = 3,19, $P < 0.001$). Crude lipid contents of all seed species except corkbark fir significantly decreased after storage. Neither the proportion of seeds infested with insect larvae ($F = 0.615$, df = 1,160, $P = 0.434$) nor energy contents ($F = 2.416$, df = 1,18, $P = 0.138$) were influenced by storage for eight months. Linoleic acid contents significantly changed with storage for eight months ($F = 319$, df = 1,12, $P < 0.001$), with a significant species/treatment interaction ($F = 172.051$, df = 3,21, $P < 0.001$). Linoleic acid contents of all seed species except corkbark fir decreased with storage.

Middens located in the transition zone had a mean (± SE) average daily internal temperature of 1.8 ± 0.1°C during the eight-month experiment, whereas that for the spruce-fir middens was 1.3 ± 0.2°C, significantly lower ($t = -3.38$, $P = 0.03$). The mean (± SE) minimum internal temperatures of the transition zone and spruce-fir middens were −2.0 ± 0.7 and −2.9 ± 1.2°C, respectively, and they were not significantly different ($t = -0.8$, $P = 0.49$). The mean (± SE) maximum internal temperature of the transition zone middens was 11.1 ± 0.9°C, whereas that for the spruce-fir middens was 11.2 ± 0.9°C and statistically equivalent ($t = 0.08$, $P = 0.90$).

Three-Month Seed/Sporocarp Comparisons

Figure 18.1 shows the nutritional compositions of control seeds and sporocarps (fig. 18.1A), as well as stored seeds and sporocarps (fig. 18.1B), during the three-month storage experiment. Control (fresh) seeds had a significantly lower ash content than the control (fresh) sporocarps collected during this same period ($t = 5.056$, df = 8, $P = 0.001$). Control seeds had significantly greater carbon ($t = 9.195$, df = 8, $P < 0.001$) and hydrogen ($t = 3.685$, df = 8, $P = 0.006$) contents. These two groups did not significantly differ in crude protein content, however ($t = 1.482$, df = 8, $P = 0.177$).

Seeds stored for three months had equivalent ash contents to sporocarps stored for the same length of time ($t = -1.778$, df = 5, $P = 0.136$). The three-month stored seeds had significantly greater carbon ($t = -5.844$,

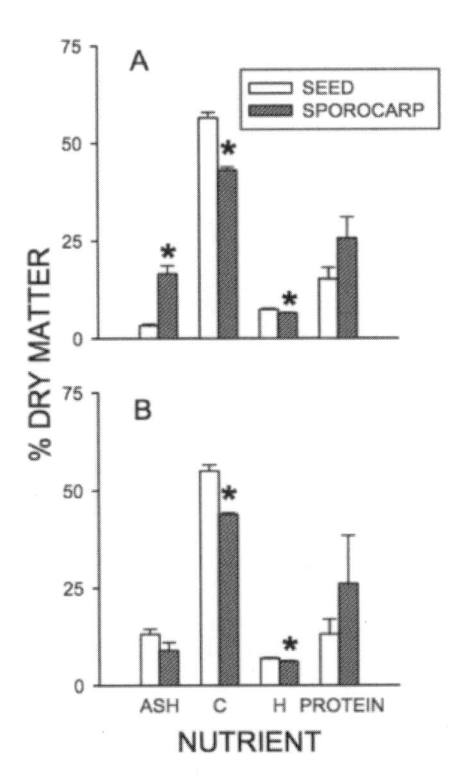

Figure 18.1. Mean (±SE) nutritional compositions of control seeds and sporocarps (A), as well as stored seeds and sporocarps (B), during the three-month storage experiments. Sporocarp means with an asterisk (*) are significantly different from the corresponding seed mean at the $P < 0.05$ level. Data for sporocarps are from Frank (in this vol.).

df = 5, P = 0.002) and hydrogen (t = –2.509, df = 5, P = 0.05) contents than sporocarps stored for three months. Both storage groups had statistically equivalent protein contents (t = 1.150, df = 5, P = 0.302), however.

Eight-Month Seed/Sporocarp Comparisons

Figure 18.2 shows the nutritional compositions of control seeds and sporocarps (fig. 18.2A), as well as stored seeds and sporocarps (fig. 18.2B), during the eight-month storage experiment. Control seeds for the eight-

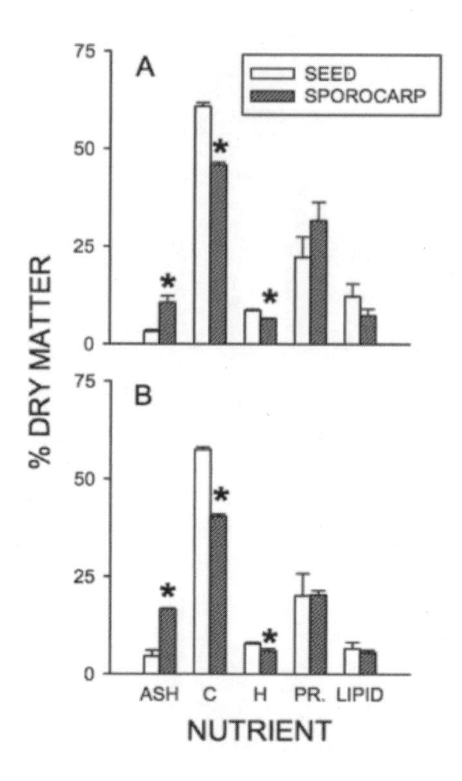

Figure 18.2. Mean (±SE) nutritional compositions of control seeds and sporocarps (A), as well as stored seeds and sporocarps (B), during the eight-month storage experiments. Sporocarp means with an asterisk (*) are significantly different from the corresponding seed mean at the $P < 0.05$ level. PR = protein content. Data for sporocarps are from Frank (2006).

month experiment had significantly lower ash contents than the control sporocarps ($t = 3.172$, df = 8, $P = 0.008$). Control seeds had significantly greater carbon ($t = -13.314$, df = 8, $P < 0.001$) and hydrogen ($t = -9.114$, df = 8, $P < 0.001$) than control sporocarps. These two control groups had equivalent crude protein ($t = 1.313$, df = 8, $P = 0.226$) and crude lipid ($t = -1.474$, df = 8, $P = 0.179$) contents. Control seeds had a mean (± SE) energy content of 25.9 ± 1.1 kJ/g ($n = 4$), which was significantly greater ($t = -5.421$, df = 8, $P = 0.001$) than that of the control sporocarps (17.9 ±0.9 kJ/g, $n = 6$).

Seeds stored for eight months had considerably lower ash contents than sporocarps stored for this period ($t = 6.673$, df = 5, $P = 0.001$). Stored seeds had greater carbon ($t = -19.759$, df = 5, $P < 0.001$) and hydrogen ($t = -5.563$, df = 5, $P = 0.003$) contents than the stored sporocarps. Both stored groups had equivalent levels of crude protein ($t = 0.042$, df = 5, $P = 0.97$) and crude lipid ($t = -0.516$, df = 5, $P = 0.63$). The stored seeds had a mean (± SE) energy content of 25.0 ± 0.6 kJ/g ($n = 4$), which was significantly greater ($t = -9.980$, df = 5, $P < 0.001$) than that of the stored sporocarps (17.0 ± 0.5 kJ/g, $n = 3$).

Discussion

All seed species increased in ash content with storage in both experiments (tables 18.1, 18.2), indicating a general loss of organic matter associated with natural hoarding. The remaining organic matter sometimes changed in nutritional composition during the storage period as well. A total of 12 different nutrient/seed species combinations other than ash content were examined in the three-month storage experiment. Seed nutritional composition remained constant during storage in only half (6) of these cases, whereas it decreased in the remaining half. A total of 28 different nutrient/seed species combinations other than ash content were investigated in the eight-month storage experiment. Stored seeds had constant nutritional compositions in only 53 percent (15) of these instances, whereas nutritional content decreased during this storage period in 46 percent (13) of the cases. Results of these two experiments therefore support my prediction that naturally stored seeds decrease in nutritional composition as frequently as they retain their nutritional composition during the caching period. In all but one of the instances where significant nutritional losses were observed, species/treatment interactions were also found. Stored corkbark fir seeds decreased in five different nutritional parameters during the eight-month storage experiment, but Engelmann spruce seeds decreased in just two nutritional factors in this experiment (table 18.2). These findings thus support my prediction that the magnitude of nutritional losses during a given storage period varies with conifer seed species.

Some of the nutritional losses observed during seed storage were sufficient to potentially affect the survival of Mt. Graham red squirrels.

Wild rodents require a dietary protein content of at least 18 percent for proper health and survival (Robbins 1993). Seed kernels of corkbark fir and white fir, however, had mean protein contents of only 13 and 8.5 percent, respectively, after eight months of storage, whereas that of white pine kernels decreased to 3.3 percent after just three months of storage. These seeds, thus, probably did not have sufficient protein contents to meet the nutritional requirements of Mt. Graham red squirrels. Rodents require a dietary linoleic acid content of at least 7 mg/g for survival (NRC 1995). Kernels of white fir, however, had a mean linoleic acid content of just 5.5 mg/g after eight months of storage (table 18.2). Mt. Graham red squirrels feeding primarily on stored white pine, corkbark fir, and white fir seeds during the late winter/early spring may thus experience nutritional stress that reduces their ability to survive this difficult period. The propensity of a food type to decrease in nutritional quality while in a food hoard is called perishability (C. Smith and Reichman 1984). This study indicates that conifer seeds vary greatly with species in perishability. Free-ranging eastern gray squirrels prefer to hoard seeds with a low perishability over those with a relatively higher perishability, apparently to maximize the useful life span of their food stores (Hadj-Chikh et al. 1996). Mt. Graham red squirrels may thus use a similar strategy when storing conifer cones. It has been proposed that food-hoarding animals may ingest the most perishable food items in their caches first, saving the least perishable items for last to better maintain a balanced diet during cache use (Gendron and Reichman 1995). Mt. Graham red squirrels may therefore have such dietary preferences when they begin to feed on their seed hoards. It would be of interest to determine both the cone-hoarding and cone-eating preferences of these squirrels.

It is important to recall that all of the seed kernels selected for nutritional analysis were free of both insect damage and fungal growth. Seeds convert stored lipids into carbohydrates as they germinate, thereby reducing their energy, carbon, and hydrogen contents (Salisbury and Ross 1985). Some of the nutritional changes observed in this seed study may therefore have been due to germination. This study suggests that insect larvae may greatly decrease the amount of seed material available to red squirrels during storage. Attack by insect larvae significantly decreased the kernel masses of all four seed species examined, with

Douglas-fir seeds experiencing the greatest mass loss/seed. Larvae found in all cones matched the description of seed maggots (*Euromyia* sp.), a common conifer seed predator in Arizona (Burns and Honkala 1990). Eastern gray squirrels prefer to ingest acorns infested with weevils immediately upon their collection, whereas they prefer to store acorns of the same species that are not infested (Steele et al. 1996). Mt. Graham red squirrels may thus minimize the losses associated with larval infestation by preferring to store only non-infested cones, and immediately ingesting all infested seeds collected. No insect larvae infested the stored fungal sporocarps examined in the previous study (Frank in this vol.). One potential advantage of hoarding sporocarps over conifer seeds may consequently be the avoidance of losses associated with insect consumption.

Control seeds from both experiments had greater ash and carbon contents than the control sporocarp groups, and greater energy contents than the eight-month sporocarp controls. Carbon, hydrogen, and energy contents of stored seeds were also greater than those of stored sporocarps. These results support my predications that (a) the nutritional content of conifer seeds is generally greater than that of fungal sporocarps, and (b) stored conifer seeds have an overall greater nutritional content than stored fungal sporocarps during winter use. Further investigation of this system will undoubtedly provided new insights into the coevolution of mammalian food hoarding and granivory.

Acknowledgments

I thank Paul Y. Young for his generous assistance with many aspects of this project. This study would not have been possible without field assistance from Vicki L. Greer, Erin Bibles, Nancy Ferguson, and J. Elaine Lowry. Helpful discussions with Reed Sanderson and Genice Froehlich greatly aided the development of this project. I am also grateful to Ellen S. Dierenfeld for the use of her equipment for the crude lipid analyses. Funding for this study was provided by the University of Arizona through the Mt. Graham Red Squirrel Study Committee, which is comprised of representatives from the Arizona Game and Fish Department, the University of Arizona, the U.S. Forest Service, and the U.S. Fish and Wildlife Service.

Literature Cited

Alexopoulos, C. J., and C. W. Mims. 1979. *Introductory mycology.* John Wiley and Sons, New York.

Andrews, J. E., and J. Ohlrogge. 1990. Fatty acid and lipid biosynthesis and degradation. Pages 339–352 in D. T. Dennis and D. H. Tarpin, eds. *Plant physiology, biochemistry, and molecular biology.* John Wiley and Sons, New York.

Brower, J. E., J. H. Zar, and C. Von Ende. 1997. *Field and laboratory methods for general ecology.* McGraw-Hill, New York.

Burns, R. M., and B. H. Honkala. 1990. *Silvics of North America*, vol. 1. U.S. Forest Service, Washington, D.C.

Christie, W. W. 1989. *Gas chromatography and lipids.* The Oily Press, Ayr, Scotland.

Cram, W. E. 1924. The red squirrel. *Journal of Mammalogy* 5:37–41.

Dice, L. R. 1921. Notes on the mammals of interior Alaska. *Journal of Mammalogy* 2:20–28.

Dobush, G. R., C. D. Ankney, and D. G. Krementa. 1985. The effect of apparatus, extraction time, and solvent type on lipid extraction of snow geese. *Canadian Journal of Zoology* 63:1917–1920.

Duke, J. A., and A. A. Atchley. 1986. Handbook of proximate analysis tables of higher plants. CRC Press, Boca Raton, Florida.

Folch, J., M. Lees, and G. Stanley. 1957. A simple method for the isolation and purification of total lipids from animal tissues. *Journal of Biological Chemistry* 226:497–507.

Froehlich, G. F., and N. S. Smith. 1990. Habitat use by the Mt. Graham red squirrel. Pages 118–125 in D. R. Krassman and N. S. Smith, eds. *Managing wildlife in southwest Arizona.* Arizona Chapter of the Wildlife Society, Phoenix.

Gendron, R. P., and O. J. Reichman. 1995. Food perishability and inventory management: A comparison of three caching strategies. *American Naturalist* 145:948–968.

Hadj-Chikh, L. Z., M. A. Steele, and P. D. Smallwood. 1996. Caching decisions by gray squirrels: A test of the handling time and perishability hypotheses. *Animal Behaviour* 52:941–948.

NRC (National Research Council). 1995. *Nutrient requirements of laboratory animals.* National Academy Press, Washington, D.C.

Paine, R. T. 1971. The measurement and application of the calorie to ecological problems. *Annual Review of Ecology and Systematics* 2:145–164.

Post, D. M. 1992. Change in nutrient content of foods stored by eastern woodrats (*Neotoma floridana*). *Journal of Mammalogy* 73:835–839.

Robbins, C. T. 1993. *Wildlife feeding and nutrition.* Academic Press, New York.

Rusch, D. A., and W. G. Reeder. 1978. Population ecology of Alberta red squirrels. *Ecology* 59:400–420.

Salisbury, F. B., and C. W. Ross. 1985. *Plant physiology.* Wadsworth Publishing, Belmont, California.

Sherry, D. F. 1985. Food storage by birds and mammals. *Advances in the Study of Behavior* 15:153–188.

Smith, A. A., and R. W. Mannan. 1994. Distinguishing characteristics of Mount Graham red squirrel midden sites. *Journal of Wildlife Management* 58:437–445.

Smith, C. C. 1968. The adaptive nature of social organization in the genus of three [*sic*] squirrels *Tamiasciurus*. *Ecological Monographs* 38:31–63.

——. 1995. The niche of diurnal tree squirrels. Pages 209–225 in C. A. Istock and R. S. Hoffman, eds. *Storm over a mountain island: Conservation biology and the Mt. Graham affair.* University of Arizona Press, Tucson.

Smith, C. C., and O. J. Reichman. 1984. The evolution of food caching by birds and mammals. *Annual Review of Ecology and Systematics* 15:329–351.

Steele, M. A., L. Z. Hadj-Chikh, and J. Hazeltine. 1996. Caching and feeding decisions by *Sciurus carolinensis*: Response to weevil-infested acorns. *Journal of Mammalogy* 77:305–314.

Vander Wall, S. B. 1990. Food hoarding in animals. University of Chicago Press, Chicago, Illinois.

Vernes, K., S. Blois, and F. Barlocher. 2004. Seasonal and yearly changes in consumption of hypogeous fungi by northern flying squirrels and red squirrels in old-growth forest, New Brunswick. *Canadian Journal of Zoology* 82:110–117.

Young, P. 1996. *Annual report of the Mt. Graham red squirrel monitoring program.* University of Arizona, Tucson.

## CHAPTER 19

The Nutritional Ecology of Fungal Sporocarp Hoarding by Mt. Graham Red Squirrels

CRAIG L. FRANK

ALTHOUGH RED SQUIRRELS are primarily granivorous rodents, the fruiting bodies (sporocarps) of mycorrhizal fungi are also ingested, at times comprising up to 95 percent of the diet (Maser et al. 1978; Currah et al. 2000; Vernes et al. 2004). The sporocarps ingested develop either below (hypogeous sporocarps) or above (epigeous sporocarps) ground. Hypogeous sporocarps are known as "truffles," whereas most epigeous sporocarps are called "mushrooms" (North et al. 1997). Sporocarps are also stored by red squirrels for winter use by placing them inside hollow downed logs or the cavities of standing trees (Dice 1921; Cram 1924; Smith 1968) and are 45–95 percent of the winter diet (Vernes et al. 2004).

Mt. Graham red squirrels ingest the sporocarps of 15 fungal species found in the Pinaleño Mountains (Young 1996). Little is known about the nutritional value of fungal sporocarps to Mt. Graham red squirrels despite their abundance in the diet. Two mushroom species (*Russula emetica* and *Cantharellus cibarius*) commonly ingested by fox squirrels had metabolizable energy contents of only 51–58 percent, whereas those of the seeds ingested were 75–94 percent (Smith 1995). Only 52 percent of the energy and 30 percent of the crude protein found in deer truffle (*Elaphomyces granulatus*) sporocarps ingested by Cascade golden-mantled ground squirrels (*Spermophilus saturatus*) could be digested and utilized (Cork and Kenagy 1989). It has been suggested that squirrels place mushrooms inside hollow logs and the cavities of trees to dry them (Moller 1983). Vander Wall (1990) suggested that air-drying mushrooms might actually conserve their nutritional value to squirrels during storage. Mushrooms stored by red squirrels quickly disintegrated upon warming after being exposed to temperatures below 0°C (Smith 1968). I thus predict that (1) the aboveground hoarding behavior of squirrels serves to reduce sporocarp moisture content, (2) this desiccation is associated with a stable sporocarp nutritional composition prior to the onset of spring freeze/thaw cycles, and (3) air-dried sporocarps stored for winter use decrease in nutritional quality during subsequent spring freeze/thawing cycles. These hypotheses were tested in field experiments involving five different sporocarp species harvested and ingested by Mt. Graham red squirrels in the Pinaleño Mountains of Arizona. The nutritional compositions of three additional sporocarp species were also determined during the late summer to further examine the nutritional limitations of mycophagy.

Methods

Study Species and Field Experiments

This study involved seven tree cavities found in the transition-zone forest on the slopes of Mt. Graham. The first field experiment detailed the effects of a three-month (93 d) storage period on sporocarp nutritional composition, whereas the second examined the influence of an eight-month (229 d) storage period. These storage periods were examined to determine the effects of air-drying on sporocarp preservation both before and after the onset of spring freeze/thaw cycles. The first experiment involved 1–3 kg each of *Amanita* spp., *Auricularia* spp., *Lactarius* spp., *Leccinum* spp., *Lycoperdon* spp., and *Russula* spp. sporocarps collected from the spruce-fir forests during September 1999. All of these species are in the fungal class Basidiomycetes, and all except *Auricularia* spp. and *Lycoperdon* spp. produce sporocarps commonly recognized as "mushrooms" (Alexopoulus and Mims 1979). All 6 species are ingested and stored by the Mt. Graham red squirrel (Young 1996). Sporocarps of only 6 species were collected because they were the only 6 of 15 species listed as ingested and stored by the Mt. Graham red squirrel by Young (1996) present at the study site during this particular year in large quantitites, and only *Auricularia* spp., *Lycoperdon* spp., and *Russula* spp. were selected for the food-hoarding experiment because sufficient quantities (3 kg each) of other sporocarps were not available. Half of the *Auricularia* spp., *Lycoperdon* spp., and *Russula* spp. sporocarps collected were immediately sealed inside 50 mL centrifuge tubes and frozen at –20°C for later analysis. The remaining *Auricularia* spp., *Lycoperdon* spp., and *Russula* spp. were used to determine the effects of hoarding by Mt. Graham red squirrels on sporocarp nutritional composition.

Four 500 g samples of each species were immediately sealed inside galvanized wire mesh (3.2 mm × 3.2 mm) envelopes (one sample/envelope) that were 8.0 cm wide by 10.0 cm long. Only sporocarps of the same species were placed inside a particular envelope. These envelopes permitted exposure of the sporocarps to the natural environment while preventing their consumption by mammals. One envelope of each species was then placed at a height of 1–2 m above ground inside the cavities of four standing quaking aspen trees with hollow main trunks to simulate hoarding by red squirrels. Mt. Graham red squirrels frequently store sporocarps in this type of tree cavity (personal observation). All four trees were located within 2 km of each other in the same study site centered at 32°41.8527' N, 109°54.3915' W, where sporocarps were collected. An Onset temperature logger (Onset, Pocasset, MA) was placed inside three of these tree cavities, adjacent to the envelopes. Each logger measured and recorded cavity air temperature at four-hour intervals throughout the experiment. The envelopes were kept inside the tree cavities until 27 December 1999. All sporocarps were then removed from the envelopes, sealed inside 50 mL plastic centrifuge tubes, and stored at –20°C.

The second field experiment involved 1–3 kg each of *Amanita* spp., *Boletus* spp., *Lactarius* spp., *Leccinum* spp., *Lycoperdon* spp., and *Russula* spp. sporocarps collected from the spruce-fir forests of the Pinaleño Mountains during August 2001, and all six species are ingested and stored by Mt. Graham red squirrels (Young 1996). Half of the *Lactarius* spp., *Leccinum* spp., and *Russula* spp. collected were immediately sealed inside 50 mL centrifuge tubes and frozen at –20°C for later analysis. The remaining *Lactarius* spp., *Leccinum* spp., and *Russula* spp. were used to determine the effects of an eight-month hoarding period by Mt. Graham red squirrels on sporocarp nutritional composition. Field experiments involving these remaining sporocarps were conducted using the methods described previously. In addition, an Onset temperature logger (Onset, Pocasset, MA) was placed below the soil surface at the base of the tree at a depth of 5 cm. The logger measured and recorded temperatures at four-hour intervals throughout the experiment to provide an estimate of the temperatures of food items buried in middens. The envelopes were kept inside the trees through 28 March 2002 to assess the effects of storage after spring freeze/thaw cycles.

Laboratory Analyses of Sporocarps

The ash, carbon, hydrogen, crude protein, crude lipid, and energy contents for each sporocarp species and treatment (control or stored) were determined by first drying samples at 60°C for 24 h in a laboratory oven. Ash contents were measured by placing 1 g samples at 500°C for 30 minutes in a muffle furnace using the techniques of Brower et al. (1997). The only chemical bonds in food that can be used by animals as an energy source are carbon-carbon and carbon-hydrogen bonds (Robbins 1993). Carbon and hydrogen contents were thus determined to estimate the amount of digestible energy in each sporocarp type. The amounts of carbon, hydrogen, and nitrogen in 1.5–2.5 mg samples of each species/treatment category were assayed using a Perkins Elmer model 2400 CHNS/O analyzer. Crude protein contents were calculated as $6.25 \times N$ content (Robbins 1993). Energy contents were determined by combusting 0.7–1.0 g samples of each species/treatment group with a Parr model 1341 calorimeter using the methods outlined by Paine (1971). Crude lipid contents were determined by petroleum ether extraction in a Soxhlet apparatus using the techniques of Dobush et al. (1985). All fresh (control) nutritional compositions were compared to each other with a one-way ANOVA (general linear models) procedure. Nutritional compositions of control and stored sporocarps were compared using a two-way ANOVA (general linear models), examining for species, treatment (control or stored), and species treatment interaction effects. If a significant species treatment interaction was found, then compositions of the control and stored groups for each species were compared with Student's t-tests. All statistical methods were performed using SYSTAT version 11 software.

Results

Three-Month Control and Stored Sporocarp Compositions

Comparisons of three-month control and stored sporocarp compositions varied within and between genera (table 19.1). The comparison of all

Table 19.1. Mean (± SE) sporocarp compositions for the control (fresh) and three-month experimental groups

Sporocarp Type†	Ash (%)	Carbon (%)	Hydrogen (%)	Crude Protein (%)	Crude Lipid (%)	Energy (kJ/g)	Moisture (% wet mass)
Amanita-C	24.6 ± 2.0[a] (4)	44.6 ± 0.5[a] (4)	6.6 ± 0.1[a] (4)	26.4 ± 1.0[a] (4)	9.0 ± 0.0[a] (4)	17.2 ±0.2[a] (4)	90.0 ± 1.0[a] (4)
Auricularia-C	11.4 ± 2.2[b] (7)	43.7 ± 0.2[a] (4)	6.4 ± 0.1[a] (4)	6.5 ± 0.3[b] (4)	5.0 ± 1.0[a] (6)	17.4 ± 0.2[a] (4)	85.2 ± 0.2[b] (10)
Auricularia-S	4.8 ± 0.4* (4)	44.6 ± 0.2 (4)	5.8 ± 0.1* (4)	8.3 ± 0.4 (4)	1.6 ± 0.1* (4)	17.6 ± 0.2 (4)	12.6 ± 1.0* (4)
Lactarius-C	11.1 ± 1.8[b] (6)	44.6 ± 0.1[a] (4)	6.7 ± 0.0[a] (4)	24.3 ± 0.2[a] (4)	4.2 ± 1.0[a] (2)	13.4 ± 2.5[b] (3)	85.2 ± 0.2[b] (5)
Leccinum-C	18.8 ± 3.8[b] (6)	39.5 ± 4.9[a] (4)	5.9 ± 0.8[a] (4)	29.8 ± 3.9[a] (4)	8.3 ± 2.1[a] (5)	18.1 ± 0.2[a] (5)	89.3 ± 0.2[b] (5)
Lycoperdon-C	15.4 ± 0.5[b] (2)	43.2 ± 0.3[a] (4)	6.7 ± 0.0[a] (4)	46.9 ± 0.3[c] (4)	4.6 ± 0.5[a] (4)	17.4 ± 0.2[a] (6)	86.3 ± 0.6[b] (10)
Lycoperdon-S	11.5 ± 1.7* (4)	43.3 ± 0.4 (4)	6.0 ± 0.1* (4)	49.8 ± 2.9 (4)	—	16.9 ± 0.2* (4)	13.3 ± 1.1* (4)
Russula-C	17.7 ± 2.6[b] (6)	43.0 ± 0.4[a] (4)	6.5 ± 0.1[a] (4)	20.3 ± 0.6[a] (4)	4.3 ±1.0[a] (4)	18.8 ± 0.2[a] (3)	89.6 ± 0.9[a] (10)
Russula-S	10.8 ± 1.2* (4)	43.7 ± 0.6 (4)	6.5 ± 0.1 (4)	20.2 ± 0.9 (4)	3.5 ± 0.6 (4)	18.1 ± 0.1* (4)	9.9 ± 1.1* (4)

*Stored [S] sporocarp mean is significantly different from the corresponding control [C] mean at the $P < 0.05$ level.

†Control [C] sporocarp means within the same experiment sharing a common lowercase letter are not significantly different from other control means at the $P < 0.05$ level; number in parentheses represents sample size (n) for that mean.

fresh (control) sporocarps collected during 1999 revealed that *Auricularia* spp. had significantly lower moisture content (F = 4.135, df = 5,39, P = 0.004) than the other five species, whereas the *Amanita* spp. sporocarps had the greatest ash content of all species collected (F = 2.996, df = 5,24, P = 0.03). All six fresh sporocarp species did not significantly differ in carbon (F = 0.855, df = 5,18, P = 0.5), hydrogen (F = 0.819, df = 5,18, P = 0.6), or crude lipid (F = 2.266, df = 5,17, P = 0.10) contents. Control *Lactarius* spp. sporocarps had a significantly lower mean energy content (F = 6.306, df = 5,22, P = 0.001) than the other five species. Control *Lycoperdon* spp. had substantially greater crude protein content than all other sporocarps, and control *Auricularia* spp. had the lowest (F = 62.32, df = 5,18, P <0.001).

Storage for 93 d significantly reduced moisture contents (F = 2986.519, df = 1,36, P <0.001), and there was a significant species treatment interaction (F = 3.958, df = 2,36, P = 0.028). The control (fresh) *Auricularia* spp. sporocarps had more than five times the moisture content (t = 20.046, df = 12, P ≤0.001) than the *Auricularia* spp. sporocarps stored for 93 d. The control *Lycoperdon* spp. sporocarps had significantly more moisture than stored *Lycoperdon* spp. (t = 58.302, df = 12, P <0.001). Likewise, control *Russula* spp. sporocarps had significantly greater moisture than stored *Russula* spp. sporocarps (t = 48.450, df = 12, P <0.001). Storage also significantly reduced (F = 9.182, df = 1,21, P = 0.006) ash contents, with no significant species treatment interaction (F = 0.215, df = 2,21, P = 0.809). Carbon contents, however, were not significantly affected (F = 4.222, df = 1,18, P = 0.06) by the 93 d storage period.

Storage significantly reduced hydrogen contents (F = 17369.118, df = 1,18, P <0.001), with significant species treatment interactions (F = 18269.632, df = 2,18, P <0.001). Control *Auricularia* spp. sporocarps had significantly more hydrogen (t = 6.303, df = 6, P = 0.001) than the stored group. The two groups of *Russula* spp. sporocarps, in contrast, did not significantly differ in hydrogen (t = −0.289, df = 6, P = 0.78) levels. Control *Lycoperdon* spp. sporocarps had greater hydrogen contents than stored *Lycoperdon* spp. sporocarps (t = 5.484, df = 6, P = 0.002). The crude lipid content of control *Auricularia* spp. sporocarps was significantly greater than that of the stored group (t = 2.693, df = 8, P = 0.027). There was an insufficient amount of stored *Lycoperdon* spp. sporocarps remaining at the end of the experiment to perform a crude lipid assay. The two groups

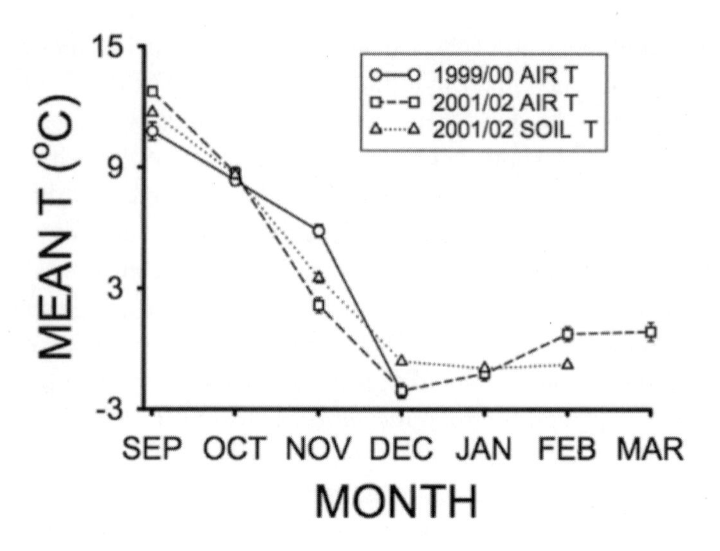

Figure 19.1. Mean (± SE) monthly temperatures of the tree cavities (air) housing the sporocarps and red squirrel middens (soil).

of *Russula* spp. sporocarps did not significantly differ in crude lipid contents as well ($t = 0.335$, df = 6, $P = 0.749$). Storage did not influence ($F = 2.054$, df = 1,18, $P = 0.17$) crude protein contents. In contrast, storage did significantly reduce energy contents ($F = 6.229$, df = 1,20, $P = 0.0021$), with no apparent species treatment interactions ($F = 0.320$, df = 2,20, $P = 0.73$). The mean (± SE) highest temperature recorded inside the tree trunk cavities was 15.9 ± 1.5°C, whereas the mean (± SE) minimum was −8.4 ± 0.8°C. Average monthly cavity temperatures were consistently below 0°C only during December 1999 (fig. 19.1).

Eight-Month Control and Stored Sporocarp Compositions

Comparisons of eight-month control and stored sporocarp compositions varied within and between genera (table 19.2). Comparing samples of all six sporocarp species collected during the late summer of 2001 revealed that *Lycoperdon* spp. had a significantly lower moisture content ($F = 24.64$, df = 5,29, $P < 0.001$) than the other five sporocarp species collected. The *Amanita* spp. sporocarps had a significantly greater ash content (table 19.2) than the fresh sporocarps of the other five species (F = 24.508, df = 5,27, P <0.001), and *Leccinum* spp. had the greatest carbon content (F = 20.886, df = 5,42, P <0.001). There were no significant differences in hydrogen content (table 19.2) among the six sporocarp species (F = 1.229, df = 5,42, P = 0.313). The *Lycoperdon* spp. sporocarps had a significantly greater crude protein content (F = 86.16, df = 5,42, P <0.001) than the other five sporocarp species. *Amanita* spp. sporocarps had a significantly greater crude lipid content (table 19.2) than all other types (F = 10.996, df = 5,16, P <0.001). Fresh *Lactarius* spp. sporocarps had a significantly lower energy content than all other species (F = 7.283, df = 5,17, P = 0.001).

Storage for eight months produced substantial reductions in moisture content (F = 72886.535, df = 1,42, P <0.001), with a significant species treatment interaction (F = 51.458, df = 2,42, P <0.001). The control (fresh) *Lactarius* spp. sporocarps had more than five times the moisture content (t = 187.476, df = 12, P <0.001) than the *Lactarius* spp. sporocarps stored for eight months. Likewise, the control *Leccinum* spp. sporocarps had significantly more moisture than the stored group (t = 271.842, df = 16, P <0.001, t = 7.247; table 19.2). Control *Russula* spp. sporocarps (table 19.2) also had significantly greater moisture than the stored *Russula* spp. sporocarps (t = 108.802, df = 14, P <0.001). Storage significantly increased ash contents (F = 103.298, df = 1,30, P <0.001) with species treatment interaction (F = 9.270, df = 2,30, P = 0.001). Stored *Lactarius* spp. had a significantly greater ash content (t = –2.479, df = 8, P = 0.038) than control *Lactarius* spp. The control *Leccinum* spp. sporocarps, similarly, had a significantly lower (t = –13.037, df = 12, P <0.001) ash content. Stored *Russula* spp. sporocarps also had a significantly greater ash content than the control sporocarps (t = –5.579, df = 10, P <0.001).

The eight-month storage period produced significantly lower carbon levels (F = 111.846, df = 1,45, P <0.001), with a species treatment interaction (F = 14.146, df = 2,45, P <0.001). Stored *Lactarius* spp. had significantly lower carbon (t = –5.286, df = 15, P <0.001; table 19.2). Control *Leccinum* spp. sporocarps had significantly more carbon (t = 7.247, df = 15, P <0.001). Likewise, control *Russula* spp. sporocarps also had a significantly greater carbon content (t = 4.573, df = 15, P <0.001) than the stored group (table 19.2). Hydrogen contents (table 19.2) also significantly decreased with storage (F = 30.133, df = 1,45, P <0.001),

Table 19.2. Mean (±SE) sporocarp compositions for the control (fresh) and eight-month experimental groups

Sporocarp Type†	Ash (%)	Carbon (%)	Hydrogen (%)	Crude Protein (%)	Crude Lipid (%)	Energy (kJ/g)	Moisture (% wet mass)
Amanita-C	17.5 ± 1.9[a] (5)	45.5 ± 0.5[a] (8)	6.5 ± 0.1[a] (8)	35.7 ± 0.4[a] (8)	14.6 ± 0.9[a] (4)	18.7 ± 0.1[a] (4)	95.5 ± 0.1[a] (8)
Boletus-C	8.7 ± 0.1[b] (6)	45.5 ± 0.8[a] (8)	6.3 ± 0.2[a] (8)	30.9 ± 2.5[a] (8)	4.1 ± 0.2[b] (4)	18.0 ± 0.4[a] (4)	94.1 ± 0.3[a] (4)
Lactarius-C	12.3 ± 0.7[c] (4)	44.9 ± 0.3[a] (8)	6.5 ± 0.1[a] (8)	22.9 ± 0.2[b] (8)	4.2 ± 1.0[b] (2)	13.4 ± 2.5[b] (3)	92.9 ± 0.3[a] (4)
Lactarius-S	16.9 ± 1.4[*] (6)	41.4 ± 0.6[*] (9)	6.1 ± 0.1[*] (9)	22.2 ± 1.0 (9)	5.4 ± 0.8 (9)	17.4 ± 0.9 (4)	13.9 ± 0.3[*] (9)
Leccinum-C	5.7 ± 0.4[b] (8)	48.9 ± 0.6[b] (8)	6.6 ± 0.1[a] (8)	23.1 ± 0.7[b] (8)	5.2 ± 0.4[b] (4)	19.4 ± 0.1[a] (4)	95.3 ± 0.1[a] (6)
Leccinum-S	16.7 ± 0.8[*] (6)	39.3 ± 1.1[*] (9)	5.9 ± 0.2[*] (9)	18.4 ± 0.4[*] (9)	6.6 ± 0.9 (9)	18.3 ± 0.2 (2)	11.6 ± 0.2[*] (22)
Lycoperdon-C	8.4 ± 0.2[b] (4)	46.2 ± 0.7[a] (8)	6.3 ± 0.1[a] (8)	53.0 ± 1.0[c] (8)	6.4 ± 0.8[b] (4)	19.8 ± 0.2[a] (4)	91.5 ± 0.3[b] (5)
Russula-C	11.0 ± 0.8[b] (6)	44.2 ± 0.3[a] (9)	6.6 ± 0.1[a] (9)	24.7 ± 1.1[b] (9)	9.3 ± 2.4[b] (4)	18.3 ± 0.1[a] (4)	92.2 ± 0.2[a] (5)
Russula-S	16.4 ± 0.6[*] (6)	40.5 ± 0.5[*] (9)	6.0 ± 0.1[*] (9)	20.4 ± 0.8[*] (9)	4.7 ± 0.9 (9)	18.0 ± 0.5 (2)	16.0 ± 0.5[*] (9)

*Stored [S] sporocarp mean is significantly different from the corresponding control [C] mean at the $P < 0.05$ level.

†Control [C] sporocarp means within the same experiment sharing a common lowercase letter are not significantly different from other control means at the $P < 0.05$ level; number in parentheses represents sample size (n) for that mean.

with no significant species treatment interaction ($F = 1.135$, df = 2,45, $P = 0.33$). Storage for eight months also produced significant reductions in crude protein ($F = 24.450$, df = 1,45, $P < 0.001$), and a significant species treatment interaction ($F = 3.974$, df = 2,45, $P = 0.026$). The control and stored *Lactarius* spp. groups did not significantly differ (table 19.2) in crude protein ($t = 0.602$, df = 15, $P = 0.556$). The control *Leccinum* spp. sporocarps had significantly more crude protein ($t = 6.057$, df = 15, $P < 0.001$), in contrast (table 19.2) to the stored group. Control *Russula* spp. sporocarps also had a significantly greater crude protein than stored sporocarps ($t = -2.962$, df = 15, $P = 0.01$). Storage for eight months did not significantly affect crude lipid ($F = 0.359$, df = 1,31, $P = 0.553$) or energy ($F = 1.898$, df = 1,31, $P = 0.178$) contents, however. The mean (± SE) temperature recorded inside one tree trunk cavity during the entire experiment was 4.1 ± 0.2° C, the minimum was –16.4°C, and the maximum was 22.7°C. Average monthly cavity (cache) temperatures were consistently below 0°C only during December 2001 and January 2002 (fig. 19.1) and consistently above 0°C during February and March 2002 (fig. 19.1), whereas the corresponding soil (midden) temperatures remained below 0°C throughout February 2002.

Discussion

The nutritional analysis and comparison of fresh sporocarps from eight different fungal species (tables 19.1, 19.2) revealed that their nutritional contents were generally uniform, but species did sometimes influence sporocarp ash, crude protein, and energy contents. Many species of mammals have a dietary preference for food items containing relatively higher levels of energy and/or crude protein (nitrogen) while foraging (Stephens and Krebs 1986). It is thus possible that red squirrels may have dietary preferences for certain sporocarp species when foraging for fresh material. The results of the sporocarp storage experiments clearly indicate that the hoarding behavior of red squirrels reduces the moisture contents of cached sporocarps even after eight months of storage, as predicted. The three-month sporocarp storage experiment demonstrated that this desiccation is initially associated with a stable nutritional composition before the onset of natural freeze/thaw cycles, as predicted. A total of 17 different nutrient/sporocarp species combinations other than moisture content were examined in the three-month storage experiment. Nutritional composition did not significantly change during storage in 11 (65 percent) of the cases.

The eight-month storage experiment supports the hypothesis that desiccation generally does not produce a stable nutritional composition once natural freeze/thaw cycles occur. A total of 18 different nutrient/sporocarp species combinations other than moisture content were examined in this experiment. In only 7 (39 percent) of these cases did nutrient content remain constant during storage. Cache (air) temperatures remained below 0°C during the fourth and fifth months of storage, but they were consistently above 0°C during the last two months of storage (fig. 19.1). Considering these two experiments together reveals that the desiccation associated with sporocarp storage conserves their nutritional composition prior to and during natural freezing, but not after natural freeze/thaw cycles have begun. Ash contents decreased during the three-month experiment, but increased after an eight-month storage period (tables 19.1, 19.2). Carbon contents were not affected during the three-month experiment, but all decreased during the eight-month experiment (tables 19.1, 19.2). Interpreting these findings together indicates that the digestible organic matter found in stored sporocarps remains constant prior to the onset of freeze/thaw cycles, but it decreases substantially after these cycles have begun.

Further consideration of fresh sporocarp nutritional compositions also suggests that there are important limitations on the use of mycophagy by Mt. Graham red squirrels. The diets of rodents must be at least 15–20 percent digestible crude protein and not less than 5 percent crude lipid for proper health and survival (Robbins 1993). Both the control and stored *Auricularia* spp. sporocarps had crude protein and lipid contents that were well below these levels, and the control *Boletus* spp. sporocarps had a crude lipid content well below 5 percent (tables 19.1, 19.2). Feeding trials with hypogeous sporocarps revealed that only about 33 percent of their crude protein contents were digestible to squirrels (Cork and Kenagy 1989), presumably because most of the nitrogen (crude protein) found in fungal sporocarps is in the form of chitin, which cannot be digested by most mammals (Robbins 1993). The crude protein digestibility of the sporocarp species included in this study is unknown. If it is also around 33 percent of crude protein content, however, then only one of the eight species examined (tables 19.1, 19.2) would have a digestible crude protein content that meets the minimum requirements (15–20 percent) of red squirrels. It therefore appears that Mt. Graham red squirrels cannot satisfy their protein requirements by consuming most sporocarp species.

Red squirrels consume fungal sporocarps year-round (Currah et al. 2000; Vernes et al. 2004), however, despite these nutritional limitations. During years of high conifer seed production, stored sporocarps represent only 45 percent of the winter diet, whereas during years of low conifer seed production they constitute 75–95 percent of red squirrel winter diets (Vernes et al. 2004). This trend suggests that red squirrels prefer to consume stored conifer seeds over stored sporocarps, and that stored sporocarp consumption increases only when there are few stored seeds available. Additional investigation of sporocarp consumption and storage by Mt. Graham red squirrels will not only provide insights for the conservation of this endangered subspecies but also contribute to our understanding of the evolution of both food hoarding and mycophagy in general.

Acknowledgments

I thank Paul Y. Young for his generous assistance with many aspects of this project. This study would not have been possible without field assistance from Vicki L. Greer, Erin Bibles, Nancy Ferguson, and J. Elaine Lowry. Helpful discussions with Reed Sanderson and Genice Froehlich greatly aided the development of this project. Funding for this study was provided by the University of Arizona through the Mt. Graham Red Squirrel Study Committee, which is comprised of representatives from the Arizona Game and Fish Department, the University of Arizona, the U.S. Forest Service, and the U.S. Fish and Wildlife Service.

Literature Cited

Alexopoulos, C. J., and C. W. Mims. 1979. *Introductory mycology.* John Wiley and Sons, New York.

Brower, J. E., J. H. Zar, and C. Von Ende. 1997. *Field and laboratory methods for general ecology.* McGraw-Hill, New York.

Cork, S. J., and G. J. Kenagy. 1989. Nutritional value of hypogeous fungus for a forest-dwelling ground squirrel. *Ecology* 70:577–586.

Cram, W. E. 1924. The red squirrel. *Journal of Mammalogy* 5:37–41.

Currah, R. S., E. A. Smreciu, T. Lehesvirta, M. Niemi, and K. W. Larsen. 2000. Fungi in the winter diets of northern flying squirrels and red squirrels in the boreal mixed-wood forest of northeastern Alberta. *Canadian Journal of Botany* 78:1514–1520.

Dice, L. R. 1921. Notes on the mammals of interior Alaska. *Journal of Mammalogy* 2:20–28.

Dobush, G. R., C. D. Ankney, and D. G. Krementa. 1985. The effect of apparatus, extraction time, and solvent type on lipid extraction of snow geese. *Canadian Journal of Zoology* 63:1917–1920.

Maser, C., J. M. Trappe, and R. A. Nussbaum. 1978. Fungal–small mammal interrelationships with emphasis on Oregon coniferous forests. *Ecology* 59:799–809.

Moller, H. 1983. Food and foraging behavior of red (*Sciurus vulgaris*) and grey (*Sciurus carolinensis*) squirrels. *Mammal Review* 13:81–98.

North, M., J. M. Trappe, and J. Franklin. 1997. Standing crop and animal consumption of fungal sporocarps in Pacific Northwest forests. *Ecology* 78:1543–1554.

Paine, R. T. 1971. The measurement and application of the calorie to ecological problems. *Annual Review of Ecology and Systematics* 2:145–164.

Robbins, C. T. 1993. *Wildlife feeding and nutrition*. Academic Press, New York.

Smith, C. C. 1968. The adaptive nature of social organization in the genus of three [*sic*] squirrels *Tamiasciurus*. *Ecological Monographs* 38:31–63.

——. 1995. The niche of diurnal tree squirrels. Pages 209–225 in C. A. Istock and R. S. Hoffman, eds. *Storm over a mountain island: Conservation biology and the Mt. Graham affair.* University of Arizona Press, Tucson.

Stephens, D. W., and J. R. Krebs. 1986. *Foraging theory.* Princeton University Press, Princeton, New Jersey.

Vander Wall, S. B. 1990. Food hoarding in animals. University of Chicago Press, Chicago.

Vernes, K., S. Blois, and F. Barlocher. 2004. Seasonal and yearly changes in consumption of hypogeous fungi by northern flying squirrels and red squirrels in old-growth forest, New Brunswick. *Canadian Journal of Zoology* 82:110–117.

Young, P. 1996. *Annual report of the Mt. Graham red squirrel monitoring program.* University of Arizona, Tucson.

## CHAPTER 20

Reproductive Ecology and Home Range Size of Red Squirrels

Do Mt. Graham Red Squirrels Fit the Pattern?

KAREN E. MUNROE, JOHN L. KOPROWSKI, AND VICKI L. GREER

UNDERSTANDING THE FACTORS that influence individual productivity within a population of conservation concern is often of utmost importance to implement recovery efforts (Beissinger and Westphal 1998; Morris et al. 2002). Creating management strategies reflecting the species' ecology and ecosystem is becoming standard practice due to increased conservation success (Beissinger and Westphal 1998; Morris et al. 2002). Recovery status, population viability analyses, and demographic assessments require specific data on productivity of a population (Morris et al. 2002; Reed et al. 2003). Therefore, it is critical to collect basic behavioral and biological data on endangered species and integrate these data into management decisions.

Red squirrels are small (200–250 g), long lived (3–5 yr), slightly dimorphic (Boutin and Larsen 1993), diurnal animals with daily bimodal peaks of activity during spring, summer, and fall (Brown 1984; Steele 1998). Natal and post-breeding dispersal are common (Boutin et al. 1993; Larsen and Boutin 1994). Red squirrels have been classified as polygamous breeders (Steele 1998). They produce one litter a year in late spring, except in mixed deciduous-conifer forests commonly found on the U.S. eastern coast, where two litters per year are possible (Steele 1998). Mating bouts have been described as involving numerous males chasing a female on her single day of estrus (Smith 1968). Red squirrels are solitary in nature, and there have been no published reports of communal nesting.

Since the Mt. Graham red squirrel was listed as endangered on 3 June 1987 (U.S. Fish and Wildlife Service 1993), several studies and thousands of hours of observation have focused on their ecology, including home range size and reproductive effort. However, these studies and data have not been compiled nor compared to other populations of North American red squirrels.

The objectives of this chapter are three-fold: (1) summarize available literature on home range size and reproductive output in North American red squirrels throughout their range; (2) integrate basic ecological data collected on Mt. Graham red squirrels in early published studies and unpublished observations; (3) compare data on Mt. Graham red squirrels with ecological data on other North American red squirrels to make predictions and informed management decisions.

Methods

Through a literature review, we obtained data on North American red squirrel home range size, dispersal distance, date of mating bout, number of individuals per mating chase, and litter size. All but one study (Becker et al. 1998) reported a single average home range size combined for the sexes. In this chapter, home range sizes were analyzed for the sexes combined because sexual dimorphism in space use is not pronounced in North American red squirrels (Larsen and Boutin 1998). Data on litter size, date of mating bout, and number of individuals in mating chases for Mt. Graham red squirrels were compiled from annual reports of the University of Arizona's Mt. Graham Red Squirrel Monitoring Program (Young et al. 1995, 1996, 1997, 1998, 1999; Koprowski et al. 2000, 2001, 2002, 2003, 2004). These annual reports synthesize data collected during four quarterly censuses and observations during other critical times over approximately 300 ha of mixed spruce-fir forest. If the mating bout for a particular female was not observed, then date of mating bout was calculated as eight weeks prior to date of litter emergence (Hamilton 1939). Linear regression of North American and Mt. Graham red squirrel litter sizes on latitude also was conducted to assess potential relationships between litter size and north-south clinal variation.

Results

Social System, Mating System, and Reproductive Biology

The red squirrel pattern. Throughout most conifer forests, North American red squirrels defend territories centered on their midden (Finley 1969) throughout the year (Smith 1968; Kemp and Keith 1970; Rusch and Reeder 1978; Gurnell 1984). This accumulation of food is considered necessary for over-winter survival (Larsen et al. 1997). Defense of a territory includes scent marking, aggressive calling, and charging an intruder (Smith 1968; Ferron 1983; Lair 1990). In mixed-conifer forests, adult red squirrels occupy permanent territories and sub-adults occupy transient territories during winter (Rusch and Reeder 1978). However, red squirrels in deciduous forests maintain overlapping territories where only nests or food caches may be defended (Layne 1954).

Males typically congregate in a female's territory during her single day of estrus and actively pursue the female in mating chases (Smith 1968). During these chases, a dominant male will pursue the female and defend her from other males with aggressive calls or direct chases (Smith 1965). The dominant male chasing closest to the female presumably obtains the highest number of initial copulations (Smith 1968). Females mate with up to seven males (Arbetan 1993), copulating multiple times with each male (Smith 1965, 1968). Arbetan (1993) reports an average of 5.23 males (range 4–7 males) per mating chase. Copulation lasts just a few seconds and occurs on the ground or in the lower branches of a tree (Smith 1968). Mating bouts are reported from February to August (Hamilton 1939; Smith 1965; Millar 1970; Dolbeer 1973; Ferron and Prescott 1977; Rusch and Reeder 1978; Koford 1982; Lair 1985; Becker 1993). Females are spontaneous ovulators (Layne 1954; Millar 1970); therefore, conception usually occurs within a few hours of mating. Mean gestation period is 33 days (Ferron and Prescott 1977; Lair 1985). Average litter size in 14 studies of North American red squirrels is 3.72 young (range of means 2.77–5.4 young, n of studies = 14) (table 20.1). Lactation occurs for eight weeks until young are weaned, although young do not disperse until a few weeks after weaning (Layne 1954; Steele 1998).

Table 20.1. Red squirrel litter sizes from literature review

Study	Mean	Range	n	Methods
Davis 1969	3.2		9	Nest young
Dolbeer 1973	3.3	2–5	16	Embryos
Ferron and Prescott 1977	5.4	4–8	7	Captive females
Humphries and Boutin 2000	2.95	1–7	583	Parturition
Kemp and Keith 1970	3.85	1–8	314	Placental scars
Lair 1985	3.7	3–6	12	Parturition
LaPierre 1986	3.6	0–7	74	Placental scars
Larsen et al. 1997	2.9	1–4	26	Parturition
Larsen and Boutin 1994	2.77	1–4	79	Parturition
Layne 1954	4.5		74	Placental scars
Millar 1970	4.4	1–6	83	Corpora lutea
Rusch and Reeder 1978	4.1		164	Placental scars
Smith 1968	3.35		24	Placental scars
Wood 1967	4.0		20	Placental scars
Mean	3.72			

Do Mt. Graham red squirrels fit the pattern? Mt. Graham red squirrels are similar to other red squirrel species. They are polygamous, territorial animals, and therefore, nest sharing is not expected. Surprisingly, Mt. Graham red squirrels have occasionally been observed to nest in pairs. From November 2002 through April 2003, three communal nesting associations were observed. On multiple occasions (n >5), an adult female that had a litter the previous summer and another adult female who did not have a litter the previous year but had a litter at least two years ago nested together (Koprowski et al. 2003). Additionally, an adult female and a sub-adult male were observed nesting together (n >5) (Koprowski et al. 2004). In addition, on one occasion a different adult female was observed to nest with a sub-adult male (Koprowski et al. 2004).

Mt. Graham red squirrel mating chases were observed from March to August with an average of 2.25 males per chase (range 1–7 males, n =

Table 20.2. Red squirrel home range sizes (ha) from literature review

Study	Mean	Range	*n*	Notes
Becker et al. 1998	0.6		~100	Females only
Boutin and Schweiger 1988	0.35	0.282–0.432	29	Trapping
Klenner 1990	0.97		18	Visual obs, Fir
Klenner 1990	0.34		37	Visual obs, Spruce
Klenner 1991	0.64		5	Telemetry
Koford 1992	0.18			Calculated
Koford 1992	0.26			Calculated
Price et al. 1986	0.38	0.278–0.456	32	Visual obs
Riege 1991	0.35		36	Visual obs
Rusch and Reeder 1978	0.35	0.28–0.42	11	Tracks
Sullivan 1990	0.88		21	Trapping/Control
Sullivan 1990	0.63		39	Trapping/Control
Wells 1987	0.56		44	Telemetry
Zirul and Fuller 1970	0.84	0.081–1.578	30	Trapping
Mean	0.53			

12). The average litter size of Mt. Graham red squirrels based on litter emergence data is 2.15 young (range 1–5 young, $n = 7$). The small litter size of Mt. Graham red squirrels is not related to latitudinal effects, as no relationship between litter size and latitude of all North American red squirrel populations was found ($F = 0.004$, df = 13, $P > 0.95$, $R^2 = 0.16$).

Home Range and Dispersal

The red squirrel pattern. Home range size of North American red squirrels averages 0.53 ha (range 0.18–0.97 ha, *n* of studies = 11) (table 20.2). Smallest home ranges were calculated from a study in mixed white spruce and lodgepole pine (*Pinus contorta*) forest (Sullivan 1990; Koford 1992). Largest home ranges were observed in mixed lodgepole pine and Douglas-fir (Klenner 1990, 1991).

A mean dispersal distance of 86.1 m (range 0–323 m, n = 67) was reported by Larsen (1993), and a mean dispersal distance of 88.6 m (range 0–323 m, n = 73) was reported by Larsen and Boutin (1994). A mean dispersal distance of 100 m (range 0–600 m, n = 8) was reported by Sun (1997). If these dispersal distances are converted to territory increments, the ratio of dispersal distance to the diameter of the home range (Larsen 1993; Larsen and Boutin 1994, 1998) varies from 2.3 (Larsen 1993), to 2.4 (Larsen and Boutin 1994), to 2.7 (Sun 1997).

Juveniles disperse from their natal territories before winter. However, short-term territory sharing by the female, with one or more of her offspring, has been observed (Price et al. 1986; Price 1992; Price and Boutin 1993). Juvenile philopatry occurred in 15 of 28 litters in two studies (Price et al. 1986; Boutin et al. 1993). Females may bequeath part or all of their territory to one or more offspring and disperse to a neighboring territory (Price 1992; Price and Boutin 1993). Territory size may also increase with additional resources acquired by the female prior to bequeathal (Boutin and Schweiger 1988). This strategy increases the ability of late-weaned juveniles to acquire a territory, therefore increasing their chances of survival and reproduction (Price and Boutin 1993).

Do Mt. Graham red squirrels fit the pattern? Froehlich (1990) reported the average Mt. Graham red squirrel home range as 3.67 ha (range 1.65–6.32 ha, n = 2). In comparison, Kreighbaum and Van Pelt (1996) reported an average home range size of 1.15 ha (range 0.03–2.0 ha, n = 7). Both reported home range sizes for Mt. Graham red squirrels are at least 200 percent larger than the average red squirrel home range and exceed the largest home ranges reported elsewhere. A mean natal dispersal distance of 584 m (range 300–1885 m, n = 2) was reported for Mt. Graham red squirrels (Kreighbaum and Van Pelt 1996). If these dispersal distances are converted to territory increments using home range size (Larsen 1993; Larsen and Boutin 1994, 1998), the ratio of average dispersal distance to territory size is 7.3 territories. Surprisingly, the ratio of dispersal distance to territory size for Mt. Graham red squirrels is at least 250 percent larger than the ratio of dispersal distance to territory size for red squirrels.

Discussion

Home Range and Dispersal

Large home range size observed for Mt. Graham red squirrels suggests forests in the Pinaleño Mountains are sub-optimal in comparison to other North American red squirrel habitat. This may be attributed to potential variations in food availability in spruce-fir and mixed-conifer forests of the Pinaleño Mountains, as well as the aridity of this habitat at the southern end of red squirrel distribution (U.S. Fish and Wildlife Service 1993). The large home range size observed for Mt. Graham red squirrels may also be an artifact of a low population density. Steury and Murray (2003) have shown territory size in North American red squirrels is inversely related to population density. Two additional studies have shown that when red squirrels are removed from their territories, relocation by neighboring animals does not occur; however, expansion or shifting of territories is observed by both sexes (Price et al. 1986; Larsen and Boutin 1995).

The great dispersal distance of juveniles in the Pinaleño Mountains also may suggest lack of available habitat. Juvenile Mt. Graham red squirrels disperse over 7 territory distances away from their natal area compared to other North American red squirrels that disperse an average of approximately 2.5 territory distances away from their natal area. This extreme natal dispersal distance may translate to decreased juvenile survival and recruitment in the Mt. Graham red squirrel population. Low recruitment of young may partially explain the small population size of Mt. Graham red squirrels.

Social System, Mating System, and Reproductive Biology

The lack of observations of mating chases in February for Mt. Graham red squirrels may be due to the limited amount of fieldwork conducted during winter or reflective of annual variability of conifer food sources (Koprowski et al. 2005). High variability of the conifer food source may have subsequent impacts on recruitment of offspring and their ability to obtain a territory. Haughland and Larsen (2004) compared mature forests to commercially thinned forests and found North American red squirrels in mature forests had a higher mean over-winter survival and success raising a juvenile to emergence. Success of these squirrels is attributed to a more consistent food source in a mature forest. Food availability increases the overall growth rate of North American red squirrel juveniles (Boutin and Larsen 1993). During periods of reduced food availability, litter size may decrease along with the overall rate of successful recruitment of young into the population. Sullivan (1990) showed that when female red squirrels were supplementally fed, they had a longer breeding season and were capable of producing a second litter.

Flexibility in onset of estrus is common in non-hibernating squirrels (Becker 1993) and may be an adaptation to variation in food and climate of conifer forests. Nutrients in spring buds of conifers might trigger estrus in females (Kemp and Keith 1970; Lair 1985). Juvenile red squirrels from litters produced later in the season are less successful at acquiring territories away from their natal area (Price et al. 1986). Females should have litters as early as possible because early-born juveniles have a better chance of obtaining a territory and consequently surviving and reproducing in the next year (Price et al. 1986). The lack of Mt. Graham red squirrel mating chases observed in February may also contribute to a smaller population size by diminishing the chances of a late-born juvenile obtaining a territory and consequentially surviving and reproducing in the next year.

The reduced number of males in Mt. Graham red squirrel mating chases may further contribute to a decrease in genetic variation in an already bottlenecked and isolated endangered species. This reduced genetic variation may be reflected in lower average litter size (Lacy et al. 1996). Unfortunately, no thorough assessment of genetic variation of Mt. Graham red squirrels has occurred.

Abundant literature on North American red squirrels throughout their range provides a useful basis for comparison; however, data on dispersal distances and mating bouts are scarce. Although the number of studies and sample sizes are limited for endangered Mt. Graham red squirrels, these results allow comparison to North American red squirrels. Furthermore, our compilation on red squirrels may be used in population viability analyses and other demographic assessments that require data on productivity of the species and endangered subspecies. Future studies on Mt. Graham red squirrels should focus on increasing sample sizes and precision of home-range and natal-dispersal studies so results may be used to facilitate informed management decisions and conservation efforts for this endangered species. Given a population of <300 Mt. Graham red squirrels and limited available habitat, we hope this literature review will contribute to future conservation efforts.

Acknowledgments

We thank four anonymous reviewers for beneficial additions to this manuscript. We also thank S. Hollands, A. Iles, S. Jensen, and T. Wilson for valuable comments on this manuscript. The University of Arizona and the Arizona Agricultural Experiment Station provided financial assistance.

Literature Cited

Arbetan, P. T. 1993. The mating system of the red squirrel, (*Tamiasciurus hudsonicus*). Master's thesis, University of Kansas, Lawrence.

Becker, C. D. 1993. Environmental cues of estrus in the North American red squirrel (*Tamiasciurus hudsonicus* Bangs). *Canadian Journal of Zoology* 71:1326–1333.

Becker, C. D., S. Boutin, and K. W. Larsen. 1998. Constraints on first reproduction in North American red squirrels. *Oikos* 81:81–92.

Beissinger, S. R., and M. I. Westphal. 1998. On the use of demographic models of population viability in endangered species management. *Journal of Wildlife Management* 62:821–841.

Boutin, S., and K. W. Larsen. 1993. Does food availability affect growth and survival of males and females differently in a promiscuous small mammal, *Tamiasciurus hudsonicus? Journal of Animal Ecology* 62:364–370.

Boutin, S., and S. Schweiger. 1988. Manipulation of intruder pressure in red squirrels (*Tamiasciurus hudsonicus*): Effects on territory size and acquisition. *Canadian Journal of Zoology* 66:2270–2274.

Boutin, S., Z. J. Tooze, and K. Price. 1993. Post-breeding dispersal of female red squirrels (*Tamiasciurus hudsonicus*): The effect of local vacancies. *Behavioral Ecology* 4:151–155.

Brown, D. E. 1984. *Arizona's tree squirrels.* Arizona Game and Fish Department, Phoenix.

Davis, D. W. 1969. The behavioral and population dynamics of the red squirrel, *Tamiasciurus hudsonicus*, in Saskatchewan. PhD diss., University of Arkansas, Little Rock.

Dolbeer, R. A. 1973. Reproduction in the red squirrel (*Tamiasciurus hudsonicus*) in Colorado. *Journal of Mammalogy* 54:536–539.

Ferron, J. 1983. Scent marking by cheek rubbing in the northern flying squirrel (*Glaucomys sabrinus*). *Canadian Journal of Zoology* 61:2377–2380.

Ferron, J., and J. Prescott. 1977. Gestation, litter size and number of litters of the red squirrel (*Tamiasciurus hudsonicus*) in Quebec. *The Canadian Field Naturalist* 91:385–391.

Finley, R. B., Jr. 1969. Cone caches and middens of *Tamiasciurus* in the Rocky Mountain region. Pages 233–273 in J. K. Jones Jr., ed. *Contributions in mammalogy.* University of Kansas Museum of Natural History Miscellaneous Publication 51.

Froehlich, G. F. 1990. Habitat use and life history of the Mt. Graham red squirrel. Master's thesis, University of Arizona, Tucson.

Gurnell, J. 1984. Home range, territoriality, caching behaviour and food supply of the red squirrel (*Tamiasciurus hudsonicus fremonti*) in a subalpine lodgepole pine forest. *Animal Behavior* 32:1119–1131.

Hamilton, W. J. 1939. Observations on the life history of the red squirrel in New York. *American Midland Naturalist* 22:732–745.

Haughland, D. L., and K. W. Larsen. 2004. Ecology of North American red squirrels across contrasting habitats: Relating natal dispersal to habitat. *Journal of Mammalogy* 85:225–236.

Humphries, M. M., and S. Boutin. 2000. The determinants of optimal litter size in free-ranging red squirrels. *Ecology* 81:2867–2877.

Kemp, G. A., and L. B. Keith. 1970. Dynamics and regulations of red squirrel (*Tamiasciurus hudsonicus*). *Ecology* 51:763–779.

Klenner, W. 1990. The effect of food abundance and territorial behaviour on population dynamics of the red squirrel. PhD diss., The University of British Columbia, Vancouver, Canada.

———. 1991. Red squirrel population dynamics. II. Settlement patterns and the response to removals. *Journal of Animal Ecology* 60:979–993.

Koford, R. 1982. Mating system of a territorial tree squirrel (*Tamiasciurus douglasii*) in California. *Journal of Mammalogy* 63:274–283.

———. 1992. Does supplemental feeding of red squirrels change population-density, movements or both? *Journal of Mammalogy* 4:930–932.

Koprowski, J. L., M. Alanen, and A. Lynch. 2005. Nowhere to run and nowhere to hide: Response of endemic Mt. Graham red squirrels to catastrophic forest damage. *Biological Conservation* 126:491–498.

Koprowski, J. L., V. L. Greer, M. I. Alanen, K. A. Hutton, and C. A. Schauffert. 2000. *Annual report of the Mt. Graham Red Squirrel Monitoring Program.* School of Natural Resources, University of Arizona, Tucson.

———. 2001. *Annual report of the Mt. Graham Red Squirrel Monitoring Program.* School of Natural Resources, University of Arizona, Tucson.

Koprowski, J. L., V. L. Greer, S. R. B. King, and S. Bertelsen. 2003. *Annual report of the Mt. Graham Red Squirrel Monitoring Program.* School of Natural Resources, University of Arizona, Tucson.

——. 2004. *Annual report of the Mt. Graham Red Squirrel Monitoring Program*. School of Natural Resources, University of Arizona, Tucson.

Koprowski, J. L., V. L. Greer, S. R. B. King, and S. Taylor. 2002. *Annual report of the Mt. Graham Red Squirrel Monitoring Program*. School of Natural Resources, University of Arizona, Tucson.

Kreighbaum, M. E., and W. E. Van Pelt. 1996. *Mount Graham red squirrel juvenile dispersal telemetry study*. Nongame and Endangered Wildlife Program Technical Report 89. Arizona Game and Fish Department, Phoenix.

Lacy, R. C., G. Alaks, and A. Walsh. 1996. Hierarchical analysis of inbreeding depression in *Peromyscus polionotus*. *Evolution* 50:2187–2200.

Lair, H. 1985. Length of gestation in the red squirrel (*Tamiasciurus hudsonicus*). *Journal of Mammalogy* 66:809–810.

——. 1990. The calls of the red squirrel: A contextual analysis of function. *Behaviour* 115:254–282.

LaPierre, L. 1986. Female red squirrel (*Tamiasciurus hudsonicus*) reproductive tracts from Fenitrothion treated and untreated forest of southeastern New-Brunswick. *Forest Chronicle* 62:233–235.

Larsen, K. W. 1993. Female reproductive success in the North American red squirrel, *Tamiasciurus hudsonicus*. PhD diss., University of Alberta, Edmonton, Canada.

Larsen, K. W., C. D. Becker, S. Boutin, and M. Blower. 1997. Effects of hoard manipulations on life history and reproductive success of female red squirrels (*Tamiasciurus hudsonicus*). *Journal of Mammalogy* 78:192–203.

Larsen, K. W., and S. Boutin. 1994. Movements, survival and settlement of red squirrel (*Tamiasciurus hudsonicus*) offspring. *Ecology* 75:214–223.

——. 1995. Exploring territory quality in the North-American red squirrel through removal experiments. *Canadian Journal of Zoology* 73:1115–1122.

——. 1998. Sex-unbiased philopatry in the North American red squirrel (*Tamiasciurus hudsonicus*). Pages 21–32 in M. A. Steele, J. F. Merritt, and D. A. Zegers, eds. *Ecology and evolutionary biology of tree squirrels*. Special Publication no. 6. Virginia Museum of Natural History, Martinsville.

Layne, J. M. 1954. The biology of the red squirrel *Tamiasciurus hudsonicus loquaz* (Bangs) in central New York. *Ecological Monographs* 24:227–267.

Millar, J. S. 1970. The breeding season and reproductive cycle of the western red squirrel. *Canadian Journal of Zoology* 48:471–473.

Morris, W. F., P. L. Bloch, B. R. Hudgens, L. C. Moyle, and J. R. Stinchcombe. 2002. Population viability analysis in endangered species recovery plans: Past use and future improvements. *Ecological Applications* 12:708–712.

Price, K. K. 1992. Territorial bequeathal by red squirrel mothers: A dynamic model. *Bulletin of Mathematical Biology* 54:335–354.

Price, K. K., and S. Boutin. 1993. Territorial bequeathal by red squirrel mothers. *Behavioral Ecology* 4:144–150.

Price, K., K. Broughton, S. Boutin, and A. R. E. Sinclair. 1986. Territory size and ownership in red squirrels: Response to removals. *Canadian Journal of Zoology* 64:1144–1147.

Reed, D. H., J. J. O'Grady, B. W. Brook, J. D. Ballou, and R. Frankham. 2003. Estimates of minimum viable population sizes for vertebrates and factors influencing those estimates. *Biological Conservation* 113:23–34.

Riege, D. A. 1991. Habitat specialization and social factors in distribution of red and gray squirrels. *Journal of Mammalogy* 72:152–162.

Rusch, D. A., and W. G. Reeder. 1978. Population ecology of Alberta red squirrels. *Ecology* 59:400–420.

Smith, C. C. 1965. Interspecific competition in the genus of tree squirrels, *Tamiasciurus*. PhD diss., University of Washington, Seattle.

——. 1968. The adaptive nature of social organization in the genus of three [*sic*] squirrels *Tamiasciurus*. *Ecological Monographs* 38:31–63.

Steele, M. S. 1998. *Tamiasciurus hudsonicus*. *Mammalian Species* 586:1–9.

Steury, T. D., and D. L. Murray. 2003. Causes and consequences of individual variation in territory size in the American red squirrel. *Oikos* 101:147–156.

Sullivan, T. P. 1990. Responses of red squirrel (*Tamiasciurus hudsonicus*) populations to supplemental food. *Journal of Mammalogy* 71:579–590.

Sun, C. 1997. Dispersal of young in red squirrels (*Tamiasciurus hudsonicus*). *American Midland Naturalist* 138:252–259.

U.S. Fish and Wildlife Service. 1993. *Mount Graham Red Squirrel* (Tamiasciurus hudsonicus grahamensis) *Recovery Plan*. Albuquerque, New Mexico.

Wells, N. M. 1987. Dynamics of space use in a red squirrel (*Tamiasciurus hudsonicus*) population. PhD diss., University of Michigan, Ann Arbor.

Wood, T. J. 1967. Ecology and population dynamics of the red squirrel (*Tamiasciurus hudsonicus*) in Wood Buffalo National Park. Master's thesis, University of Saskatchewan, Saskatoon, Canada.

Young, P. J., V. L. Greer, E. L. Bibles, N. Ferguson, J. E. Lowry, and K. D. Morse. 1996. *Annual report of the Mt. Graham Red Squirrel Monitoring Program*. Steward Observatory, University of Arizona, Tucson.

Young, P. J., V. L. Greer, E. L. Bibles, N. Ferguson, J. E. Lowry, and E. A. Point. 1997. *Annual report of the Mt. Graham Red Squirrel Monitoring Program*. Steward Observatory, University of Arizona, Tucson.

——. 1998. *Annual report of the Mt. Graham Red Squirrel Monitoring Program*. Steward Observatory, University of Arizona, Tucson.

Young, P. J., V. L. Greer, E. L. Bibles, N. Ferguson, J. E. Lowry, and C. Schauffert. 1999. *Annual report of the Mt. Graham Red Squirrel Monitoring Program*. Steward Observatory, University of Arizona, Tucson.

Young, P. J., V. L. Greer, W. E. Kreisel Jr., J. E. Lowry, K. D. Morse, and J. A. Reed. 1995. *Annual report of the Mt. Graham Red Squirrel Monitoring Program*. Steward Observatory, University of Arizona, Tucson.

Zirul, D. L., and W. A. Fuller. 1970. Winter fluctuations in size of home range of the red squirrel (*Tamiasciurus hudsonicus*). *Transactions of the 35th North American Wildlife and Natural Reservation Conference* 35:115–127.

Part VI

Risk and Disturbance

> Vision is not enough; it must be combined with venture. It is not enough to stare up the steps; we must step up the stairs.
> —Vaclav Havel

A SMALL POPULATION of any species that is restricted to a single locality, which is experiencing natural and anthropogenic change, raises the concern of many interested in the science or practice of conservation. The endangered Mt. Graham red squirrel, with populations averaging around 300 animals in <10,000 ha of potential habitat, is no different. While exact risks of extinction are unknown, such conditions justifiably are cause for concern. One of the major risks associated with a single population is the increased risk of extinction due to catastrophic events. In addition to natural successional changes in the high-elevation forests of Mt. Graham, a diversity of human-related changes of varying scales also continues to occur from use and development to wildfire and insect infestations to exotic species and climate change.

In Part VI, we review the many challenges that face Mt. Graham red squirrels due to a variety of natural and anthropogenically induced disturbances. Sarah King and John Koprowski assess the impact of fire and detail the slow recovery of areas burned by the Clark Peak fire. Ann Lynch examines the important role of insect damage as a force shaping the future of the Pinaleño Mountains. The potential negative impacts of the introduction of Abert's squirrels by the Arizona Game and Fish

Department in the early 1940s are examined empirically by Tom Morrell and others as they describe subtle species differences in nest-tree selection. Andrew Edelman and John Koprowski follow with a review of potential mechanisms for the impact of introduced Abert's squirrels on the rare native red squirrel. Finally, Elaine Harding and others synthesize the plethora of challenges and produce a mathematical risk assessment that not only highlights the explicit need for additional data but also concludes that fire poses a major threat to the long-term persistence of the population.

CHAPTER 21

Effect of Human and Non-human Disturbance on Mt. Graham Red Squirrels

Sarah R. B. King and John L. Koprowski

The Mt. Graham red squirrel is endemic to the Pinaleño Mountains of southeastern Arizona and is vulnerable to catastrophes; currently, there are fewer than 300 animals (Arizona Game and Fish Department 2004). Previously, fire was considered the most severe threat to the squirrel population. However, recent insect outbreaks have become potentially more destructive in terms of reducing squirrel habitat and providing dead trees as fuel for a potentially devastating fire.

Mushrooms are an important food source for squirrels. The protein in fungi is comparable to that in spruce and pine seeds (12 to 27 percent; Miller and Halls 1969), and they are energetically less costly to eat (Grönwall and Pehrson 1984). In addition, mushrooms have high levels of other nutrients and easily digestible cell walls (Grönwall and Pehrson 1984). In Finland, mushrooms accounted for an average of 11 percent of the stomach contents of Eurasian red squirrels over all seasons, with 80 percent to 92 percent of the contents composed of conifer seeds (Sulkava and Nyholm 1987). Hypogeous fungi are most commonly eaten by small mammals, forming 88 percent of spores in the digestive tract of more than 400 North American rodents of different species examined by Maser, Trappe, and Nussbaum (1978). Red squirrels have been observed eating hypogeous fungi, particularly truffles, but they also have a specialized method of collecting and storing epigeous species (Sulkava and Nyholm 1987). The structure of epigeous fungi is conducive to air drying and caching (Maser, Trappe, and Nussbaum 1978); Eurasian red squirrels as well as red squirrels in North America have been observed to collect mushrooms and dry them on twigs or logs before storing them in their middens (Buller 1920; Cram 1924; Smith 1968; Sulkava and Nyholm 1987; Lurz and South 1998).

Mushrooms appear to provide a supplemental food through the summer and are most important for squirrels when stored and eaten through the winter (Smith 1968). Between 14 percent and 57 percent of stored mushrooms were eaten each month from October to December in Siberia (Sulkava and Nyholm 1987). However, 10 percent to 16 percent of stores were still left in April, so it is likely that in the late spring stored mushrooms become a crucial food source as the supply of cones in the midden becomes depleted (Smith 1968). In England, four species of mushroom were observed cached in trees: *Colybia maculata* and three species of *Russula* (Lurz and South 1998). In Siberia, mainly *Boletus* and *Nematoloma* species were stored (Sulkava and Nyholm 1987). Very little work has been done on this behavior in red squirrels, but *Russula* species (Buller 1920) and *Rhizopagon* species (Maser and Maser 1988) are stored by these squirrels.

Mycorrhizal fungi are important for forest health in addition to providing a food source for many small mammals. Pinaceae depend on mycorrhizae for nutrient and water uptake from the soil, and they can enhance host resistance to pathogens (Maser and Maser 1988). Hypogeous fungi are dependent on small mammals to distribute their spores (Maser, Trappe, and Ure 1978; States and Wettstein 1998). In Oregon, chipmunks were observed to inoculate clearcuts with mycorrhiza-forming spores (Maser, Trappe, and Nussbaum 1978), and in the same way, small mammals could help to regenerate the flora of burned areas of Mt. Graham.

Over the last century, there were two serious fires on Mt. Graham: the Clark Peak fire of 1996, which encompassed 2,718 ha over 15 days (Froehlich 1996), and the Nuttall Complex fire, which covered 11,898 ha in 2004 (Koprowski et al. 2006). Some trees on Mt. Graham date back to the 1300s, and tree-ring evidence shows that prior to 1872 fires tended to occur about every 8 years, with fires in which 75 percent of trees were scarred occurring every 70 years on average (Grissino-Mayer et al. 1995). A similar pattern was observed on Mica Mountain, Arizona (Baisan and Swetnam 1990), and in Yellowstone National Park (Despain and Romme 1991). Following a fire, a typical pattern of succession occurs in vegetation communities. Herbaceous plants generally dominate for the first years, before pine seedlings establish, and canopy closure occurs 30 to 50 years after the fire (Despain and Romme 1991). The growth of trees and paucity of ground-layer vegetation means the forest is able to resist fire quite well until 200 to 400 years after a stand-replacing event (Despain and Romme 1991). During this period, tree mortality begins to occur and fuels become abundant on the forest floor. This stage of forest succession, such as exists on Mt. Graham at the present time, ignites readily and burns intensely, resulting in a new stand-replacing fire.

Krefting and Ahlgren (1974) found that species diversity of vegetation did not change after a fire, but percent ground cover did differ, particularly abundance of herbs. Seeds of some herb species will be present in soil before a burn, and fire creates conditions conducive to germination. Consumption and dispersal of these seeds by small mammals may also influence the pattern of vegetation change (Krefting and Ahlgren 1974). Although fire can induce some species of mushroom to fruit when a substantial portion of the litter is destroyed, the number and kinds of fleshy fungi decline (Miller and Halls 1969). Thus, for the Mt. Graham red squirrel, a stand-replacing fire will not only remove their main source of food and shelter but also may reduce the amount of mushrooms, an important supplemental food.

This study aimed to examine impacts of disturbances such as fire and insect infestation on the vegetation and mushrooms used as food by Mt. Graham red squirrels.

Study Area

The University of Arizona's Mount Graham Red Squirrel Monitoring Program (MGRSMP) study area in the Pinaleño Mountains covers 311 ha, ranging in elevation from 2,830 m to 3,257 m. The transitional forest (TR) habitat type (106 ha), between 2,830 m and 3,050 m, consists of forest dominated by corkbark fir (52 percent), Engelmann spruce (16 percent), and Douglas-fir (14 percent). Above 3,050 m, the spruce-fir (SF) habitat type on the study area (205 ha) consists almost entirely of Engelmann spruce (61 percent) and corkbark fir (38 percent)(Hutton et al. 2003).

The Clark Peak fire started on Mt. Graham by undetermined human cause on 24 April 1996. By May 8, it was contained within a perimeter of 2,718 ha. The fire created a mosaic of burned and unburned forest of which 1,200 ha burned at a high or moderate intensity (Froehlich 1996). Potential red squirrel habitat on Mt. Graham has been divided into habitat management zones (HMZ) (U.S. Fish and Wildlife Service 1993). Of the 879 ha considered best for red squirrels (HMZ 1 and 2), 143 ha (16 percent) were burned at a high intensity and 17 ha (2 percent) were burned at a moderate intensity (Froehlich 1996).

Insect damage first began affecting large areas of spruce-fir forest in 1992 after blowdown from an ice storm provided dead timber habitat for increasing bark beetle populations (Young et al. 1999). Between 1996 and 1998, a geometrid looper moth defoliated Engelmann spruce and corkbark firs, and by 1998, several species of bark beetle had increased in trees weakened by the moth. In 1999, an introduced spruce aphid was observed in Engelmann spruce trees around High, Hawk, and Emerald Peaks (Young et al. 1999), and it has now spread farther.

Methods

Data Collection

In 1994, 35 transects were placed randomly over the entire study area, with another added on the SF in 1995. Transects were 50 m long and 1 m wide in each cardinal direction. Mushrooms were sampled within the 1 m wide strip along the transect, and vegetation was sampled along the transect line. Nine transects were destroyed or damaged by the Clark Peak fire in 1996 (seven in the TR and four in the SF habitat) and were replaced by three transects in the TR and one in the SF habitat on non-burned areas. Eleven transects in the SF habitat have been affected by a suite of insect infestations since 1996. Most transects in the SF are now likely affected by insect damage, but the nine transects outside the natural outbreak areas were less damaged.

Mushroom Data

From 1994 to 2002, all transects were checked once every two weeks during the field season for the presence of mushrooms. Only mushroom genera indicated in the literature or observed by MGRSMP biologists as being used by squirrels were collected. The number of mushrooms and total wet and dry weight were obtained for each genus on each transect. Mushrooms of 13 genera were collected in 1994 and 1995; from 1996,

3 additional genera (*Auricularia*, *Clitocybe*, and gastroid species) were collected (table 21.1).

Dry weight biomass and number of mushrooms found in each year and in each habitat were compared using chi-square tests, Mann-Whitney U tests, and Kruskal-Wallis non-parametric ANOVA tests because the data were not normal. Diversity of mushrooms across the years and habitats was examined using the Shannon-Weiner diversity index. The effect of weather on the biomass of mushrooms gathered was examined using Spearman's rank correlation. Mann-Whitney U tests and the Shannon-Weiner diversity index were used to explore the effect of insects on mushrooms in the SF habitat.

Vegetation Data

The effect of fire on the vegetation was examined by comparing the data from transects in 1995 that were burned in the 1996 Clark Peak fire with data from the same transects in 1997. To examine the vegetation on these transects, the ground cover substrate was recorded every 50 cm along the transect line in each cardinal direction. A chi-squared test was used to compare vegetation types pre- and post-fire. Between 1999 and 2001, ground cover in burned transects was recorded by noting the presence of any species in five 1 m^2 quadrats along the transect line. These data were analyzed using the Shannon-Weiner diversity index, and any change in vegetation over the years was explored with a Kruskal-Wallis test.

Results

Abundance of Mushroom Species Eaten by Red Squirrels on Mt. Graham

Only three fruiting bodies (0.2 g dry weight) of *Pholiota* were found on the burned part of the SF habitat in September 1996. Because no other mushrooms were found, these burned transects were not checked after 1996. Therefore, all mushroom data presented represents unburned transects only.

All of the 15 mushroom species used by the Mt. Graham red squirrel were found in SF and TR habitats between 1994 and 2002, but not all

Table 21.1. Biomass (g/ha) of mushroom genera found on transects on spruce-fir and transitional forest habitats of Mt. Graham between 1994 and 2002

Blank spaces indicate that genus was not collected in that year. Genera in bold print contain mycorrhizal species.

Genus	1994	1995	1996	1997	1998	1999	2000	2001	2002	Mean
Spruce-fir habitat										
Amanita	734	141	1138	340	266	123	158	197	0	344.1
Auricularia			2	5	2	23	127	46	121	46.7
Boletus	362	56	188	444	788	0	8	97	0	216.0
Clavaria	17	66	1	70	6	0	0	0	0	17.7
Clitocybe			37	306	243	257	150	1075	19	298.2
Cortinarius	1667	534	2500	1024	396	516	147	419	37	804.4
Gastroid sp.			381	5	0	6	0	0	0	56.1
Hydnum	343	7	943	211	294	18	0	0	0	201.8
Lactarius	155	0	563	548	247	57	2	5	3	175.5
Leccinum	3124	237	3037	904	275	8	0	40	0	847.2
Lycoperdon	62	0	322	967	961	1621	1157	1018	215	702.6
Pholiota	67	3	58	45	0	0	53	0	0	25.2
Ramaria	1272	572	1477	473	357	216	117	10	0	499.4
Russula	5204	643	3970	1881	2694	658	300	1129	0	1831.0
Suillus	158	0	180	19	77	2	0	1	0	48.5
Total	13164	2261	14796	7241	6607	3506	2219	4040	395	

Transitional forest habitat										
Amanita	67	0	76	297	68	125	241	560	84	168.5
Auricularia			273	313	69	659	1670	589	1174	678.2
Boletus	142	168	0	189	0	0	69	67	0	70.5
Clavaria	0	0	0	0	0	0	0	0	0	0.0
Clitocybe			296	622	141	1113	423	460	169	460.6
Cortinarius	506	683	596	795	108	522	730	1936	521	710.8
Gastroid sp.			0	0	0	0	0	0	0	0.0
Hydnum	0	0	0	0	0	73	0	123	0	21.9
Lactarius	18	13	335	269	24	364	25	472	64	176.0
Leccinum	419	0	261	59	0	0	0	214	0	106.0
Lycoperdon	140	78	839	1807	105	423	583	818	90	542.6
Pholiota	72	17	39	2	0	0	156	0	113	44.2
Ramaria	22	520	67	19	99	15	169	4	0	101.5
Russula	2095	475	1686	794	1766	675	573	4713	67	1427.0
Suillus	149	27	29	42	63	12	0	439	0	84.6
Total	3630	1981	4496	5209	2443	3982	4638	10394	2283	

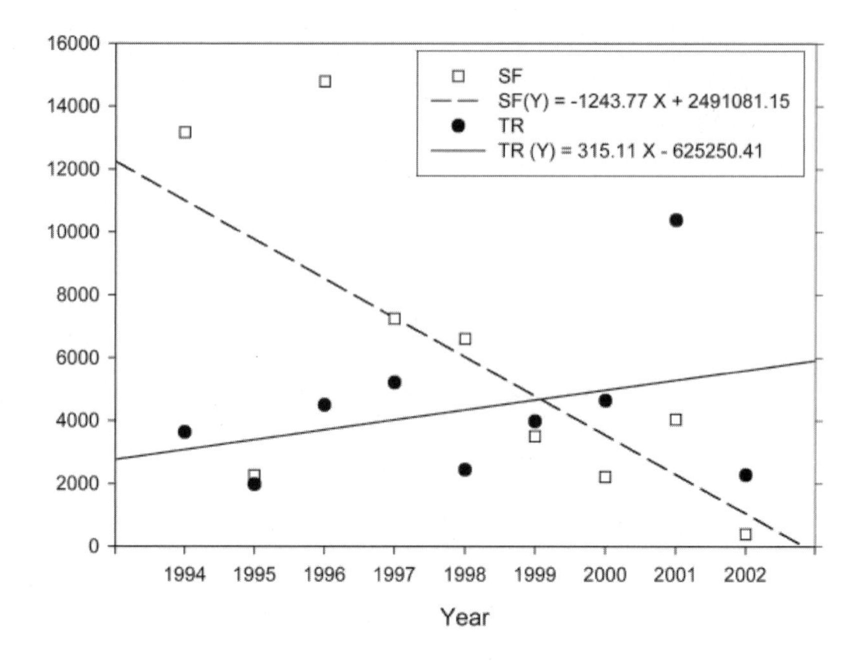

Figure 21.1. Results of a regression of the biomass of mushrooms on the spruce-fir (SF) and transitional forest (TR) habitats of Mt. Graham against time. The biomass of mushrooms is decreasing on the SF (r^2 = 0.46, P = 0.0436), but not on the TR habitat (r^2 = 0.12, P = 0.3707).

species were found in every year (table 21.1). Two genera found in SF habitat were not seen in TR habitat: *Clavaria* and gastroid species. Over all genera, there was no significant difference in biomass between the SF and TR habitats (Mann-Whitney U = 131, n = 15, P = 0.45). Among the years, biomass of mushrooms collected differed in the SF (Kruskal-Wallis H = 36.03, df = 8, P = <0.0001), but not TR habitat (H = 8.13, df = 8, P = 0.421). Non-parametric multiple comparison tests showed that median biomass in the SF was significantly lower in 2002 than in any other year. This trend remains when biomass of mushrooms in each habitat is regressed against time (fig. 21.1). In the SF habitat, biomass decreased over time, but not in the TR habitat.

No overall difference exists between the diversity (Student's t = 0.43, df = 16, $P = 0.677$), species richness ($t = 1.03$, df = 16, $P = 0.322$), or evenness of the two habitats ($t = -0.12$, df = 16, $P = 0.906$). Despite the lack of significant inter-year variation (SF habitat: $X^2 = 6.88$, df = 8, $P \geq 0.5$; TR habitat: $X^2 = 2.26$, df = 8, $P \geq 0.95$), the species richness was lower in both habitats in 2002 with a concurrent drop in H'.

In the SF habitat, more *Lycoperdon* fruiting bodies (median = 1484 ha^{-1}, IQR = 237–1918) were collected than any other genus (Kruskal-Wallis $H = 57.45$, df = 14, $P \leq 0.0001$), but *Russula* dominated in biomass (median = 1129 g/ha, IQR = 471–3332). In the TR habitat, *Auricularia* were the most abundant fruiting bodies (median = 3357 ha^{-1}, IQR = 233–8867; $H = 86.27$, df = 14, $P \leq 0.0001$), but *Russula* were again most abundant by biomass (median = 1427 g/ha, IQR = 524–1930). There was a difference in the abundance of the most common species over the years they were collected, and in all cases, the trend was for a decline from 1994 to 2002 (table 21.1). Of particular concern is the decline in mycorrhizal species (Arora 1986): *Leccinum* in particular have become very rare in both the SF and TR habitats. In 2002, no *Amanita*, *Ramaria*, or *Russula* mushrooms were found in the SF, and no *Ramaria* were found in the TR (table 21.1).

Effects of Weather on Mushroom Abundance

The earliest date in the year mushrooms were found and collected was 15 June in 1998, and the latest was 16 October in both 1995 and 1996. The greatest biomass of mushrooms was collected from both SF and TR sites in August (SF: $H = 21.90$, df = 4, $P \leq 0.0001$; TR: $H = 15.86$, df = 4, $P = 0.003$), but in the SF habitat a large biomass of mushrooms was also found in September. In both sites, mushrooms were rare in June.

In the SF habitat, the only significant correlations between weather variables and number and biomass of mushrooms were between rainfall and the number and biomass of mushrooms (number: Spearman's $r_s = 0.40$, $n = 34$, $P = 0.0094$; biomass: $r_s = 0.37$, $n = 34$, $P = 0.0158$) and the average relative humidity and biomass ($r_s = 0.51$, $n = 13$, $P = 0.037$; no significance with number: $r_s = 0.22$, $n = 13$, $P = 0.2295$). There was no significant relationship with temperature (number: $r_s = -0.15$, $n = 36$, $P = 0.1835$; biomass: $r_s = -0.17$, $n = 36$, $P = 0.1622$). However, in the TR

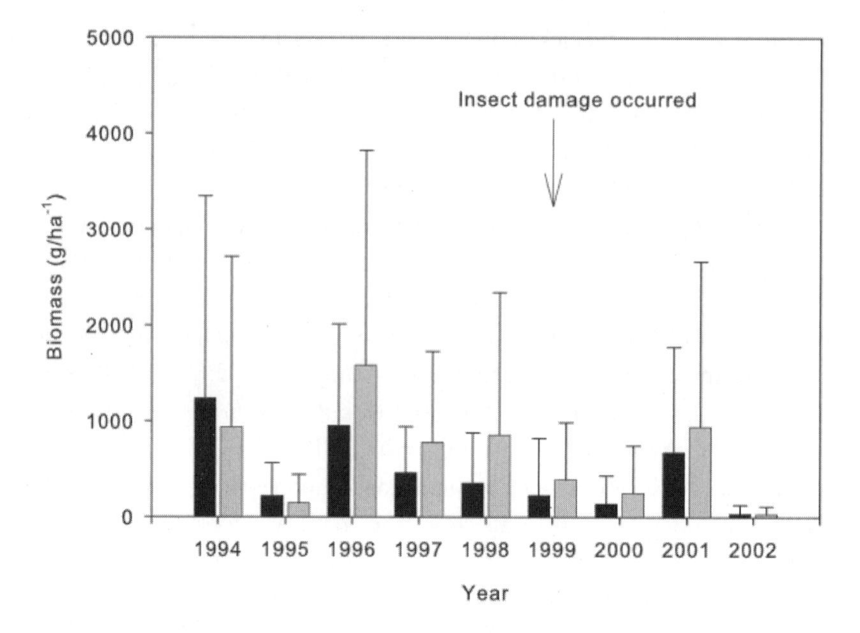

Figure 21.2. Difference in biomass between damaged (black columns) and undamaged (grey columns) transects on the spruce-fir area of Mt. Graham. There was significantly less biomass on the damaged transects ($X^2 = 6907.087$, df = 14, $P \geq 0.0001$).

habitat, the mean temperature and biomass were also negatively related ($r_s = -0.40$, $n = 34$, $P = 0.0102$; no significance for number: $r_s = -0.25$, $n = 34$, $P = 0.0809$), while rainfall and relative humidity were positively related with both biomass and number of mushrooms (rainfall: biomass $r_s = 0.49$, $n = 36$, $P = 0.0013$, number $r_s = 0.69$, $n = 36$, $P \leq 0.0001$; relative humidity: biomass $r_s = 0.75$, $n = 32$, $P \leq 0.0001$, number $r_s = 0.79$, $n = 32$, $P \leq 0.0001$). The average snow depth and average snow cover were not correlated with annual biomass or number of mushrooms in either SF (snow depth: biomass $r_s = -0.18$, $n = 9$, $P = 0.3067$, number $r_s = -0.37$, $n = 9$, $P = 0.1563$; snow cover: biomass $r_s = 0.14$, $n = 8$, $P = 0.376$, number $r_s = 0.18$, $n = 8$, $P = 0.3517$) or TR habitat (snow depth: biomass $r_s = -0.08$, $n = 9$, $P = 0.405$, number $r_s = -0.45$, $n = 9$, $P = 0.1063$; snow cover: biomass $r_s = 0.50$, $n = 8$, $P = 0.1081$, number $r_s = -0.06$, $n = 8$, $P = 0.4201$).

Effects of Insect Damage in the Spruce-Fir Forest on Mushrooms

Insect damage of trees has occurred on 10 transects in the SF habitat, with 11 being less damaged. Mushroom biomass was greater before the infestation (pre-1998) than after (post-1999) (U = 3030, P = 0.0004). Total biomass also differed between damaged and relatively less damaged transects (fig. 21.2). No significant difference was detected between these damage classes in diversity (H') (paired t-test t = −0.57, df = 17, P = 0.575), evenness (J') (t = −0.89, df = 17, P = 0.393), or species richness (S) (t = −0.22, df = 17, P = 0.832).

Although sample sizes were small (table 21.2), diversity (H') (t = −3.00, df = 8, P = 0.03) and evenness (J') (t = −3.26, df = 8, P = 0.023) differed on damaged transects before and after the infestation, but there was no significant difference in the species richness (S) (t = 1.95, df = 8, P = 0.10).

Habitat Capability Model: Ground Cover Vegetation Pre– and Post–Clark Peak Fire

Composition of the ground cover differed pre- and post-burn (X^2 = 1150.010, df = 7, P = <0.0001). More bare ground was present before (66 percent) than after the burn (26 percent). After the fire, the ground was covered by various understory vegetation: fern (3 percent), grass (11 percent), and herb (2 percent). Species increased after the fire to 8 percent, 21 percent, and 6 percent, respectively. *Carex* species also appeared. Tree or tree parts increased after the fire (17 percent before to 25 percent after the fire), and litter from the downed and dead trees also was more abundant.

Diversity of Vegetation Following the Clark Peak Fire

Twice as many species of plants were found on the burned parts of the SF habitat as on the burned areas in the TR habitat. However, within those species, relative diversity and evenness were similar between the two sites (table 21.3). On both sites, some species were more abundant than others (SF: H = 105.25, df = 20, $P \leq 0.0001$; TR: H = 313.09, df = 41, $P \leq 0.0001$). In the SF habitat, *Carex* spp., *Laennecia scheideana* (pineland marshtail),

Table 21.2. Diversity of transects before (pre-1999) and after the insect infestation in the spruce-fir habitat

H' = Shannon-Wiener diversity value, S = species richness, H_{max} = log$S$, J' = evenness (H'/H_{max})

		Pre-infestation					Post-infestation			
		1994	1995	1996	1997	1998	1999	2000	2001	2002
Damaged transects	S	11	8	13	14	11	8	9	11	5
	H'	0.72	0.75	0.93	0.95	0.80	0.47	0.65	0.73	0.41
	H_{max}	1.04	0.90	1.11	1.15	1.04	0.90	0.95	1.04	0.7
	J'	0.69	0.83	0.84	0.83	0.77	0.52	0.69	0.70	0.58
Undamaged transects	S	12	5	14	14	13	12	9	10	4
	H'	0.67	0.54	0.81	0.88	0.70	0.72	0.69	0.64	0.35
	H_{max}	1.08	0.70	1.15	1.15	1.11	1.08	0.95	1.00	0.60
	J'	0.62	0.77	0.70	0.77	0.63	0.67	0.72	0.64	0.58

and *Senecio wootonii* (Wooton's ragwort) were the most abundant species. These plants were also among the most common in the TR habitat, along with moss species and quaking aspen. In neither habitat did species abundance differ over the three years of collection (SF: $H = 1.53$, df = 2, $P = 0.465$; TR: $H = 0.69$, df = 2, $P = 0.708$).

Effect of Habitat Damage on the Red Squirrel Population

The Clark Peak fire burned 2,718 ha in the Pinaleño range. Sixty-six ha were in the environs of the study area, with 59 ha (19 percent) of the study area being burned. Insect damage covered a minimum of 240 ha of spruce-fir forest, 170 ha of which is on the study area (55 percent). Of the 427 middens that are located on the study area, 33 (7 percent) were burned and 143 (33 percent) were on the area damaged by insects. Most of these 175 middens are not currently habitable by squirrels (June 2003); thus, 41 percent of all middens on the study area were lost. Effects of the 2004 Nuttall Complex fire on the Mt. Graham red squirrels can be found in Koprowski et al. (2006) and are not dealt with here.

Table 21.3. Shannon-Weiner diversity index results for each year following the Clark Peak fire of 1996 in the spruce-fir (SF) and transitional forest (TR) habitats of Mt. Graham

S = number of species present, H' = Shannon-Weiner diversity value ($0 - H_{max}$), H_{max} = logS, J' = evenness (H'/H_{max}).

	1999		2000		2001	
	TR	SF	TR	SF	TR	SF
S	42	20	40	20	38	19
H'	1.30	1.11	1.30	1.08	1.31	1.11
H_{max}	1.62	1.30	1.60	1.30	1.58	1.28
J'	0.80	0.86	0.82	0.83	0.83	0.87

Conclusions

Insect and fire damage undoubtedly has affected the squirrel population on Mt. Graham. These two disturbances have meant that not only cover and cone food but also mushrooms are reduced, thus depriving squirrels of an important food source from late summer through late winter. The two different areas in this study had similar diversity and species abundance of mushrooms, but until 1999 the SF habitat had a greater biomass of mushrooms than the TR. The insect damage is responsible for the decline in the abundance of mushrooms since the first infestation, but a general trend of decline on both habitats may also be due to recent years of drought. These results point to the dynamic nature of an ecosystem faced by endangered Mt. Graham red squirrels as a montane isolate.

Studies of several populations show that *Russula* mushrooms are most selected by squirrels (e.g., Buller 1920; Lurz and South 1998). In both habitats on Mt. Graham, these mushrooms were the most abundant by biomass and would have provided an important food source, especially in late winter when cone stores are low. The ground cover increased after the fire and likely provided conditions for mushrooms to recover. The higher number of plant species found in the SF habitat than in the TR habitat could have been related to insect damage; if the trees do not recover their foliage, herbaceous species are not shaded out.

The insect damage has had a more insidious effect on the habitat than the fire damage and ultimately may prove to be more of a threat to the squirrel population, although fire is still a possible danger. The 1996 Clark Peak fire burned fiercely, but in patches, and it is possible that the squirrels could move away from the burn. A search of the burned areas following the fire detected no dead red squirrels, but a dead and charred Abert's squirrel was found (V. Greer, personal communication). Over the next 200 years, vegetation will regenerate (Despain and Romme 1991), and squirrels can recolonize the area. Squirrels, like other small mammals (Maser, Trappe, and Nussbaum 1978), act as dispersal agents of spores following consumption of fungi (States and Wettstein 1998). The mycorrhizae of fungi enhance seedling survival and the regeneration of the forest, and States and Wettstein (1998) postulate that squirrel distribution and abundance may be linked to the availability of fungi. Therefore, by traversing the burned areas, the squirrels are likely to be inoculating the forest with mycorrhizae and so promoting the health of the future trees, as well as inadvertently providing a food source.

The insect damage likely caused a third of all middens on the study area to be abandoned by 1999. Although young trees are growing in the infested area, the spruce aphids and the beetles are still present; thus, the squirrel habitat continues to be reduced as the trees are attacked as they reach maturity. The squirrels that live in infested areas are likely to move when conditions become critical. However, maintaining a bad territory is often preferred to moving to an unfamiliar area (Larsen and Boutin 1994, 1995), and moving to a new territory increases the risk of predation (Van Vuren et al. 1997). Consequently, squirrels remaining in poor habitat are likely to produce fewer young (Wauters et al. 2001), and reduced resources will result in slower growth and higher mortality of juveniles (Boutin and Larsen 1993). Squirrels are also likely to be more susceptible to predation in the open environment created by the insect damage (Stuart-Smith and Boutin 1995).

It remains to be seen how the Mt. Graham red squirrel population will cope with the insect damage. In the past 7,000 to 10,000 years of isolation (van Devender and Spaulding 1979), the squirrels have coped with stand-replacing fires and bark-beetle infestations, but the non-native aphids are a new threat. The persistence of the population appears to depend on the ability of the Mt. Graham red squirrel to handle such challenges yet again.

Acknowledgments

We would like to thank Vicki Greer and Sadie Bertelsen for their help during the formation of this manuscript. We also thank the present and past biologists of the MGRSMP for collecting all the data presented here. Funding was provided by the University of Arizona, Arizona Game and Fish Department, and the U.S. Forest Service. Research was conducted under permits from the Coronado National Forest, Arizona Game and Fish Department, U.S. Fish and Wildlife Service, and the University of Arizona's Institutional Animal Care and Use Committee.

Literature Cited

Arizona Game and Fish Department. 2004. Mount Graham red squirrel fall 2004 count. Press release. Phoenix.

Arora, D. 1986. *Mushrooms demystified.* Ten Speed Press, Berkeley, California.

Baisan, C. H., and T. W. Swetnam. 1990. Fire history on a desert mountain range: Rincon Mountain Wilderness, Arizona, U.S.A. *Canadian Journal of Forestry Research* 20:1559–1569.

Boutin, S., and K. W. Larsen. 1993. Does food availability affect growth and survival of males and females differently in a promiscuous small mammal, *Tamiasciurus hudsonicus? Journal of Animal Ecology* 62:364–370.

Buller, A. H. R. 1920. The red squirrel of North America as a mycophagist. *Transactions of the British Mycological Society* 6:355–362.

Cram, W. E. 1924. The red squirrel. *Journal of Mammalogy* 5:37–47.

Despain, D. G., and W. H. Romme. 1991. Ecology and management of high- intensity fires in Yellowstone National Park. Pages 43–58 in S. M. Herman, ed. *High-intensity fire in wildlands: Management challenges and options: Proceedings, Tall Timbers Fire Ecology Conference, Tall Timbers Research Station, May 18–21, 1989, Tallahassee, Florida.* Tall Timbers Research Station, Tallahassee, Florida.

Froehlich, G. 1996. *Biological assessment and evaluation for Clark Peak fire suppression and rehabilitation.* U.S. Forest Service, Safford Ranger District, Safford, Arizona.

Grissino-Mayer, H. D., C. H. Baisan, and T. W. Swetnam. 1995. Fire history in the Pinaleño Mountains of southeastern Arizona: Effect of human-related disturbances. Pages 399–407 in L. F. DeBano, P. F. Ffolliott, A. Ortega-Rubio, G. J. Gottfried, R. H. Hamre, and C. B. Edminster, technical coordinators. *Biodiversity and management of the Madrean Archipelago: The sky islands in southwestern United States and northwestern Mexico: September 19-23, 1994, Tucson, Arizona*. General Technical Report RM-GTR-264. U.S. Dept. of Agriculture, Forest Service, Rocky Mountain Forest and Range Experiment Station, Fort Collins, Colorado.

Grönwall, O., and Å. Pehrson. 1984. Nutrient content in fungi as a primary food supply of the red squirrel *Sciurus vulgaris* L. *Oecologia* 64:230–231.

Hutton, K. A., J. L. Koprowski, V. L. Greer, M. I. Alanen, C. A. Schauffert, and P. J. Young. 2003. Use of mixed conifer and spruce-fir forests by an introduced population of Abert's squirrels (*Sciurus aberti*). *The Southwestern Naturalist* 48(2):257–260.

Koprowski, J. L., K. M. Leonard, C. A. Zugmeyer, and J. L. Jolley. 2006. Direct effects of fire on endangered Mt. Graham red squirrels. *Southwestern Naturalist* 51:59–63.

Krefting, L. W., and C. E. Ahlgren. 1974. Small mammals and vegetation changes after fire in a mixed conifer-hardwood forest. *Ecology* 55:1391–1398.

Larsen, K. W., and S. Boutin. 1994. Movements, survival, and settlement of red squirrel *Tamiasciurus hudsonicus* offspring. *Ecology* 75(1):214–223.

——. 1995. Exploring territory quality in the North American red squirrel through removal experiments. *Canadian Journal of Zoology* 73(6):1115–1122.

Lurz, P. W. W., and A. B. South. 1998. Cached fungi in non-native conifer forests and their importance for red squirrels (*Sciurus vulgaris* L.). *Journal of Zoology London* 246:471–477.

Maser, C., and Z. Maser. 1988. Interactions among squirrels, mycorrhizal fungi, and coniferous forests in Oregon. *Great Basin Naturalist* 48:358–369.

Maser, C., J. M. Trappe, and R. A. Nussbaum. 1978. Fungal–small mammal interrelationships with emphasis on Oregon coniferous forests. *Ecology* 59(4):799–809.

Maser, C., J. M. Trappe, and D. C. Ure. 1978. Implications of small mammal mycophagy to the management of western coniferous forests. *Transactions of the North American Wildlife and Natural Resources Conference* 43:78–88.

Miller, H. A., and L. K. Halls. 1969. *Fleshy fungi commonly eaten by southern wildlife*. Research Paper SO-49. U.S. Dept. of Agriculture, Forest Service, Southern Forest Experiment Station, New Orleans, Louisiana.

Smith, M. C. 1968. Red squirrel responses to spruce cone failure in interior Alaska. *Journal of Wildlife Management* 32:305–317.

States, J. S., and P. J. Wettstein. 1998. Food habits and evolutionary relationships of the tassel-eared squirrel (*Sciurus aberti*). Pages 185–194 in M. A. Steele, J. F. Merritt, and D. A. Zegers, eds. *Ecology and evolutionary biology of tree squirrels*. Special Publication Number 6. Virginia Museum of Natural History, Martinsville.

Stuart-Smith, A. K., and S. Boutin. 1995. Behavioral differences between surviving and depredated juvenile red squirrels. *Ecoscience* 2(1):34–40.

Sulkava, S., and E. S. Nyholm. 1987. Mushroom stores as winter food of the red squirrel, *Sciurus vulgaris*, in northern Finland. *Aquilo Serie Zoologica* 25:1–8.

U.S. Fish and Wildlife Service. 1993. *Mount Graham Red Squirrel* (Tamiasciurus hudsonicus grahamensis) *Recovery Plan*. Albuquerque, New Mexico.

van Devender, T. R., and W. G. Spaulding. 1979. Development of vegetation and climate in the southwestern United States. *Science* 204:701–710.

Van Vuren, D., A. J. Kuenzi, I. Loredo, A. L. Leider, and M. L. Morrison. 1997. Translocation as a nonlethal alternative for managing California ground squirrels. *Journal of Wildlife Management* 61(2):351–359.

Wauters, L. A., J. Gurnell, D. Preatoni, and G. Tosi. 2001. Effects of spatial variation in food availability on spacing behaviour and demography of Eurasian red squirrels. *Ecography* 24:525–538.

Young, P. J., V. L. Greer, E. L. Bibles, N. Ferguson, J. E. Lowry, and E. A. Point. 1999. *The Mt. Graham Monitoring Program 1998 annual report.* University of Arizona, Tucson.

CHAPTER 22

Spruce Aphid, *Elatobium abietinum* (Walker)

Life History and Damage to Engelmann Spruce in the Pinaleño Mountains, Arizona[1]

ANN M. LYNCH

SPRUCE APHID IS AN EXOTIC INSECT recently introduced to the Pinaleño Mountains. It feeds on dormant Engelmann spruce, and possible effects include tree-growth suppression, tree mortality, and reduction in seed and cone production. Potential longer-term effects include changes in forest structure and species composition—primarily through reduction in Engelmann spruce dominance in spruce-fir and mixed-conifer forests. These effects would impact Mt. Graham red squirrel food resources, cache sites, and general habitat requirements.

Spruce aphid is folivorous on spruce species, with known infestations previously restricted to coastal areas (Koot 1991). Historically, economically important damage has occurred on Sitka spruce (*Picea sitchensis*), white spruce (*P. glauca*), and Norway spruce (*P. abies*) in areas with mild maritime winter climates: Britain, Iceland, Norway, Denmark, southern Sweden, northwestern Germany, Holland, New Zealand, and the northwestern North America coast (Bejer-Petersen 1962; Day et al. 1998). Spruce aphid was first reported in North America in 1916 in British Columbia (Koot and Ruth 1971; Carter and Halldórsson 1998), probably originating from European infestations (Bejer-Petersen 1962; Carter and Halldórsson 1998). The insect was found in Santa Fe, New Mexico, in 1976, although no wildland outbreak occurred in the Southwest until 1988, in the White Mountains of Arizona (U.S. Forest Service 1997; Lynch 2003, 2004). Numerous outbreaks of this insect have recently occurred on Engelmann spruce and Colorado blue spruce (*P. pungens*) in the subhumid mountains of the southwestern United States. Defoliation episodes, sometimes exceeding 11,000 ha, have occurred in the White Mountains, Pinaleño Mountains, and San Francisco Peaks of Arizona, and in the Mogollon Mountains and Sacramento Mountains of New Mexico (U.S. Forest Service 1997, 2001; Lynch 2004).

Spruce aphid was first found in the Pinaleño Mountains near High Peak in early winter 1999 by staff with the University of Arizona's Mt. Graham Red Squirrel Monitoring Program (MGRSMP). The majority of the Engelmann spruce on the mountain was severely defoliated by spring 2000. In the White Mountains, severe defoliation resulted in 24–41 percent mortality of Engelmann spruce (Lynch 2004). The extent and severity of spruce aphid damage in such a short period in the Pinaleño Mountains raised concerns about the insect's potential effects on Mt. Graham red squirrel habitat. Additionally, the timing of defoliation raised questions about the seasonal history of this insect in high-elevation southwestern ecosystems. Previous damaging outbreaks of this insect have been in maritime environments in the spring and early summer, with only infrequent and mild population increases being recorded in the autumn (Carter and Halldórsson 1998). Almost complete defoliation of many Engelmann spruce stands, and the nearly undetectable levels of aphid populations by April, indicated that the epizootic event occurred prior to spring in the Pinaleño Mountains.

Extensive European literature (see Day et al. 1998) describes spruce aphid maritime populations as increasing on dormant spruce in late winter through late spring, with occasional modest increases in the autumn. The aphids suck sap from host needle phloem cells, attacking the most recently produced needles after the foliage matures and hardens. Needles die and fall from the tree.

The insect's ability to over-winter is facilitated through supercooling ability, but aphid mortality does occur from ice nucleation (Carter 1972; Powell 1974), and populations are limited by cold winter temperatures. Aphids reared in cool conditions acquire cold-hardiness (Day and Kidd 1998), which varies seasonally and is lost in the spring (Carter 1972; Day and Kidd 1998). Spruce aphid epidemics have been restricted to areas with maritime climates, which moderates winter temperatures (Ohnesorge 1961; Bejer-Petersen 1962; Carter and Halldórsson 1998). Developing outbreaks in Britain are checked by sudden late spring frost (-11°C) after mild winters (Carter 1972). Epidemics in Denmark correspond to years with all monthly mean temperatures above 0°C (Bejer-Petersen 1962) and in Britain to regions where ambient temperatures do not fall to −8°C (Carter 1972). Some mild winters in Britain are not followed by outbreaks, and this is thought to be due to ambient temperatures falling below a limiting level at some point (Carter 1972; Day and Crute 1990), to an abrupt onset of cold conditions (Carter 1972), or to overcompensating density dependence (Day and Kidd 1998). The coldest temperature recorded for aphid survival is −23°C (Carter 1972). Using weather station data, Ohnesorge (1961, as summarized by Bejer-Petersen 1962) found that temperatures below −10°C to −14°C prevented or diminished outbreaks, with 0.5°C the lowest monthly mean in winters before an outbreak. Powell and Parry (1976) reported that in Scotland over-wintering populations are noticeably reduced when ambient temperatures fall below −7°C. In the North American Pacific Northwest, springtime populations collapse when temperatures fall below −5°C or −6°C (D. Overhulser and M. Schultz, personal communication). Temperatures often fall below these population limits in southwestern high-elevation forests, raising additional questions about the insect's life history.

Research was initiated in 2000 to determine the life history and potential frequency of outbreaks of the insect and the severity of damage to Engelmann spruce in Mt. Graham red squirrel habitats in the Pinaleño Mountains.

Methods

Seasonal Life History

Aphid populations were sampled 15 times between August 2000 and July 2001 to determine the seasonal abundance of the species in southwestern high-elevation ecosystems. On each sample date, one branch encompassing the 1996–2000 foliage was collected from 10 different trees and bagged individually at each of two sites, one on Emerald Peak and one near the Vatican Advanced Technology Telescope. Branches and trees were haphazardly selected each sample date within a 1 ha area. Branches were taken to the laboratory, where aphids were washed off in a lukewarm, weak saltwater bath. Straight-edged artists' brushes were used to gently brush tenacious aphids from the upper- and under-sides of the needles and twigs. When aphids were numerous, coffee filters in fast-flow funnels were used to filter them from the water bath. Aphids were placed in 85 percent ethyl alcohol and counted later. The distribution of aphids by foliage cohort was not obtained. Aphid count data is reported here as the mean number of aptera (wingless forms) and alates (winged forms) per branch.

Contemporary temperatures in the Pinaleño spruce-fir forest, and population increase and decline with respect to temperature, were assessed using temperature data from two sources. Temperature data for 1995–2000 were obtained from the MGRSMP weather station located in an opening on Emerald Peak. Temperature data for 2000–2003 were obtained from Onset optic temperature data loggers approximately 0.4 km away in the forest. In both cases, the ambient temperature was recorded at 2–3 m above ground. The MGRSMP weather station recorded minimum and maximum diurnal temperatures, from which a diurnal midpoint temperature was computed. The optic data loggers recorded temperature every 15 minutes, and mean daily temperatures were computed from 96 observations. Humidity and rainfall affect spruce aphid when rime ice is formed (Carter 1972; Carter and Nichols 1989; Carter and Halldórsson 1998), but the effect of snow is not known. In maritime populations, freezing of individual aphids of instars II through adult occurs when ice nucleation begins in the needles in which the aphid stylets are inserted (Powell 1974). During several sample dates over the 2002–2003 autumn–winter (data not reported here), aphids were observed to be alive and well when branches were covered by snow. Rather than damaging the aphids, snow may insulate them from nighttime low temperatures.

Tree Defoliation and Mortality

Temporary and permanent impact assessment plots were established and assessed on a grid for the area of the Pinaleños above 2,600 m. The primary objective was to determine if defoliation and mortality in the Pinaleños will be more or less severe than in the White Mountains of Arizona, as reported by Lynch (2004). Forty-three plots of 0.0314 ha (10 m radius) were measured in summer 2000 on a grid pattern above 2,600 m, including 34 temporary and 9 permanent plots. In 2001, the plots with few Engelmann spruce were expanded to 0.126 ha (20 m radius), and one additional permanent plot was added in the mixed-conifer zone. All live trees 10 cm or larger diameter at breast height (dbh) were assessed for species and diameter. All Engelmann spruce 10 cm dbh or larger were assessed for spruce dwarf mistletoe (*Arceuthobium microcarpum*) infection using the six-point Hawksworth DMR system (Hawksworth 1977), attack by bark beetles (spruce beetle and spruce ips [*Ips hunteri* Swaine], both Coleoptera: Scolytidae), and defoliation by spruce aphid. A defoliation index (DI) was computed as the sum of three crown-third ratings—where each crown-third was rated as 0, 1, 2, or 3—by 33 percent defoliation classes (an index of nine indicates that each crown third was 67–100 percent defoliated) (Lynch 2004). Trees or plots with defoliation indices of 0–3, 4–7, and 8–9 were considered lightly, moderately, and severely defoliated, respectively. DI class 7 trees were considered moderately rather than severely defoliated because Lynch (2004) found that significant mortality after spruce aphid defoliation occurred only in trees with DI 8 and 9.

Eight permanent plots were compared with nearby plots of the same size containing occupied Mt. Graham red squirrel middens. Midden plots were of the same size (10 m radii), approximately centered on the midden, and assessed in the same manner as non-midden plots. Statistical comparison of midden and non-midden plots utilized paired-comparison t-tests.

It should be noted that defoliation caused by a sap-sucking insect such as spruce aphid is not the same as defoliation from a defoliator such as western spruce budworm (*Choristoneura occidentalis* Freeman [Lepidoptera: Tortricidae]). The needle replenishes and subsequently loses fluid removed by the aphid. Eventually, the needle may die from necrosis or dehydration. Therefore, defoliation estimates made here are not directly comparable to similar levels of defoliation from a leaf-chewing insect.

Results

Aphid Populations and Temperatures

Aphids were found from early August 2000 through late March 2001, and again in July 2001, though population densities were very low after November 2000 (table 22.1). Aphid densities were high in autumn, from sometime in July through October 2001, and increased rapidly between

Table 22.1. Mean number of aphids collected on the 1996–2000 foliage of 10 branches at Emerald Peak and near the Vatican telescope over the 2000–2001 winter*

	Emerald Peak			Vatican		
Date	Aptera	Alates	Percent alates	Aptera	Alates	Percent alates
7 Jul 00	present	present	unknown			
5 Aug 00	57.0	0.0	0.0	0.0	0.0	
24 Aug 00	49.6	1.2	2.3	0.0	0.0	
21 Sep 00	50.6	1.8	3.4	1.5	0.0	0.0
11 Oct 00	318.9	8.1	2.5	7.3	0.3	4.1
5 Nov 00	8.7	0.0	0.0	4.0	0.4	0.1
22 Nov 00	2.4	0.0	0.0	0.3	0.0	0.0
13 Dec 00	0.8	0.0	0.0	0.0	0.0	
3 Jan 01	0.1	0.0	0.0	0.0	0.0	
1 Feb 01	0.3	0.0	0.0	0.0	0.0	
7 Mar 01	0.0	0.2	100.0	0.1	0.0	0.0
29 Mar 01	0.0	0.4	100.0	0.0	0.0	
12 Apr 01	0.0	0.0		0.0	0.0	
8 May 01	0.0	0.0		0.0	0.0	
28 Jun 01	0.0	0.0		0.0	0.0	
24 Jul 01	0.3	0.0	0.0	0.3	0.0	0.0

* Observations on 7 July 2000 are unquantified observations from a reconnaissance trip.

late September and mid-October (table 22.1; fig. 22.1). Populations were undetectable from April through June 2001, and barely detectable in July 2001 (table 22.1). Alates were found in autumn and spring. Population densities followed the same seasonal trends at both sample sites, but levels were much lower at the Vatican telescope site (table 22.1).

Aphid population density increased to detectable levels prior to the first sample date in early August 2000 (table 22.1; fig. 22.1). Population density was fairly stable from 5 August until 21 September, when mean daily temperatures were between 7°C and 14°C and minimum daily temperatures were above 2.7°C. Aphid population density increased 530

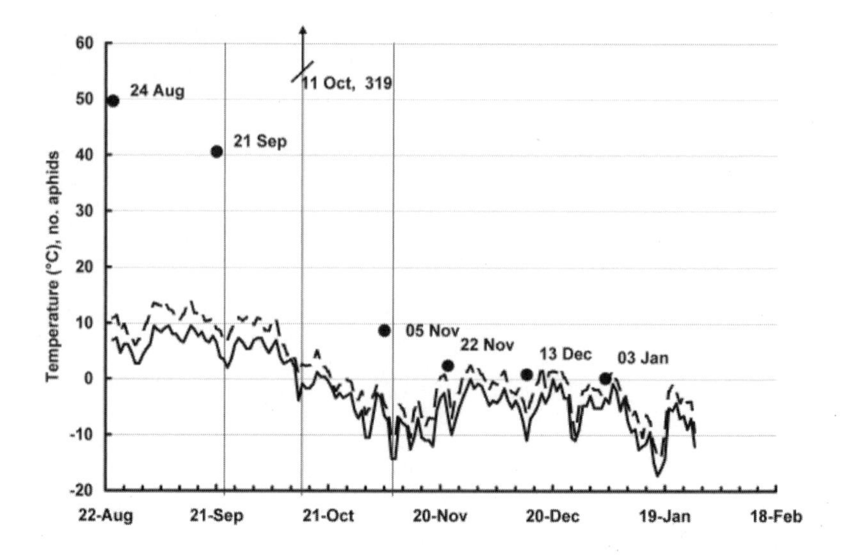

Figure 22.1. Minimum (solid line) and mean (dashed line) daily temperatures, and mean number of aphids per branch (dots, each computed as the mean of 10 branches including 1996–2000 foliage) at Emerald Peak, August 2000 through May 2001. Thin vertical lines are for guidance to the axis position.

percent in the 21-day period between 21 September and 11 October, when mean daily temperatures were somewhat cooler, between 7 and 10°C, with minimum temperatures never falling below 2°C (fig. 22.1). After the 2001 summer, aphid population density had increased to only barely detectable levels by the end of July 2001, much lower than by early August in the previous year (table 22.1).

Aphid population density decreased 97 percent between 11 October and 5 November 2000 (fig. 22.1). During the intervening 25-day period, minimum temperatures fell to –3.8°C, –7.0°C, and –10.5°C, and daily mean temperatures were below 0°C for 16 days. It is not known whether one of the cold events, or the extended cold period, caused the population to collapse. Populations were not quantified in 1999, the first year the insect was found in the Pinaleños, but populations were very high at least through December (collections were made for genetic assessment), when the mean daily temperature was –3.3°C and the minimum temperature was −12.1°C. The October and November 1999 minimum temperatures were −4.6°C and −8.7°C, respectively. Defoliation from the 1999–2000 episode was more severe and extensive than from the 2000–2001 episode, so aphid populations were either of higher density or present at damaging densities for a longer period of time during 1999–2000. Also, I have observed populations at damaging densities surviving temperatures as low as −13.6°C in the Pinaleños, during the 2002–2003 winter, and −15°C in the San Francisco Peaks, during the 2001–2002 winter.

Significant defoliation was observed in 2000, 2001, and 2003 from spruce aphid populations in the preceding autumn–winter periods of 1999–2000, 2000–2001, and 2002–2003, respectively. Monthly mean temperatures fell below 0°C three to five months each winter, and ambient temperatures fell below −12°C every autumn–winter period from 1996–1997 through 2002–2003. Spruce aphid population densities were not quantified in the Pinaleño Mountains the first winter that the species was present, but infested branch samples collected on Emerald Peak for identification purposes in November and December 1999 indicated that temperatures fell to at least −4.6°C and possibly below −10°C without terminating the outbreak.

The majority of aphids collected were aptera (table 22.1). A small portion of the collections from late August to November was male alates, and the collections in March at Emerald Peak were entirely dispersal females.

Tree Defoliation and Mortality

Fifteen plots contained no Engelmann spruce 10 cm or larger dbh and were not evaluated further. The 29 remaining plots (19 temporary and 10 permanent) include 16 habitat types from corkbark fir, Douglas-fir, Engelmann spruce, and white fir series (Stuever and Hayden 1997). Site physiography varies from riparian to rocky ridge tops, ranging in slope from 3 to 72 percent and in elevation from 2,728 to 3,226 m, and encompassing the full range of aspect. Engelmann spruce density (trees 20 cm or larger) ranges from 32 to 795 trees per ha, and total tree density ranges from 64 to 954 trees per ha. Other tree species present include corkbark fir, white fir, Douglas-fir, ponderosa pine, southwestern white pine, quaking aspen, and willow (*Salix* sp.). The 10 stands in the permanent

Table 22.2. Character of the 10 stands sampled from 2000 through 2002 for spruce aphid defoliation and mortality

Habitat type abbreviations: Abco = White fir, Abla = Corkbark fir, Erex = Forest fleabane [*Erigeron eximius*], Pien = Engelmann spruce, Psme = Douglas-fir, Rone = New Mexico locust [*Robinia neomexicana*]

Name	Aspect (deg)	Slope (%)	Elev (m)	Habitat Type	Physiography	Density (tph)	Engelmann Spruce Dominance (%)	Engelmann Spruce dbh ± SD (cm)
Shingle Mill	320	27	2826	Abla/moss	mid-slope, rocky soil	72	36	28 ± 14
Lower Webb Pk	116	20	2961	Abla/moss	lower slope	863	100	36 ± 12
Bible Camp	12	53	2889	Abla/moss	steep mid- to upper-slope	401	37	17 ± 6
Soldier Creek	168	25	2910	Pien/Erex	gentle mid-slope	64	10	33 ± 10
Rock Pile	50	6	2947	Abla/riparian	wetland	223	50	33 ± 9
Access Rd	204	30	3140	Pien/moss	upper slope	863	96	31 ± 7
Plain View	188	30	3115	Pien/moss	mid-slope, deep soil	382	100	30 ± 9
Crazy Horse	310	3	2890	Psme/Rone	ridge top, shallow soil	401	49	20 ± 3
Heliograph Peak	250	36	2907	Pien/Erex	mid-slope below steep slope	159	19	46 ± 19
Grant Hill Trailhead	150	10	2800	Abco/Carex	gentle drainage hollow	374	75	22 ± 14

sample were a representative sample of the total sample, representing a wide range of density, dominance by Engelmann spruce, physiography, and Engelmann spruce size (table 22.2).

Defoliation from spruce aphid was moderate to severe (DI ≥ 5) on 55 percent of the total sample. Little or no defoliation occurred from the 1999–2000 episode on plots on, near, or east of the ridge that runs from Mt. Graham down towards Shannon Park. However, all plots in the permanent sample have subsequently had at least moderate defoliation from spruce aphid at least once (table 22.3). This indicates that aphids did not disperse from the original establishment area (probably near Highwater Cienega) to the east side of the mountain range with enough time for populations to increase and cause defoliation the first year. Mean plot defoliation in 2000 of plots within the 1999–2000 outbreak area was 7.1 ± 2.7 DI.

Mean mortality on plots was 21.2 percent ± 21.1 percent by 2002 (table 22.3). Many plots were defoliated two years in a row (table 22.3). Plots with a mean defoliation index in either 2000 or 2001 (or both) of 8 or greater incurred, on average, 32 percent mortality by 2002, while plots with mean defoliation indices of 7 or less each year incurred 5 percent mortality (table 22.3). The two plots with no mortality by 2002, which incurred no defoliation in the 1999–2000 episode, were located on the eastern side of the study area, as described earlier.

Of the trees that died, 10 percent died from aphid defoliation without additional attack from bark beetles or pre-existing severe mistletoe infection. Fourteen percent had pre-existing severe mistletoe infection, and 75 percent were attacked by bark beetles after being severely defoliated (DI ≥ 8) by spruce aphid. Fifty percent of trees with heavy defoliation (DI ≥ 8) and severe dwarf mistletoe (DMR ≥ 5) were dead by 2002. Much greater incidence of bark beetle attack in heavily defoliated trees was found in the Pinaleños than was found in the White Mountains, undoubtedly due to the fact that a spruce beetle epidemic was underway when spruce aphid first established in this mountain range (see Discussion). No tree in the sample died from spruce beetle attack that was not first severely defoliated by spruce aphid.

Paired-comparisons showed that midden plots were slightly less defoliated in 2000 than were non-midden plots (DI of 4.2 ± 3.7 vs. 5.1 ± 4.3, $t = 1.88$, df = 7, $P = 0.10$). Compared to non-midden plots, midden plots

Table 22.3. History of spruce aphid defoliation and mortality from 2000 through 2002

Note that defoliation in a given year results from aphid populations in the preceding autumn and winter.

	Mean defoliation index (0 = none; 9 = complete)			Percent mortality of trees alive in 2000		
Site name	2000	2001	2002	2001	2002	Total
Shingle Mill	9.0	7.3	2.0	0.0	16.6	16.6
Lower Webb Pk	5.3	6.9	2.7	0.0	5.9	5.9
Bible Camp	9.0	8.9	2.0	28.6	28.5	57.1
Soldier Creek	9.0	7.4	3.5	12.5	0.0	12.5
Rock Pile	2.4	4.8	2.7	0.0	14.3	14.3
Access Rd	8.1	7.6	1.6	0.0	47.1	47.1
Plain View	0.0	8.4	2.0	0.0	47.1	47.1
Crazy Horse	0.0	6.0	3.0	0.0	0.0	0.0
Heliograph Peak	0.0	3.6	3.3	0.0	0.0	0.0
Grant Hill Trailhead	M	8.0	2.6	0.0	11.1	11.1

had greater density (482 vs. 345 trees per hectare [tph]) and dominance (53.1 percent vs. 42.2 percent, by density) by Engelmann spruce. The Engelmann spruce were also larger on the midden plots (35.5 vs. 32 cm mean dbh).

Discussion

In its native range in continental Europe, the spruce aphid life cycle is holocyclic, with parthenogenic population increases in autumn and spring, winged male and female forms, and sexual forms in autumn that produce an over-wintering egg stage (von Scheller 1963; Lampel 1968; Carter and Austarå 1994; Carter and Halldórsson 1998). The presence of autumn males in Arizona indicates that the life cycle is either holocyclic or paracyclic (both egg and parthenogenic forms over-winter). Winged females would account for rapid dispersal of the species throughout the mountain range; they rarely occur in maritime environments where the life cycle is usually limited to the parthenogenic apterous form. Sampling over several winters will be necessary to determine if the life cycle in the Southwest is holocyclic or paracyclic. Although eggs have not yet been found, the presence of males indicates that eggs are produced in Arizona. Very little is known about the egg stage, but since it is an over-wintering stage in continental Europe, it is probably very cold hardy.

Aphid populations in the Pinaleños appear to be more cold hardy than populations in maritime Europe and the Pacific Northwest. In maritime climates, temperatures below –10°C to –14°C prevent or diminish subsequent spring outbreaks, and outbreaks do not develop after a monthly mean of 0.5°C occurs in any winter month (Bejer-Petersen 1962; Day and Kidd 1998). Maritime populations collapse between –5°C and –7°C, and development of individual aphids occurs only above 5°C (Day and Crute 1990). Temperatures below these levels are reached in the Pinaleños almost every October, November, and December, and aphid populations continue to increase.

There may be considerable inter-annual variability in autumn–winter population dynamics. Aphid population densities in early autumn 2000 and 2001 were very different (table 22.1; fig. 22.1). Additionally, densities were quite high in late December 1999 and early January 2000, prior to this investigation, but very low in December 2000 and January 2001. Spring population dynamics in maritime Europe are influenced by the density of the pre-existing aphid population, and by temperatures during the previous winter (Day and Kidd 1998). Data here are not adequate to model population dynamics, but autumn temperatures show considerable variation, especially with respect to minimum temperatures. October minimum temperatures have ranged from –5°C to –14°C, which would undoubtedly affect spruce aphid population dynamics.

Even though the aphid sampling methodology was fairly crude, it is adequate to demonstrate that mean daily temperatures between 7°C and 10°C, and minimum temperatures above 2°C, are suitable for aphid population increase. However, aside from demonstrating that population increase occurs under late autumn environmental conditions typical of the Pinaleño Mountains, which are colder than temperatures that limit spruce aphid density in maritime environments, these data do not establish a direct relationship between temperature and population dynamics. Population growth and decline is associated with fecundity, which is strongly influenced by nutritional factors in the phloem fluid, as influenced by dormancy (Day 1984; Fisher and Dixon 1986; Day and Kidd 1998). Engelmann spruce dormancy would progress at different rates from year to year.

At some point, below-freezing temperatures limit aphid population growth and/or cause population decline. Several different cold events preceded a population crash in the Pinaleño Mountains during the course of this study. These events included temperatures falling to –3.8°C to –10.5°C, and mean and minimum diurnal temperatures remaining below 0°C for 16 days. However, damaging populations survived temperatures as low as –13.6°C the next year and –15°C in northern Arizona. This indicates some variability in the cold-hardiness of aphid populations, possibly associated with conditioning. Probably a sharp decline into cold temperatures is more damaging to aphid populations than a gradual decline. An extended cold period probably results in low developmental rates, starvation, and a low birth rate, such that dying individuals are not replaced.

Defoliation in the Pinaleño Mountains was more severe than what was seen in the White Mountains by Lynch (2004). Mortality levels are consistent with, or more severe than, those observed in the White Mountains, where 24–41 percent mortality was observed four years after a single defoliation episode in heavily defoliated trees (Lynch 2004). Here, mean plot mortality reached 21 percent by 2002 and 32 percent on severely defoliated plots within two years. These differences may be due to more severe defoliation, consecutive years of defoliation on some plots, or to spruce beetle activity in defoliated trees. Spruce beetle activity was high in the Pinaleño Mountains when spruce aphid was introduced, but has been low in the White Mountains. Based on patterns seen in the White Mountains, additional mortality from the 1999–2000 and 2000–2001 episodes may have continued through 2005 in the Pinaleño Mountains.

At the time that spruce aphid was first found in the Pinaleño Mountains, the spruce-fir forest was in the middle of outbreaks of both spruce beetle and western balsam bark beetle, both of which originated and developed slowly after damage from ice and snow in the winter of 1992–1993. About 160 ha of this area were subsequently severely defoliated in 1996–1999 by *Nepytia janetae* Rindge (Lepidoptera:Geometridae), and the beetle populations increased rapidly in the defoliated trees (Lynch et al. in press). Spruce beetle is an aggressive species, and during outbreaks, it kills most mature spruce over extensive areas of mature and over-mature spruce or spruce-fir forests, especially the larger trees (Furniss and Carolin 1977). Much greater incidence of bark beetle attack in heavily defoliated trees was found in the Pinaleño Mountains than was seen in the White Mountains, undoubtedly due to the coincidence of outbreaks. No trees on the study plots died from spruce beetle attack that were not first severely defoliated by spruce aphid. Spruce beetle hazard is high in dense stands composed primarily of large spruce (Schmid and Frye 1976, 1977), but defoliated trees in this study were attacked by spruce beetle in stands not dominated by large spruce, such as near the Bible Camp. Unfortunately, there were too few trees with light levels of defoliation (there were only 17 individual trees with DI $\leq$ 7) to definitively conclude that defoliation predisposes trees to spruce beetle attack.

The high level of mortality seen in trees that were both severely defoliated by spruce aphid and severely infected with spruce dwarf mistletoe (50 percent by 2002) is consistent with findings from the White Mountains, where almost 70 percent of such trees died within four years of aphid attack (Lynch 2004). Because the overall levels of defoliation were so high in the Pinaleños, site and vegetation factors could not be assessed for association with defoliation, other than to conclude that Engelmann spruce in the Pinaleño Mountains is susceptible to spruce aphid attack in the mixed-conifer, transition, and spruce-fir zones. The only other factor associated with increased likelihood of defoliation and mortality in the White Mountains was position in the lower-stand canopy (not tree size directly) (Lynch 2004). Future risk-assessment efforts should be directed towards weather factors rather than stand and site character.

Plots containing active Mt. Graham red squirrel middens in 2000 were slightly less severely defoliated than paired non-midden plots. Although I have seen Mt. Graham red squirrels feeding on spruce aphids, it is difficult to think that such activity would significantly reduce aphid populations on the entire plot. More likely, the less severe defoliation reflects an absence of smaller Engelmann spruce on the plots, which generally are the ones most severely defoliated (Lynch 2004). Average size of Engelmann spruce is larger on midden plots.

Conclusions

The population dynamics information described here from the Pinaleño Mountains, repeated outbreaks two years in a row, as well as the frequency of outbreaks in the White Mountains, indicates that autumn and possibly winter temperatures are frequently suitable for spruce aphid population increase in the Pinaleño Mountains. Aphid populations in the southwestern United States are clearly surviving, increasing, and causing severe defoliation during periods when diurnal minimums are below –5°C, the point where ice nucleation occurs in Sitka spruce–attached aphids, and below –8°C to –14°C, the range of ambient temperatures below which subsequent epidemics are thought to be limited. Temperatures seldom fall below –15°C, and in some years do not fall below –10°C until December, while aphid population increase may begin as early as July. There is evidence that autumn spruce aphid outbreaks follow dry winter–spring periods, but the mechanisms are unknown (Lynch 2003). Further research is necessary to establish the temperature regimes that terminate autumn–winter spruce aphid outbreaks, and that limit and favor population increases in southwestern high-elevation environments. Such information will allow better estimation of outbreak frequency and severity.

There is little hope that spruce aphid will be eliminated from the ecosystem due to a severe cold event, as the species persisted in the White Mountains after December 1990 when temperatures fell to –25°C (NOAA 1990).

Unfortunately, if weather patterns seen in recent years continue, it appears that spruce aphid outbreaks will occur with some frequency. The severity of damage seen in both the Pinaleño Mountains and the White Mountains indicates that, given frequent outbreaks, Engelmann spruce populations will be diminished in all habitat types in which it occurs. Additional research is needed to determine the effect of aphid feeding on Engelmann spruce seed and cone production, seed viability, and established regeneration. Spruce aphid drastically reduces height growth of other spruce species (Carter 1977; Warrington and Whittaker 1990; Seaby and Mowat 1993; Thomas and Miller 1994; Straw et al. 1998), and similar growth loss effects would affect the ability of Engelmann spruce to compete with other species. Increased defoliation and mortality of the lower-canopy layers, seen in the White Mountains, indicates that regeneration size classes will be impacted the most by this insect. Though research reported here does not address impact to small trees, spruce aphid has heavily damaged Engelmann spruce seedlings and saplings in the Pinaleño Mountains.

The combined short-term effects of beetle, looper, and aphid outbreaks in the spruce-fir, transition, and mixed-conifer forests include an abundance of standing dead timber of all size classes, increased amounts of coarse woody debris, an abundance of desiccating and dead needles distributed throughout the forest canopy, and dramatically altered sunlight regimes within the forest. In stands with a large Engelmann spruce component, airflow and temperatures on the site have also changed. The abundance of dead and dying needles from repeated spruce aphid outbreaks represents a canopy ladder of fine fuels, affecting the risk and vulnerability to wildfire.

The combined effects of outbreaks of the multiple insect species have devastated the spruce-fir forest and pure spruce stands and individual Engelmann spruce in the mixed-conifer forest in the Pinaleño Mountains. Spruce beetle, western balsam bark beetle, and N. *janetae* drastically altered the character of the highest-elevation forest before spruce aphid established. Spruce aphid defoliation caused additional mortality to Engelmann spruce in both the spruce-fir and mixed-conifer forests (table 22.3), and will in all likelihood continue to cause mortality. The catastrophic insect-related damage was rapidly followed by a precipitous decline in the Mt. Graham red squirrel population (Koprowski et al. 2005). Severe outbreaks of bark beetles are catastrophic, but natural, disturbances in spruce and spruce-fir forests (Furniss and Carolin 1977). The geometrid is a native insect, and it has probably incurred outbreaks in the past as well (Lynch et al. in press). The Mt. Graham red squirrel, and red squirrel populations in other southwestern forests, have obviously survived such disturbances in the past, including catastrophic wildfire (Grissino-Mayer et al. 1995). However, spruce aphid is a new disturbance agent in the high-elevation forest, and the future cannot be assured by historical success.

Engelmann spruce is highly susceptible to fire damage (Fischer and Bradley 1987; Grissino-Mayer et al. 1995), but risk is somewhat mitigated by the relatively moist and cool habitat, except during very dry years

(Zwolinski 1996). Anthropogenic influences over the last century, especially fire exclusion, have produced dense, multi-storied canopies in the mixed-conifer forests that are now much more prone to catastrophic wildfire (Baisan and Swetnam 1995; Grissino-Mayer et al. 1995; Allen 1996; Touchan et al. 1996; Zwolinski 1996). Prior to European settlement, the relatively high frequency of surface fires and the heterogeneous nature of the mixed-conifer forest contributed to the formation of a buffer zone below the highly flammable spruce-fir forest, promoting long-term stability in the spruce-fir forest. The integrity of this buffer has been reduced by fire exclusion practices (Grissino-Mayer et al. 1995). Fine fuels produced by aphid feeding throughout the forest canopy may increase fire risk and hazard. Fire hazard will also be altered due to the significant amounts of mortality in the mature canopy from the combined effects of beetle, looper, and aphid outbreaks, and in the lower-canopy layers from spruce aphid. Wildfire concerns are heightened by an apparent association between aphid outbreaks and warm, dry winter weather (Lynch 2003), and the compromised ability of the lower-elevation forests to buffer the high-elevation forests from encroaching wildfires.

The time frames of the studies reported here, as well as those discussed from the White Mountains, are too short to predict the ultimate fate of Engelmann spruce in the ecosystem. However, in all likelihood, Engelmann spruce populations will be diminished in the future forest. Aphid-caused spruce mortality and altered seed production could have long-term effects on stand structure, species composition, and Mt. Graham red squirrel food resources and habitat.

Acknowledgments

I am grateful to Genice Froehlich for reconnaissance and aphid sampling assistance, to Jill Wilson and Michelle Frank for helpful suggestions, and to Stephanie Jentsch, Brian Nowicki, and Jerry Snow for counting aphids and for assistance with the defoliation and mortality surveys. Funding for this study was provided by the University of Arizona through the Mt. Graham Red Squirrel Study Committee, which is comprised of representatives from the Arizona Game and Fish Department, the University of Arizona, the U.S. Forest Service, and the U.S. Fish and Wildlife Service; and by the U.S. Forest Service, Rocky Mountain Research Station.

Note

[1] This manuscript was written and prepared by a U.S. government employee on official time, and therefore it is in the public domain and not subject to copyright.

Literature Cited

Allen, L. S. 1996. Ecological role of fire in the Madrean Province. Pages 5–10 in P. F. Ffolliott, L. F. DeBano, M. B. Baker, G. J. Gottfried, G. Solis-Garza, C. B. Edminster, D. G. Neary, L. S. Allen, and R. H. Hamre, technical coordinators. *Effects of fire on Madrean province ecosystems: A symposium proceedings.* General Technical Report RM-GTR-289. U.S. Dept. of Agriculture, Forest Service, Rocky Mountain Forest and Range Experiment Station, Fort Collins, Colorado.

Baisan, C. H., and T. W. Swetnam. 1995. Management implications of historical fire occurrence patterns in remote mountains of southwestern New Mexico and northern Mexico. Pages 153–156 in J. K. Brown, R. W. Mutch, C. W. Spoon, and R. H. Wakimoto, technical coordinators. *Proceedings: Symposium on Fire in Wilderness and Park Management: Missoula, MT, March 30–April 1, 1993.* General Technical Report INT-320. U.S. Dept. of Agriculture, Forest Service, Intermountain Research Station, Ogden, Utah.

Bejer-Petersen, B. 1962. Peak years and regulation of numbers in the aphid *Neomyzaphis abietina* Walker. *Oikos* 13:155–168.

Carter, C., and G. Halldórsson. 1998. Origins and background to the green spruce aphid in Europe. Pages 1–10 in Day, K. R., G. Halldórsson, S. Harding, and N. A. Straw, eds. *The green spruce aphid in Western Europe: Ecology, status, impacts and prospects for management.* Forestry Commission Technical Paper 24. Forestry Commission, London.

Carter, C. I. 1972. *Winter temperatures and survival of the green spruce aphid.* Forest Record 84. Forestry Commission, London.

——. 1977. *Impact of green spruce aphid on growth: Can a tree forget its past?* Forestry Commission Research and Development Paper 116. Forestry Commission, London.

Carter, C. I., and Ø. Austarå. 1994. The occurrence of males, oviparous females, and eggs within anholocyclic populations of the green spruce aphid *Elatobium abietinum* (Walker) (Hemiptera: Aphididae). *Norwegian Journal of Entomology* 41:53–58.

Carter, C. I., and J. F. A. Nichols. 1989. *The green spruce aphid and Sitka spruce provenances in Britain.* Forestry Commission Occasional Paper 19. London.

Day, K. R. 1984. The growth and decline of a population of the spruce aphid, *Elatobium abietinum* during a three year study, and the changing pattern of fecundity, recruitment, and alary polymorphism in a Northern Ireland forest. *Oecologia* 64:118–124.

Day, K. R., and S. Crute. 1990. The abundance of spruce aphids under the influence of oceanic climate. Pages 25–33 in A. D. Watt, S. R. Leather, M. D. Hunter, and N. A.

C. Kidd, eds. *Population dynamics of forest insects.* Intercept, Andover, Hampshire, England.

Day, K.R., G. Halldórsson, S. Harding, and N.A. Straw, eds. 1998. *The green spruce aphid in Western Europe: Ecology, status, impacts and prospects for management.* Forestry Commission Technical Paper 24. Forestry Commission, London.

Day, K.R., and N.C. Kidd. 1998. Green spruce aphid population dynamics: Effects of climate, weather, and regulation. Pages 41–52 in Day, K. R., G. Halldórsson, S. Harding, and N. A. Straw, eds. *The green spruce aphid in Western Europe: Ecology, status, impacts, and prospects for management.* Forestry Commission Technical Paper 24. London.

Fischer, W.C., and A.F. Bradley. 1987. *Fire ecology of western Montana forest habitat types.* General Technical Report INT-223. U.S. Dept. of Agriculture, Forest Service, Intermountain, Ogden, Utah.

Fisher, M., and A.F.G. Dixon. 1986. Role of photoperiod in the timing of dispersal in the green spruce aphid *Elatobium abietinum. Journal of Animal Ecology* 55:657–667.

Furniss, R. L., and V. M. Carolin. 1977. *Western forest insects.* U.S. Forest Service Miscellaneous Publication Number 1339. Washington, D.C.

Grissino-Mayer, H.D., C.H. Baisan, and T.W. Swetnam. 1995. Fire history in the Pinaleño Mountains of southeastern Arizona: Effects of human-related disturbances. Pages 399–407 in L.F. DeBano, G.J. Gottfried, R.H. Hamre, C.B. Edminster, P.F. Ffolliott, and A. Ortega-Rubio, technical coordinators. *Biodiversity and management of the Madrean Archipelago: The sky islands of southwestern United States and northwestern Mexico: September 19–23, 1994, Tucson, Arizona.* General Technical Report RM-GTR-264. U.S. Dept. of Agriculture, Forest Service, Rocky Mountain Forest and Range Experiment Station, Fort Collins, Colorado.

Hawksworth, F. G. 1977. *The 6–class dwarf mistletoe rating system.* General Technical Report RM-48. U.S. Dept. of Agriculture, Forest Service, Fort Collins, Colorado.

Koot, H.P. 1991. *Spruce Aphid.* Forestry Canada, Forest Insect and Disease Survey, Pacific Forestry Centre, Forest Pest Leaflet No. 16, 4p.

Koot, H.P., and D.S. Ruth. 1971. *Spruce aphid in British Columbia.* Forest Insect and Disease Survey Pest Leaflet No. 16. Department of Fisheries and Forestry, Canadian Forestry Service, Victoria, British Columbia.

Koprowski, J.L., M.I. Alanen, and A.M. Lynch. 2005. Nowhere to run and nowhere to hide: Response of endemic Mt. Graham red squirrels to catastrophic forest damage. *Biological Conservation* 126:491–498.

Lampel, G. 1968. *Die Biologie des Blattlaus-Generationwechsels.* Gustav Fischer Verlag, Jena, Germany.

Lynch, A.M. 2003. Spruce aphid in high elevation habitats in the Southwest U.S. Pages 60–63 in S.L.C. Fosbroke and K.W. Gottschalk, eds. *Proceedings: U.S. Department of Agriculture Interagency Research Forum on Gypsy Moth and Other Invasive Species: January 15–18, 2002, Annapolis, Maryland.* General Technical Report NE-300. U.S. Dept. of Agriculture, Forest Service, Northeastern Research Station, Newton Square, Pennsylvania.

——. 2004. Fate and characteristics of *Picea* damaged by *Elatobium abietinum* (Walker) (Homoptera: Aphididae) in the White Mountains of Arizona. *Western North American Naturalist* 64:7–17.

Lynch, A. M., R. A. Fitzgibbon, and T. J. Rogers. In press. Observations on the life history of *Nepytia janetae* in spruce-fir and mixed-conifer forests in Arizona and New Mexico. *The Southwestern Entomologist.*

NOAA. 1990. Climatological data, Arizona. National and Oceanic and Atmospheric Administration, National Climate Data Center, 94. Asheville, North Carolina.

Ohnesorge, B. 1961. Wann sind Schäden durch die Sitkalaus zu erwarten? *Allg. Forstzeitschrift* 16:408–410.

Powell, W. 1974. Supercooling and low-temperature survival of the green spruce aphid *Elatobium abietinum. Annals of Applied Biology* 78:27–37.

Powell, W., and W. H. Parry. 1976. Effects of temperature on overwintering populations of the green spruce aphid *Elatobium abietinum. Annals of Applied Biology* 82:209–219.

Schmid, J. M., and R. H. Frye. 1976. *Stand ratings for spruce beetles.* Research Note RM-309. U.S. Dept. of Agriculture, Forest Service, Rocky Mountain Forest and Range Experiment Station, Fort Collins, Colorado.

——. 1977. *Spruce beetle in the Rockies.* General Technical Report RM-49. U.S. Dept. of Agriculture, Forest Service, Rocky Mountain Forest and Range Experiment Station, Fort Collins, Colorado.

Seaby, D. A., and D. J. Mowat. 1993. Growth changes in 20-year-old Sitka spruce *Picea sitchensis* after attack by the green spruce aphid *Elatobium abietinum. Forestry* 66:371–379.

Straw, N. A., G. Halldórsson, and T. Benedikz. 1998. Damage sustained by individual trees: Empirical studies on the impact of the green spruce aphid. Pages 15–31 in Day, K. R., G. Halldórsson, S. Harding, and N. A. Straw, eds. *The green spruce aphid in Western Europe: Ecology, status, impacts and prospects for management.* Forestry Commission Technical Paper 24. Forestry Commission, London.

Stuever, M., and J. Hayden, compilers. 1997. *Plant associations of Arizona and New Mexico: Forests,* vol. 1. U.S. Dept. of Agriculture, Forest Service, Southwestern Region, Albuquerque, New Mexico.

Thomas, R. C., and H. G. Miller. 1994. The interaction of green spruce aphid and fertilizer applications on the growth of Sitka spruce. *Forestry* 67:329–341.

Touchan, R., C. D. Allen, and T. W. Swetnam. 1996. Fire history and climatic patterns in ponderosa pine and mixed-conifer forests of the Jemez Mountains, northern New Mexico. Pages 33–46 in C. D. Allen, technical ed. *Fire effects in southwestern forests: Proceedings of the Second La Mesa Fire Symposium, 29–31 March 1994, Los Alamos, New Mexico.* General Technical Report RM-GTR-286. U.S. Dept. of Agriculture, Forest Service, Rocky Mountain Forest and Range Experiment Station, Fort Collins, Colorado.

U.S. Forest Service. 1997. *Forest insect and disease conditions in the Southwestern Region, 1996.* Report R3–97–1. Southwestern Region, Albuquerque, New Mexico.

——. 2001. *Forest insect and disease conditions in the Southwestern Region, 2000.* R3–00–1. Southwestern Region, Albuquerque, New Mexico.

von Scheller, H. D. 1963. Zur Biologie und Schadwirkung der Nadelholzspinnmilbe *Oligonychus ununguis* Jacobi und der Sitkafichtenlaus *Liosomaphis abietina* Walker (Hom. Aphid.). *Zeitschrift für Angewandte Entomologie* 51:258–284.

Warrington, S., and J. B. Whittaker. 1990. Interactions between Sitka spruce, the green spruce aphid, sulphur dioxide pollution, and drought. *Environmental Pollution* 65:363–370.

Zwolinski, M. J. 1996. Effects of fire on montane forest ecosystems. Pages 55–63 in P. F. Ffolliott, L. F. DeBano, M. B. Baker, G. J. Gottfried, G. Solis-Garza, C. B. Edminster, D. G. Neary, L. S. Allen, and R. H. Hamre, technical coordinators. *Effects of fire on Madrean Province ecosystems: A symposium proceedings.* General Technical Report RM-GTR-289. U.S. Dept. of Agriculture, Forest Service, Rocky Mountain Forest and Range Experiment Station, Fort Collins, Colorado.

CHAPTER 23

Nest-Site Characteristics of Sympatric Mt. Graham Red Squirrels and Abert's Squirrels in the Pinaleño Mountains

THOMAS E. MORRELL, ERIC A. POINT, AND JAMES C. DEVOS JR.

ABERT'S SQUIRRELS were introduced to the Pinaleño Mountains in the early 1940s by the Arizona Game and Fish Department (Davis and Brown 1989) and are now sympatric with Mt. Graham red squirrels (MGRS). The extent of competition between this introduced species and MGRS is not known, but may be an important factor in MGRS ecology (U.S. Fish and Wildlife Service 1993). Additionally, Abert's squirrels use forest types in the Pinaleño Mountains that are not presently occupied by MGRS, including ponderosa pine and pine-oak–dominated forests at lower elevations. However, where Abert's squirrels occur at higher elevations in the Pinaleño Mountains, ponderosa pine is not a dominant overstory species and therefore may be considered somewhat uncharacteristic Abert's squirrel habitat.

Nest structures are important to squirrels because they serve as nurseries, as shelter from inclement weather, as places for resting, and as protection from predators (Keith 1965; Farentinos 1972; Rothwell 1979; Wauters and Dhondt 1990; Koprowski 1991). Additionally, nests can provide insulation important in maintaining thermal regulation (Brown 1984).

Although specific nest-site habitat characteristics of Abert's squirrels and other red squirrel subspecies have been studied elsewhere in the United States (Ferner 1974; Rothwell 1979; Yahner 1980; Hall 1981; Vahle and Patton 1983; Halloram and Bekoff 1994), little is known about their nesting requirements in the Pinaleño Mountains (but see Smith and Mannan 1994 for midden characteristics). Moreover, the Mount

Graham Red Squirrel Recovery Plan (U.S. Fish and Wildlife Service 1993) does not report nest-site habitat characteristics of MGRS. Provided with this information, resource managers could develop and initiate forest management practices that enhance MGRS nesting habitat.

Our objectives were to quantitatively describe and compare nest-tree and nest-plot habitat characteristics of sympatric MGRS and Abert's squirrels in the Pinaleño Mountains.

Methods

We quantified nest-site habitat characteristics of MGRS and Abert's squirrels during June–October 1995 and 1996. Squirrel nests were located by initiating searches at known MGRS middens previously located during annual surveys (Arizona Game and Fish Department, Nongame Branch, Phoenix, unpublished data). Middens used in our study were randomly selected from all known midden sites. Mt. Graham red squirrels observed at or near middens were monitored until their nests were located. Randomly selected midden sites also were used as starting locations for Abert's squirrel nest searches. Starting at a midden, we moved in ever-increasing concentric circles until we located the nearest Abert's squirrel nest.

Active squirrel nests were verified by using a combination ≥2 of the following criteria: (1) a squirrel was observed entering a tree cavity or bolus nest and did not emerge for ≥30 minutes, (2) conifer needles remaining on twigs were observed in a bolus nest or protruding from a cavity, (3) a squirrel was observed carrying nesting material to a cavity, bolus nest, or hole, (4) a squirrel was observed <10 m from a nest on >1 occasion separated by at least two days, (5) evidence of recent feeding, such as teeth marks on fungi, scattered green cone scales, or cached green cones, was observed <10 m from a nest structure, or (6) juvenile squirrels were observed <10 m from a nest.

Information on nest-site habitat characteristics was collected at nest trees and within 10 m radius circular plots (0.03 ha) centered on the nest trees (nest plot) in each of two forest types that were classified by elevation and tree species composition (Stromberg and Patton 1991; Smith and Mannan 1994). Stands above 3,110 m were classified as spruce-fir forest, whereas stands below 3,110 m were classified as transition-zone forest.

Nest-Tree Characteristics

At each nest tree, we recorded nest type (bolus or cavity), tree species, diameter at breast height (dbh), tree height, nest height, percent nest height relative to tree height, nest aspect relative to the bole of the tree (bolus nests only), number of trees with crowns interlocking the nest tree's crown, distance to nearest opening in the forest canopy, and distance to nearest active MGRS midden. Nest trees were placed into one of four dbh size classes (10.0–30.2, 30.3–45.5, 45.6–61.0, and >61.1 cm dbh). Nest aspect measurements were placed into one of four categories: north (316°–45°), east (46°–135°), south (136°–225°), and west (226°–315°). We defined an opening in the forest canopy as a point under the canopy where a concave spherical densiometer reading was <50 percent, or where all four center squares of a concave spherical densiometer were devoid of cover (Smith and Mannan 1994).

We assigned snags to one of five decay categories (from Thomas et al. 1979, and Smith and Mannan 1994): (1) bole pointed with branches remaining and bark intact, (2) bole pointed with broken branches and patchy bark, (3) bole pointed with branches and bark absent, (4) top of bole broken with bark and branches absent, and (5) bottom of bole broken with bark and branches absent. We did not measure characteristics of nests occurring in holes in the ground.

Nest-Plot Characteristics

Variables measured within nest plots included percent slope, aspect, percent canopy cover, basal area (m^2/ha), log volume (m^3/ha), snag density occurring in two size classes (small = 10.0–45.5 cm dbh; large ≥ 45.6 cm dbh), live-tree density by species (trees >10 cm dbh/ha), and live-tree density occurring in five size classes (≤9.9, 10.0–30.2, 30.3–45.5, 45.6–61.0, and ≥ 61.1 cm dbh/ha). Nest-plot aspect was categorized using the same procedure as for nest aspect. Percent canopy cover was measured at four locations in each plot using a concave spherical densiometer (as modified by Strickler 1959) placed 5 m from the nest tree. The first location was determined using a random compass azimuth from the nest tree, and subsequent measurements were taken at 90° intervals. The mean of the four readings was used for analysis. Log volume within the nest plot was calculated by summing the cubic area of all logs ≥ 20 cm in diameter at the smallest end that occurred in the plot.

Statistical Analyses

We used principal components analysis (PCA) to explore covariance patterns in 24 habitat variables at and around squirrel nest sites in the entire study area. Components (axes) were evaluated if they met the broken-stick criterion (Jackson 1993). Observed eigenvalues that exceeded expected eigenvalues using the broken-stick distribution were considered meaningful and were retained for interpretation (McGarigal et al. 2000).

Kruskal-Wallis tests were used on continuous variables to test for nest-tree differences among MGRS bolus nests, MGRS cavity nests, and Abert's squirrel bolus nests in the spruce-fir and transition-zone forests. Mann-Whitney U tests were used to test for differences between squirrel species when Kruskal-Wallis tests were significantly different ($P < 0.005$). We also used Mann-Whitney U tests on continuous nest-plot variables to test for differences between squirrel species. G-tests of independence were used to test if squirrel species nest placement was independent of tree species, dbh size class, and aspect (Zar 1996). We used Rayleigh tests to determine if bolus-nest aspect and nest plots were randomly distributed for each squirrel species (Zar 1996). Log-likelihood ratio goodness-of-fit tests were used to determine if nest-plot aspect of both species of squirrel in each forest type occurred in proportions similar to those found at random sites. Proportions of random site aspects were generated from findings reported by Smith and Mannan (1994) who measured aspect at 58 random sites in the spruce-fir forest and 143 random sites in the transition-zone forest.

Because of distinct structural differences between live trees and snags, we did not pool MGRS nest types for analyses of nest-tree characteristics. Because MGRS cavity and bolus nests often occurred together within a nest plot, we pooled all MGRS nest types for plot analyses. However, only one nest within a plot was used for analyses (the first observed active nest). Because we conducted a large number of comparisons, we reduced the likelihood of committing a Type I error by using $P \leq 0.005$ to test for significance on all of our tests.

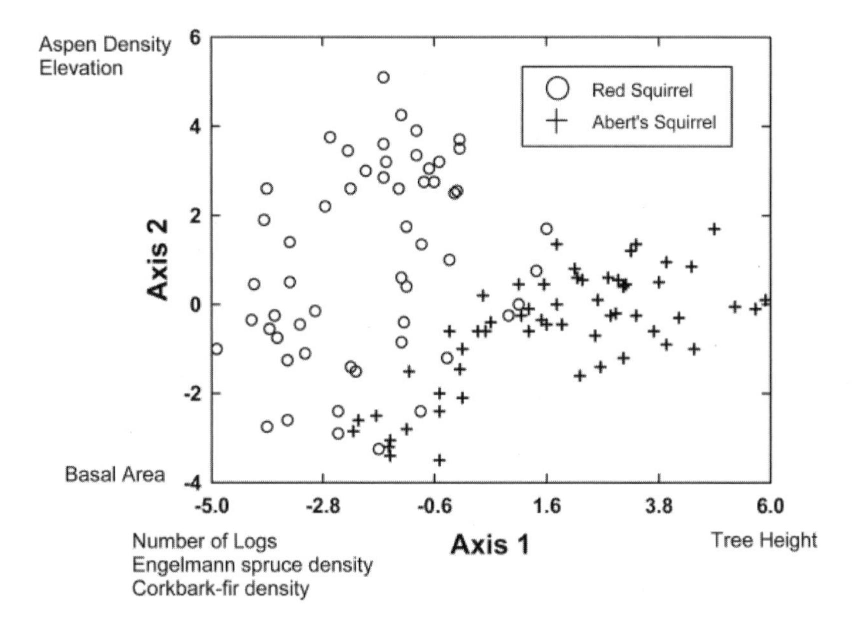

Figure 23.1. Ordination of 116 squirrel nests on the first two principal components (axes) obtained from PCA of 24 habitat variables in spruce-fir and transition forests, Pinaleño Mountains, Arizona, June to October 1995 and 1996. Variables that provided the highest and lowest component loadings are provided on their respective axes.

Results

We quantified nest-site habitat characteristics at 60 MGRS nests and 60 Abert's squirrel nests. Forty (67 percent) MGRS nests were located in cavities in trees, 16 (27 percent) were bolus nests, and 4 (7 percent) were ground nests. All Abert's squirrel nests were bolus nests. Principal components analysis of 24 habitat variables measured at 116 nest sites generated two components that met the broken-stick criterion and accounted for 38 percent of the total variation in the sample (fig. 23.1). Principal component analysis resulted in MGRS occurring largely on the left of ordination space and Abert's squirrels occurring on the lower right of ordination space. The first component explained 25 percent of the variation with the variables of number of logs, Engelmann spruce

Table 23.1. Mean (± SE) and range of bolus and cavity nest-tree characteristics of Mt. Graham red squirrels and Abert's squirrels in spruce-fir and transition-zone forests, Pinaleño Mountains, Arizona, from June to October 1995 and 1996

P-values were derived using Kruskal-Wallis tests. Means with different letters within rows indicate significant differences ($P < 0.005$) between Mt. Graham red squirrel bolus nests, cavity nests, and Abert's squirrel nests (Mann-Whitney *U* test).

	Mt. Graham Red Squirrel						Abert's squirrel			
	Bolus (n = 7, 9)[a]			Cavity (n = 10, 30)			Bolus (n = 13, 47)			
Variable	Mean	SE	Range	Mean	SE	Range	Mean	SE	Range	*P*
Spruce-Fir										
dbh (cm)	40.6	6.3	19.8–63.3	57.1	3.7	37.4–73.3	42.7	4.4	22.4–71.9	0.026
Nest height (m)	8.1A	0.7	6.0–10.3	8.1A	0.9	4.7–13.3	12.7B	0.7	9.7–18.0	<0.001
Nest-tree height (m)	17.0	1.8	9.0–21.0	13.9	2.1	6.7–26.9	17.1	1.3	11.0–28.8	0.230
% nest height	49.8A	4.5	29.5–68.8	63.6AB	5.6	35.2–87.8	76.4B	3.7	49.0–96.9	0.005
No. interlocking trees	3.9	0.6	1–6	2.2	0.5	0–6	4.2	0.5	0–7	0.032
Nearest opening (m)	5.0	0.9	2.2–7.8	6.9	1.3	1.0–13.2	5.9	0.7	2.0–10.1	0.579
Nearest midden (m)	6.6A	3.0	1–24	2.4A	1.4	0–12.2	63.7B	12.5	6.3–150	<0.001
Transition-Zone										
dbh (cm)	49.5	8.9	20.6–103.5	56.4	3.6	26.2–101.4	72.6	4.6	28.3–207.0	0.009
Nest height (m)	12.8AB	1.9	5.6–20.6	7.3A	0.6	2.0–16.6	15.4B	0.6	8.8–24.7	<0.001
Nest-tree height (m)	24.4A	2.6	13.6–39.6	15.8B	1.3	5.0–29.6	24.2A	0.8	11.9–36.7	<0.001
% nest height	52.5A	6.3	25.6–83.7	52.1A	4.4	12.8–100.0	65.9B	2.1	32.4–92.0	0.006
No. interlocking trees	4.9A	0.6	1–7	1.7B	0.2	0–3	4.4A	0.4	0–13	<0.001
Nearest opening (m)	10.1	3.7	1–38.0	7.4	1.1	1.0–28.0	7.2	0.8	0.3–20.0	0.891
Nearest midden (m)	18.6A	3.9	0–35.0	6.1B	1.5	0–31.0	212.6C	50.9	14.5–1,651	<0.001

[a]Sample sizes in the spruce-fir and transition-zone forests.

density, corkbark fir density, and tree height providing the highest loadings. The second component explained 13 percent of the variation with the variables of elevation, basal area, and aspen density contributing to the highest loadings.

Spruce-Fir Forest

Seventeen MGRS and 13 Abert's squirrel nests were located in spruce-fir. Seven (41 percent) MGRS nests were bolus nests and 10 (59 percent) were cavity nests. Both squirrel species placed bolus nests in three tree species, and the frequency of tree species used by MGRS and Abert's squirrels did not differ (bolus nests: $G = 3.131$, $df = 2$, $P = 0.209$; cavity nests: $G = 0.209$, $df = 2$, $P = 0.253$). Four of 7 (57 percent) MGRS bolus nests were observed in corkbark fir trees, whereas 9 of 13 (69 percent) Abert's squirrel nests were observed in Engelmann spruce trees. All MGRS cavity nests were found in Engelmann spruce trees. Except for 1 cavity nest in a live tree, all other MGRS cavity nests were located in snags. Fifty-five percent of the cavity nests we observed occurred in snags that were absent of limbs and bark and had a broken bole. All Abert's squirrel nests were constructed in live trees.

Three of 13 variables describing nest-tree characteristics differed between MGRS and Abert's squirrel bolus nests. MGRS bolus nests were closer to the ground, had lower percent nest height, and were closer to the nearest midden than Abert's squirrel nests (table 23.1). Six of 7 (86 percent) MGRS bolus nests were <10 m above ground. In contrast, 12 of 13 (92 percent) Abert's squirrel bolus nests were located >10 m above ground. Most MGRS bolus (71 percent) and cavity (80 percent) nests were within 5 m of the nearest midden; however, the closest Abert's squirrel nest occurred 6.3 m from a midden (table 23.1).

Frequency of MGRS and Abert's squirrel nests occurring in four dbh size class categories did not differ (bolus nests: $G = 0.806$, $df = 2$, $P = 0.848$; cavity nests: $G = 7.517$, $df = 2$, $P = 0.057$). Trees >61.1 cm dbh comprised 0.6 percent of all plot trees, yet 14 percent, 60 percent, and 31 percent of MGRS bolus, cavity, and Abert's squirrel nest trees were >61.1 cm dbh, respectively, and were the largest tree in the nest plot. Ninety percent of MGRS cavity nests occurred in trees >45.6 cm dbh. Frequency of MGRS and Abert's squirrel bolus nests occurring in four aspect categories did not differ, and bolus nests of both squirrel species were randomly distributed with respect to the bole (MGRS: $z = 1.149$, $P = 0.359$; Abert's: $z = 0.579$, $P = 0.594$). Nonetheless, 38 percent of Abert's squirrel nests had a northern aspect, whereas no MGRS bolus nests were found in this aspect category.

Total density of live trees was greater at MGRS nest plots than at Abert's squirrel nest plots (table 23.2). Engelmann spruce and corkbark fir trees were the most abundant plot trees for both squirrel species (table 23.3). In contrast, Douglas-fir trees occurred in only two (12 percent) MGRS nest plots and one (8 percent) Abert's squirrel nest plot, and southwestern white pine was not observed in MGRS nest plots and occurred in only one (8 percent) Abert's squirrel nest plot (table 23.3). Additionally, there was no difference between MGRS and Abert's squirrel nest plots occurring in four plot aspect categories ($G = 1.967$, df = 3, $P = 0.579$), and the aspects of nest plots of both squirrels were randomly distributed (MGRS: $z = 2.189$, $P = 0.174$; Abert's: $z = 0.138$, $P = 0.906$). Similarly, nest plots of both squirrel species were distributed by aspect in proportions similar to random sites.

Transition-Zone Forest

Forty-three MGRS and 47 Abert's squirrel nests were located in the transition-zone forest. Nine (21 percent) MGRS nests were bolus nests, 30 (70 percent) were cavity nests, and 4 (9 percent) were ground nests. Three MGRS ground nests were below ground and 1 was located in a downed aspen log. Both squirrel species nested in five tree species, and there was no difference in the frequency of tree species used for bolus nests ($G = 13.2$, df = 4, $P = 0.013$). However, MGRS cavity nest and Abert's squirrel bolus nest placement was dependent on tree species ($G = 55.8$, df = 5, $P \leq 0.001$). All Abert's squirrel nests were observed in live trees, mostly Douglas-fir (62 percent). Twenty-one of 30 (70 percent) MGRS cavity nests were observed in snags, of which 67 percent ($n = 14$) were aspen. MGRS cavity nests were also observed in Douglas-fir ($n = 4$), Engelmann spruce ($n = 2$), and southwestern white pine ($n = 1$) snags. Seventy-six percent ($n = 16$) of the snags used for cavity nests were absent of branches and bark, and had a broken bole. Ten (30 percent) MGRS cavity nests were constructed in living trees: 9 in aspen and 1 in a southwestern white pine.

Table 23.2. Mean (± SE) and range of nest plot habitat characteristics of Mt. Graham red squirrels and Abert's squirrels in spruce-fir forest, Pinaleño Mountains, Arizona, from June to October 1995 and 1996

	Mt. Graham red squirrel (n = 17)			Abert's squirrel (n = 13)				
Variable	Mean	SE	Range	Mean	SE	Range	Mann-Whitney U	P
% slope	19.2	2.0	8.0–35.0	16.5	1.8	8.0–29.0	127.5	0.475
% canopy cover	78.0	1.5	65.3–88.8	79.5	1.9	67.3–91.4	98.0	0.601
Basal area (m^2/ha)	64.5	5.3	31.7–127.0	53.3	3.8	32.5–80.1	135.0	0.144
No. small snags/ha (0.0–45.5 cm dbh)	205.8	41.3	0.0–700.0	191.7	40.0	0.0–500.0	104.0	0.929
No. large snags/ha ≥45.6 cm dbh)	31.3	8.3	0.0–133.3	36.1	11.9	0.0–100.0	99.0	0.888
Log volume (m^3/ha)	269.3	50.6	12.6–713.4	313.8	63.6	25.6–751.0	89.0	0.565
No. live trees/ha by dbh size class								
≤9.9 cm	2,133.3	276.3	900.0–5,399.9	1,315.4	193.2	500.0–2,733.3	170.5	0.012
10–30.2 cm	1,023.5	56.5	600.0–1,600.0	861.1	101.8	133.3–1,400.0	130.0	0.214
30.3–45.5 cm	156.8	23.3	33.3–333.3	177.8	25.4	66.7–333.3	89.0	0.562
45.6–61.0 cm	41.2	8.8	0.0–100.0	16.7	6.5	0.0–66.7	141.0	0.066
≥61.1 cm	19.6	7.6	0.0–100.0	11.1	4.7	0.0–33.3	110.0	0.673
Total no. live trees/ha	3,374.4	275.3	1,933.3–6,366.6	2,382.1	213.1	1,566.6–4,199.9	165.5	0.005

Table 23.3. Mean (± SE) and range of tree density (trees/ha) by species at Mt. Graham red squirrel and Abert's squirrel nest plots in spruce-fir and transition-zone forests, Pinaleño Mountains, Arizona, from June to October 1995 and 1996

	Mt. Graham red squirrel (n = 17, 43)[a]			Abert's squirrel (n = 13, 47)				
Tree species	Mean	SE	Range	Mean	SE	Range	Mann-Whitney U	P
Spruce-fir								
Corkbark fir	531.4	57.5	66.7–1,000.0	600.0	117.3	100.0–1,443.3	91.0	0.906
Douglas-fir	3.9	3.9	0.0–66.7	2.5	2.5	0.0–33.3	112.0	0.885
Engelmann spruce	694.1	59.2	166.7–1,133.3	500.0	108.7	33.3–1,400.0	147.0	0.046
Southwestern white pine	0.0	0.0	0.0	2.5	2.5	0.0–33.3	-	-
Transition-zone								
Aspen	110.1	20.2	0.0–533.3	50.3	14.1	0.0–433.3	1,391.5	0.001
Corkbark fir	368.2	55.0	0.0–1,166.7	145.4	32.7	0.0–900.0	1,442.5	<0.001
Douglas-fir	82.9	13.6	0.0–433.3	113.5	14.1	0.0–433.3	830.0	0.139
Engelmann spruce	190.7	30.4	0.0–833.3	87.9	20.5	0.0–666.7	1,445.0	<0.001
Locust (*Robinia* spp.)	0.7	0.7	0.0–33.3	4.9	3.6	0.0–166.7	969.0	0.348
Maple (*Acer* spp.)	16.3	9.0	0.0–333.3	8.5	5.0	0.0–166.7	1,043.0	0.614
Ponderosa pine	0.0	0.0	0.0	19.1	5.9	0.0–166.7	-	-
Southwestern white pine	24.0	6.1	0.0–133.0	146.1	31.6	0.0–1,066.7	621.0	0.001
White fir	62.8	20.7	0.0–666.7	90.8	23.5	0.0–933.3	840.0	0.138

[a]Sample sizes in the spruce-fir and transition-zone forests.

Table 23.4. Mean (± SE) and range of nest plot habitat characteristics of Mt. Graham red squirrels and Abert's squirrels in transition-zone forest, Pinaleño Mountains, Arizona from June to October 1995 and 1996.

	Mt. Graham red squirrel ($n = 43$)			Abert's squirrel ($n = 47$)				
Variable	Mean	SE	Range	Mean	SE	Range	Mann-Whitney *U*	*P*
% slope	23.2	1.6	5.0–44.0	26.3	1.8	4.0–55.0	888.5	0.324
% canopy cover	79.5	1.3	54.3–94.0	77.7	1.4	53.5–95.5	1,121.5	0.370
Basal area (m^2/ha)	78.7	5.1	26.2–181.1	61.9	3.4	25.6–115.6	1,335.5	0.010
No. small snags/ha (10.0–45.5 cm dbh)	199.9	23.5	0.0–666.7	157.4	18.8	0–533.3	1,117.5	0.184
No. large snags/ha ≥ 45.6 cm dbh)	12.4	4.3	0.0–166.7	17.0	3.6	0.0–100.0	887.0	0.228
Log volume (m^3/ha)	212.0	21.7	0.0–673.2	141.0	23.6	0.0–634.4	1,389.0	0.002
No. live trees/ha by dbh size class								
≤ 9.9 cm	1,237.9	177.7	95.5–5,400.0	1,100.0	162.1	66.7–7,000.0	1,048.0	0.762
10.0–30.2 cm	604.7	71.0	66.7–2,974.8	447.5	37.9	0.0–1,166.7	1,235.5	0.069
30.3–45.5 cm	197.7	16.8	33.3–500.0	118.4	10.2	0.0–333.3	1,452.5	<0.001
45.6–61.0 cm	68.9	9.2	0.0–266.7	48.9	7.8	0.0–266.7	1,225.0	0.076
≥ 61.1 cm	34.1	6.0	0.0–133.3	43.3	6.5	0.0–200.0	873.0	0.247
Total no. live trees/ha	**2,143.3**	**208.7**	**600.0–6,266.7**	**1,758.1**	**58.4**	**300.0–7,266.7**	**1,183.5**	**0.162**

Four variables describing nest-tree characteristics in the transition-zone forest differed between MGRS and Abert's squirrels. MGRS bolus and cavity nests had lower percent nest height and were closer to the nearest MGRS midden than were Abert's squirrel nests (table 23.1). Additionally, MGRS cavity nests were in shorter trees and had fewer interlocking branches than Abert's squirrel nest trees. Thirty-three percent and 80 percent of MGRS bolus and cavity nests were <10 m above the ground, respectively, whereas 91 percent of Abert's squirrel nests were >10 m above the ground. Although most (67 percent) MGRS cavity nests were <5 m from the nearest midden, only one of nine (11 percent) MGRS bolus nest and no Abert's nests were within 5 m of the nearest midden.

Frequency of MGRS and Abert's squirrel nests occurring in four dbh size class categories did not differ (bolus nests: $G = 6.93$, $df = 3$, $P = 0.074$; cavity nests: $G = 10.19$, $df = 3$, $P = 0.017$). However, 28 of 47 (59 percent) Abert's squirrel nests occurred in trees ≥ 61.1 cm dbh, and 47 percent of their nest trees were the largest-diameter tree in the nest plot. In contrast, 20 percent and 23 percent of MGRS bolus and cavity nest trees were the largest-diameter tree within the nest plot. Only 2 percent of all plot trees in the transition-zone forest were ≥ 61.1 cm dbh. Frequency of MGRS and Abert's squirrel bolus nests occurring in four aspect categories did not differ ($G = 2.491$, $df = 3$, $P = 0.476$), and bolus nests of both squirrel species were randomly distributed with respect to the bole of the tree (MGRS: $z = 2.626$, $P = 0.070$; Abert's: $z = 3.547$, $P = 0.317$).

Six MGRS nest plot variables differed from Abert's squirrel nest plots in the transition-zone forest. MGRS nest plots had greater log volume and density of live trees in the size class 30.3–45.5 cm dbh than Abert's squirrel nest plots (table 23.4). Additionally, significantly greater tree densities of aspen, corkbark fir, and Engelmann spruce trees occurred in MGRS nest plots than in Abert's squirrel nest plots (table 23.3), whereas the density of southwestern white pine was greater in Abert's squirrel nest plots than MGRS nest plots.

Frequency of nest plots occurring in four aspect categories did not differ between squirrel species ($G = 5.222$, $df = 3$, $P = 0.156$). However, whereas the aspects of Abert's squirrel nest plots were randomly distributed ($z = 2.751$, $P = 0.486$), MGRS nest plots were not random in their distribution ($z = 7.180$, $P < 0.001$). Only 7 percent ($n = 3$) of MGRS nest plots had a south facing slope. Mean aspect of MGRS nest plots was 357°.

Plots of both squirrel species were distributed by aspect in proportions similar to random sites.

Discussion

Mt. Graham red squirrel nests were located lower in the nest tree, had lower percent nest height, and were closer to middens than Abert's squirrel nests. Lower nest placement close to middens by MGRS may facilitate quick midden access for food storage or acquisition. Nest placement near middens would also facilitate defense of this habitat component and enable MGRS squirrels to quickly return to their nests if predators are nearby. In contrast, because Abert's squirrels typically feed in the upper canopy where they consume the inner bark of the terminal branches (States and Wettstein 1998) and seeds where cones are produced (Brown 1984), it may be more efficient for them to place nests relatively higher. Nest height placement differences between the squirrel species may represent a form of habitat partitioning in the Pinaleño Mountains that reduces interspecific competition for nest sites.

Red squirrel investigations conducted elsewhere have reported that bolus nests were more commonly used than cavity nests, and in some studies, cavities were not used by red squirrels (Klugh 1927; Vahle 1978). Thus, the Pinaleño Mountains are somewhat unique with respect to the frequency of cavity use by red squirrels. However, in New York, Layne (1954) found that cavity nests were used more than bolus nests. In Minnesota, Sun (1989) reported that females nursing young were located in cavity nests more often than bolus nests. Where available, cavities may be preferred by MGRS because cavity nests (1) are more likely to be resistant to the effects of temperature (Froehlich and Smith 1990), precipitation, and wind (Wauters and Dhondt 1990; Edwards and Guynn 1995), (2) require less maintenance and last longer than bolus nests (Wauters and Dhondt 1990), (3) can be used for cone and mushroom storage (Froehlich and Smith 1990), (4) offer greater protection from predators (Sun 1989; Froehlich and Smith 1990), and (5) may be used as nurseries (Klugh 1927; Hamilton 1939; Sun 1989).

We believe the extensive use of cavity nests by MGRS also may be related to the age of the forest. Smith and Mannan (1994) obtained core samples from MGRS midden sites and reported that more than 75 percent of the dominant trees in the spruce-fir forest were 150–300 years old, and 61 percent of the dominant trees in the transition-zone forest were >150 years old. Similarly, Stromberg and Patton (1991) determined that both the spruce-fir and transition-zone forests contained trees >230 years old. Therefore, in some areas on the Pinaleño Mountains, trees have probably had more time to develop fungal infections and/or be used by primary cavity-nesting birds than in managed forests that have fewer or no old-growth characteristics. We observed proportionately more MGRS cavity nests in the transition-zone forest than in the spruce-fir forest, possibly because of aspen occurrence in this forest type. Aspen are fast-growing, short-lived trees that can reach large sizes (Preston 1976), and they are susceptible to trunk rot from fungus infection (DeByle and Winokur 1985).

Seventeen of 30 snags (57 percent) used by MGRS occurred in snags without branches or bark and broken tops. Snags with these characteristics may have been used more because they have reached a point of decay that promotes heartwood rot. Additionally, broken top snags also provide increased protection from some ground predators because the nest often is higher than in snags where the bottom of the bole has broken.

Only one nest plot variable, total live-tree density, differed between squirrel species in the spruce-fir forest, with greater densities occurring in MGRS nest plots. Nest plots of both species were dominated by corkbark fir and Engelmann spruce. Although not statistically significant, Abert's squirrel nest plots had somewhat lower densities of Engelmann spruce and higher densities of corkbark fir than MGRS nest plots.

Because red squirrels often forage on the ground and retrieve cones from middens buried by snow, log volume may be an important habitat component for MGRS by providing travel pathways. Additionally, downed logs provide a medium for growth of fungi, a red squirrel food source (U.S. Fish and Wildlife Service 1993). Mean log volumes at MGRS nest plots in our study (spruce-fir = 269.3 m^3/ha; transition-zone = 212.0 m^3/ha) were considerably less than mean log volumes reported by Smith and Mannan (1994) at spruce-fir (359.9 m^3/ha) and transition-zone forest midden sites (331.3 m^3/ha). Mt. Graham red squirrel middens are often placed on or adjacent to logs, perhaps enhancing storage qualities of a midden site (Smith and Mannan 1994), and may be more important for MGRS midden placement than nest-site selection. Nonetheless, we found mean log volume at MGRS nest sites to be considerably higher than at random sites described by Smith and Mannan (1994) who reported mean log volumes of 112.1 m^3/ha and 145.7 m^3/ha in the spruce-fir and transition-zone forests, respectively. This suggests that logs also are important components of MGRS nest sites.

Mt. Graham red squirrels in the transition-zone forest selected nest sites in areas with greater tree density (although only one size class was statistically different), consisting of smaller trees, whereas Abert's squirrels selected nest sites in slightly more open areas consisting of fewer but larger trees. Smith and Mannan (1994) suggested that MGRS midden sites were located in areas with unusually thick foliage, dense canopies, and high stem density, in part because these areas provided cool moist microclimates that might preserve the integrity of stored cones.

In the transition-zone forest, MGRS nest plots had higher densities of aspen, corkbark fir, and Engelmann spruce trees than Abert's squirrel nest plots. In contrast, Abert's squirrel nest plots had a higher density of southwestern white pine than MGRS nest plots. Greater aspen densities at MGRS nest sites are likely related to the availability and selection of nesting cavities in aspen stands. Differences in tree species densities in nest plots also may be related to the forage requirements/preferences of the squirrels. Abert's squirrels are dependent on pine trees for over-winter survival (e.g., Rasmussen 1941; Keith 1965; Patton 1975; Brown 1984), and in both forest types, they selected nest sites in areas with higher southwestern white pine densities. Additionally, they selected sites with higher ponderosa pine densities than MGRS sites in the transition-zone forest.

Both squirrel species often used large-diameter (>45.6 cm dbh) trees for nesting in the spruce-fir and transition-zone forests. Larger trees may indicate a greater foliage volume that may provide more protection from the effects of wind and sun (Holbo and Childs 1987). Moreover, larger snags may provide increased food-storage space and greater insulation from the elements than smaller snags.

We likely underestimated the use of ground nests by MGRS because nests are cryptic (Mayfield 1948), and squirrels may have avoided using these when observers were present. On two occasions, we observed a MGRS using a ground nest in addition to a cavity nest.

To what extent habitat partitioning exists as a function of interspecific interactions between MGRS and Abert's squirrels is unknown and was beyond the scope of this study. However, habitat partitioning in dimensions similar to those reported here (differences in nest type, nest height, and forest stand structure) has been reported as a mechanism that facilitates sympatry in fox and gray squirrel populations (Edwards and Guynn 1995; Edwards et al. 1998). Additionally, observed variation in some habitat variables, such as density of Engelmann spruce, corkbark fir, aspen, number of logs, and basal area were important in separating the two squirrel species along ecological gradients in ordination space. Just how often and to what extent territorial MGRS exclude Abert's squirrels from suitable nest sites is unclear, and the density and distribution of Abert's squirrels in the Pinaleño Mountains are not currently known. We recommend that future research in the Pinaleño Mountains investigate these questions.

Management Implications

Habitat suitability for the MGRS depends on the ability of the forest to produce suitable nesting trees and reliable and adequate conifer cone crops for food, as well as microclimate conditions suitable for storage of closed cones (U.S. Fish and Wildlife Service 1993). Additionally, because of the importance of cavities and snags as nest substrates for MGRS, forest stands on the Pinaleño Mountains should be managed to provide an abundance of old and standing dead trees. Specifically, managing for the recruitment, growth, and decadence of Engelmann spruce trees in the spruce-fir forest, and aspen in the transition-zone forest, may increase the potential for suitable MGRS nest sites. We recommend that snag use and availability be investigated on the Pinaleño Mountains to better determine if there is a need for such management.

The importance of decadent downed woody debris in providing squirrels runways was noted in our study, and its importance in enhancing the growth of dietary fungi, thereby influencing red squirrel distribution, has been noted in the Pacific Northwest (Carey et al. 1999). Thus, practices such as snag removal and firewood log removal should not be permitted in areas where MGRS might occur.

Other studies have found lower densities of red squirrels in forest stands that have been thinned compared to areas with unthinned stands

(Medin 1986; Sullivan and Moses 1986; Sullivan et al. 1996). Moreover, Smith and Mannan (1994) found MGRS middens in stands with greater foliage density and tree density than random sites. In our study, we observed MGRS nest sites in areas that had overall greater tree densities in the spruce-fir forest and in stands with greater tree densities in the dbh size-class of 30.3–45.5 cm than at Abert's nest sites. Thus, we recommend that stands where MGRS currently exist, as well as adjacent areas, be managed to provide high tree densities, closed-canopy forest stands, and large-diameter trees and snags.

Acknowledgments

We thank Paul Young and Reed Sanderson, the University of Arizona, for their advice and support during the course of this project. We are indebted to Genice Froehich, U.S. Forest Service, who provided logistical support. The University of Arizona's Mt. Graham Red Squirrel Monitoring Team supplied important information on red squirrel nest locations, and we are indebted. We would also like to thank Melissa Kreighbaum and Carl Russworm, Arizona Game and Fish Department, for their assistance. Stan Boutin provided helpful editorial comments.

Literature Cited

Brown, D. E. 1984. *Arizona's tree squirrels.* Arizona Game and Fish Department, Phoenix.

Carey, A. B., J. Kershner, B. Biswell, and L. Dominguez de Toledo. 1999. Ecological scale and forest development: Squirrels, dietary fungi, and vascular plants in managed and unmanaged forests. *Wildlife Monographs* 142:1–71.

Davis, R., and D. E. Brown. 1989. Documentation of the transplanting of Abert's squirrels. *Southwestern Naturalist* 33:490–492.

DeByle, N. V., and R. P. Winokur. 1985. *Aspen: Ecology and management in the western United States.* General Technical Report RM-119. U.S. Dept. of Agriculture, Forest Service, Rocky Mountain Forest and Range Experiment Station, Fort Collins, Colorado.

Edwards, J., and D. C. Guynn Jr. 1995. Nest characteristics of sympatric populations of fox and gray squirrels. *Journal of Wildlife Management* 59:103–110.

Edwards, J., D. G. Heckel, and D. C. Guynn Jr. 1998. Niche overlap in sympatric populations of fox and gray squirrels. *Journal of Wildlife Management* 62:354–363.

Farentinos, R. C. 1972. Nest of the tassel-eared squirrel. *Journal of Mammalogy* 53:902–903.

Ferner, J. W. 1974. Habitat relationships of *Tamiasciurus hudsonicus* and *Sciurus aberti* in the Rocky Mountains. *Southwestern Naturalist* 18:470–473.

Froehlich, G. F., and N. S. Smith. 1990. *Natural history and reproductive performance of the Mount Graham red squirrel.* Final report. Arizona Cooperative Fish and Wildlife Research Unit, Tucson.

Hall, J. G. 1981. A field study of the Kaibab squirrel in Grand Canyon National Park. *Wildlife Monographs* 75:6–54.

Halloram, M., and M. Bekoff. 1994. Nesting behavior of Abert's squirrels (*Sciurus aberti*). *Ethology* 97:236–248.

Hamilton, W. J., Jr. 1939. Observations on the life history of the red squirrel in New York. *American Midland Naturalist* 22:732–745.

Holbo, H. R., and S. W. Childs. 1987. Summertime radiation balances of clearcut and shelterwood slopes in southwest Oregon. *Forest Science* 33:504–516.

Jackson, D. A. 1993. Stopping rules in PCA: A comparison of heuristical and statistical approaches. *Ecology* 74:2205–2214.

Keith, J. O. 1965. The Abert's squirrel and its dependence on ponderosa pine. *Ecology* 46:150–163.

Klugh, A. B. 1927. Ecology of the red squirrel. *Journal of Mammalogy* 8:1–32.

Koprowski, J. L. 1991. The evolution of sociality in tree squirrels: The comparative behavioral ecology of fox squirrels and eastern gray squirrels. PhD diss., University of Kansas, Lawrence.

Layne, J. N. 1954. The biology of the red squirrel, *Tamiasciurus hudsonicus loquax* (Bangs), in central New York. *Ecological Monographs* 24:227–267.

Mayfield, H. 1948. Red squirrel nesting on the ground. *Journal of Mammalogy* 24:186.

McGarigal, K., S. Cushman, and S. Stafford. 2000. *Multivariate statistics for wildlife and ecology research.* Springer-Verlag, New York.

Medin, D. E. 1986. The impact of logging on red squirrels in an Idaho conifer forest. *Western Journal of Applied Forestry* 1:73–76.

Patton, D. R. 1975. *Abert's squirrel cover requirements in southwestern ponderosa pine.* Research Paper RM-145. U.S. Dept. of Agriculture, Forest Service, Rocky Mountain Forest and Range Experiment Station, Fort Collins, Colorado.

Preston, R. J. 1976. *North American trees.* The Iowa State University Press, Ames.

Rasmussen, D. I. 1941. Biotic communities of the Kaibab Plateau, Arizona. *Ecological Monographs* 11:229–275.

Rothwell, R. 1979. Nest sites of red squirrels (*Tamiasciurus hudsonicus*) in the Laramie Range of southeastern Wyoming. *Journal of Mammalogy* 60:404–405.

Smith, A. A., and R. W. Mannan. 1994. Distinguishing characteristics of Mount Graham red squirrel midden sites. *Journal of Wildlife Management* 58:437–445.

States, J. S., and P. J. Wettstein. 1998. Food habits and evolutionary relationships of the tassel-eared squirrel (*Sciurus aberti*). Pages 185–194 in M. A. Steele, J. F. Merritt, and D. A. Zegers, eds. *Ecology and evolutionary biology of tree squirrels.* Special Publication No. 6. Virginia Museum of Natural History.

Strickler, G. S. 1959. *Use of the densiometer to estimate density of forest canopy on permanent sample plots.* Research Note PNW-180. U.S. Dept. of Agriculture, Forest Service, Pacific Northwest Forest and Range Experiment Station, Portland, Oregon.

Stromberg, J. C., and D. T. Patton. 1991. Dynamics of the spruce-fir forests on the Pinaleño Mountains, Graham Co., Arizona. *Southwest Naturalist* 36:37–48.

Sullivan, T. P., W. Klenner, and P. K. Diggle. 1996. Response of red squirrels and feeding damage to variable stand density in young lodgepole pine forest. *Ecological Applications* 6:1124–1134.

Sullivan, T. P., and R. A. Moses. 1986. Red squirrel populations in natural and managed stands of lodgepole pine. *Journal of Wildlife Management* 50:595–601.

Sun, C. 1989. Young dispersal and the seasonal changes of territorial behavior in red squirrels (*Tamiasciurus hudsonicus*). Master's thesis, University of Wisconsin, Madison.

Thomas, J. W., R. G. Anderson, C. Maser, and E. L. Bull. 1979. Snags. Pages 60–70 in J. W. Thomas, ed. *Wildlife habitats in managed forests: The Blue Mountains of Oregon and Washington.* U.S. Forest Service. Agriculture Handbook No. 533. U.S. Government Printing Office, Washington, D.C.

U.S. Fish and Wildlife Service. 1993. *Mount Graham Red Squirrel* (Tamiasciurus hudsonicus grahamensis) *Recovery Plan*. Albuquerque, New Mexico.

Vahle, J. R. 1978. Red squirrel use of southwestern mixed coniferous habitat. Master's thesis, Arizona State University, Tempe.

Vahle, J. R., and D. R. Patton. 1983. Red squirrel cover requirements in Arizona mixed conifer forests. *Journal of Forestry* 81:14–15.

Wauters, L. A., and A. A. Dhondt. 1990. Nest-use by red squirrels. *Mammalia* 54: 377–389.

Yahner, R. H. 1980. Burrow system used by red squirrels. *American Midland Naturalist* 103:409–411.

Zar, J. H. 1996. *Biostatistical analysis.* Prentice-Hall Inc., Englewood Cliffs, New Jersey.

CHAPTER 24

Introduced Abert's Squirrels in the Pinaleño Mountains

A Review of Their Natural History and Potential Impacts on the Red Squirrel

ANDREW J. EDELMAN AND JOHN L. KOPROWSKI

THE ACCIDENTAL AND INTENTIONAL INTRODUCTION of non-native species into ecosystems can have substantial negative impacts on endemic biodiversity. Successfully introduced species can negatively affect native species directly through resource competition and predation, and indirectly through habitat alteration, hybridization, and spread of diseases and parasites (Manchester and Bullock 2000). In the sky islands of the southwestern United States, one of the most successfully translocated species is the Abert's squirrel (*Sciurus aberti*). This tree squirrel was successfully introduced to 10 montane areas in Arizona and New Mexico for hunting purposes and has subsequently colonized 3 additional sites (Davis and Brown 1988, 1989). Prior to translocations, Abert's squirrels were not present in these areas, due to local extinctions or lack of dispersal corridors (Brown 1984; Davis and Brown 1989). Several sky islands already contained native tree squirrel species prior to Abert's squirrel introductions (Hoffmeister 1986). Introduced Abert's squirrels have been implicated in the decline of the Arizona gray squirrel (*Sciurus arizonensis catalinae*) in the Santa Catalina Mountains and the Mt. Graham red squirrel in the Pinaleño Mountains of Arizona (Lange 1960; Minckley 1968).

The Mt. Graham red squirrel is a federally endangered subspecies (U.S. Fish and Wildlife Service 1993). Abert's squirrels are believed to affect Mt. Graham red squirrels through resource competition, but the exact mechanism is unknown (Spicer 1985). Most research on Abert's squirrels has been conducted on populations in ponderosa pine forests that lack syntopic tree squirrel species (Nash and Seaman 1977; Brown

1984). In the Pinaleño Mountains, introduced Abert's squirrels are syntopic with Mt. Graham red squirrels in mixed-conifer and spruce-fir forests (Hutton et al. 2003). The potential impacts of introduced Abert's squirrels on endangered Mt. Graham red squirrels are largely unstudied. To promote and direct future research and management of Abert's squirrels in the Pinaleño Mountains, we review in this chapter Abert's squirrel ecology and history in the Pinaleño Mountains, possible mechanisms that allow syntopy with Mt. Graham red squirrels, and potential impacts on Mt. Graham red squirrels.

Ecology of Abert's Squirrels

Six subspecies of Abert's squirrel naturally occur in montane areas (elev. 1,800–3,000 m) of Arizona, Colorado, New Mexico, Utah, and Wyoming, in the United States and Chihuahua and Durango in Mexico (Nash and Seaman 1977). This large-bodied squirrel (600–900 g) is also called the tassel-eared squirrel, due to its large ears that are adorned with tufts during winter (Steele and Koprowski 2001). Typically, Abert's squirrels have a white venter, slate-gray dorsum with a russet patch on the back, white eye ring, and a gray bushy tail fringed with white (fig. 24.1). Partial or complete melanic forms commonly occur in populations from the Mogollon Plateau of Arizona and New Mexico to the northern extent of the species (Brown 1984).

Abert's squirrels are typically found in forests dominated by ponderosa pine, which has contributed to their depiction as ponderosa pine obligates. Many studies have documented the Abert's squirrel's specialized use of ponderosa pine for food, cover, and nests (Keith 1965; Pederson et al. 1976; Hall 1981; Snyder 1993; Halloran and Bekoff 1994; States and Wettstein 1998). The extent of the Abert's squirrel's dependence on ponderosa pine has been compared to highly specialized mammalian herbivores, including the koala (*Phascolarctos cinereus*) and giant panda (*Ailuropoda melanoleuca*) (Murphy and Linhart 1999). Anecdotal information, however, suggests that the Abert's squirrel's use of ponderosa pine may be more facultative than obligate. Individuals have been observed in oak-dominated woodland (Baker and Greer 1962), piñon-pine woodland (Reynolds 1966), mixed-conifer forest (Findley et al. 1975; Hall 1981; Polechla 2005), spruce-fir forest (J. Allen 1895; Pedersen et al.

Figure 24.1. Adult Abert's squirrel from the Pinaleño Mountains, Arizona. Photograph by J. L. Koprowski.

1976), and alpine tundra (Cooper 1987). Our review of Abert's squirrel diet and nesting ecology also indicates more plasticity than is generally acknowledged.

Diet of Abert's squirrels varies throughout the year. Food is rarely cached (Keith 1965), unlike most other North American tree squirrels (Steele and Koprowski 2001). Hypogeous fungi are the dominant food in summer and fall (Stephenson 1975; Kotter 1981). Apical buds and inner bark of ponderosa pine are eaten extensively in winter (Farentinos et al. 1981). Use of inner bark with low amounts of digestible carbohydrates, proteins, and fats (Snyder 1992) during winter is probably due to low levels of other seasonally abundant foods (e.g., fungi, seeds, and acorns) and restricted access to ground foraging due to snow cover (Stephenson 1975; Kotter 1981). Staminate cones and seeds of ponderosa pines and acorns of Gambel oak (*Quercus gambelii*) are eaten when available (Stephenson 1975; Forbes 1997). Abert's squirrels are also known to eat seeds and inner bark from piñon pine and Douglas-fir as well as Utah juniper (*Juniperus osteosperma*) seeds and dwarf mistletoe berries and tissue (Reynolds 1966; Ratcliff et al. 1975; Stephenson 1975; Pederson et al. 1987; Soderquist 1987). Epigeous fungi, grasses, forbs, berries, insects, ponderosa pine needles, and bone are consumed in small amounts (Pederson et al. 1976; Stephenson 1975; Kotter 1981).

Spherical nests constructed by Abert's squirrels, known as dreys or bolus nests, typically are composed of twigs and are located close to the trunk in the upper canopy of ponderosa pines (Farentinos 1972a; Halloran and Bekoff 1994). Dreys can also occur in piñon pine (Hoffmeister 1971). Cavity nests are occasionally found in ponderosa pine, Gambel oak, quaking aspen, and Fremont cottonwood (*Populus fremontii*) (Keith 1965; Patton and Green 1970; Brown 1984; Halloran and Bekoff 1994). Artificial nest boxes are used when available (Pederson et al. 1978). Nest trees are larger than the majority of surrounding trees, have branches interdigitated with several other trees, and are located in closed stands (Halloran and Bekoff 1994).

Abert's Squirrels in the Pinaleño Mountains

Between September 1941 and May 1943, 69 Abert's squirrels (38 male, 31 female) were translocated to the Pinaleño Mountains by the Arizona

Game and Fish Department from the Fort Valley Experimental Forest near Flagstaff, Arizona (Davis and Brown 1988). The source location indicates that the subspecies introduced was likely *S. a. aberti*, which inhabits the south rim of the Grand Canyon and Mogollon Plateau (Hoffmeister 1986). There is no evidence that Abert's squirrels naturally existed in the Pinaleño Mountains before the 1940s introductions. After introduction, Abert's squirrels were first officially reported in mixed-conifer forests in 1952 (Hoffmeister 1956) and spruce-fir forests in 1967 (Minckley 1968). Both Hoffmeister (1956) and Minckley (1968) reported that Abert's squirrels were abundant throughout the Pinaleño Mountains, whereas red squirrels were scarce. Both authors also speculated that because Abert's squirrels occupied the same habitat as red squirrels, they were causing the red squirrel decline through resource competition. Mixed-conifer forests in the southwestern United States (elev. 2,500–3,050 m) contain varying numbers of Engelmann spruce, corkbark fir, southwestern white pine, Douglas-fir, white fir, ponderosa pine, and quaking aspen. Spruce-fir forests (elev. 3,050–3,250 m) consist solely of Engelmann spruce and corkbark fir (Stromberg and Patten 1991). Abert's squirrels were also observed to use Gambel oaks in riparian areas at lower elevations in the Pinaleño Mountains (Brown 1984).

Although no formal attempts at estimating Abert's squirrel densities in the Pinaleño Mountains have been conducted, nest surveys (Morrell et al. in this vol.) and published observations (Hutton et al. 2003) indicate that Abert's squirrels are common throughout the mixed-conifer and spruce-fir forests. From 1989 to 2002, the University of Arizona's Mt. Graham Red Squirrel Monitoring Program opportunistically observed Abert's squirrels 324 times in mixed-conifer forests and 174 times in spruce-fir forests of the Pinaleño Mountains during 41,000 field hours (Hutton et al. 2003). Over half of the Abert's squirrels observed in both forest types were within 30 m of a red squirrel midden. Squirrels engaged in normal behaviors such as feeding, foraging, and building nests. Juveniles and reproductive adults were observed in both forest types, and two mating chases were observed. Within the study areas, ponderosa pine accounted for only 0.01 percent and 0 percent of tree compositions in mixed-conifer and spruce-fir forests, respectively (Hutton et al. 2003).

During the course of a radiotelemetry study, we captured 47 individual Abert's squirrels (table 24.1) in the mixed-conifer forest of the Pinaleño

Table 24.1. Age class and reproductive condition of marked male and female Abert's squirrels

Live-trapped from September 2001 to July 2003 in mixed-conifer forests of the Pinaleño Mountains, Arizona

		Male		Female	
Age Class	N	*n*	Descended Testes	*n*	Signs of Lactation
Adult	35	17	14	18	14
Young of the year	12	4	2	8	-

Mountains (same study area as Hutton et al. 2003) from September 2001 to July 2003 (A. J. Edelman and J. L. Koprowski, unpublished data; for methods see Edelman 2004). More adults than juveniles were captured (ratio = 2.92:1, $\chi^2 = 11.23$, df = 1, $P = 0.0008$). We detected no difference between number of female and male adult squirrels (ratio = 1.06:1, $\chi^2 = 0.03$, df = 1, $P = 0.866$) or juveniles captured (ratio = 2:1, $\chi^2 = 1.33$, df = 1, $P = 0.248$). About 78 percent of adult females showed signs of recent or past lactation (i.e., elongated and pigmented teats) during at least one capture. Among males, 82 percent of adults and 50 percent of juveniles had descended testes during at least one capture (table 24.1). Mating chases involving marked squirrels were recorded in or around the study area; one mating chase was observed in May 2002 and seven in May 2003 (A. J. Edelman, personal observation). Four of five radio-collared adult females produced litters during 2003. Upon juvenile emergence from the natal nest, all four females had three offspring each (A. J. Edelman, personal observation).

We measured 129 drey trees used by Abert's squirrels in the mixed-conifer forest of the Pinaleño Mountains. About 2 percent of dreys were found in ponderosa pine, whereas most were found in Douglas-fir, Engelmann spruce, corkbark fir, and southwestern white pine (Edelman and Koprowski 2005a). Only 5 percent of conifer food items were taken from ponderosa pine during our observations of Abert's squirrels in the mixed-conifer forest of the Pinaleño Mountains (Edelman and Koprowski 2005b). Douglas-fir and southwestern white pine were heavily utilized for seeds and buds during the spring and summer, and Engelmann spruce inner bark was fed on during winter (Edelman and Koprowski 2005a). Home ranges of Abert's squirrels were smaller in mixed-conifer forests of the Pinaleño Mountains than ponderosa pine forests, possibly due to higher food resource abundance in the former (Edelman and Koprowski 2006b).

From the large number of observations and live-trapping records, it is apparent that since shortly after introduction, Abert's squirrels in mixed-conifer and spruce-fir forests of the Pinaleño Mountains have existed in close proximity to Mt. Graham red squirrels. We still do not know what mechanisms allow a presumed obligate of ponderosa pine to survive in these forests types that contain extremely low levels of ponderosa pine.

Mechanisms of Abert's and Red Squirrel Syntopy

The absence of Abert's squirrels in other forest types is given as evidence for the dependence of this species on ponderosa pine (Davis and Brown 1989). However, the use of mixed-conifer and spruce-fir forests by the population in the Pinaleño Mountains since shortly after being translocated does not support the Abert's squirrel's strict dependence on ponderosa pine. We suggest that the Abert's squirrel's fundamental niche is likely broader than previously believed. Below, we list several alternative explanations why Abert's and red squirrels are only syntopic in the Pinaleño Mountains. If possible, we have provided evidence for or against each explanation.

Paucity of Range Data

Abert's squirrels may occupy mixed-conifer and spruce-fir forests in their natural range, but have not been documented due to a lack of intensive surveys. As reviewed above, there are occasional observations within their natural range of individual Abert's squirrels occurring in non–ponderosa pine forest. Mixed-conifer and spruce-fir forests are denser than ponderosa pine forests, which could hinder detection of Abert's squirrels. However, it seems unlikely that Abert's squirrels would have gone undetected in the more heavily recreated areas (e.g., San Francisco Mountains) that contain mixed-conifer and spruce-fir forests. In the Pinaleño Mountains, Abert's squirrels are secretive, but are regularly seen crossing roads or vocalizing.

Red Squirrel Decline

Abert's squirrels may be competitively excluded from mixed-conifer and spruce-fir forests where healthy populations of red squirrels are present. The decline of the Mt. Graham red squirrel may have allowed Abert's squirrels to invade mixed-conifer and spruce-fir forests in the Pinaleño Mountains.

Character Displacement

Populations of Abert's and red squirrels that naturally co-occur may have evolved character displacement (behavioral, ecological, and morphological) to reduce interspecific competition. Mt. Graham red squirrels have been isolated from other tree squirrel species for at least 10,000 years (Lomolino et al. 1989). Due to a relaxation of competition, adaptations (e.g., territorial or cone-storing behavior) that could allow red squirrels to exclude or outcompete Abert's squirrels may be reduced (Minckley 1968). Mt. Graham red squirrels are reported to be more quiet and secretive than red squirrels in the White Mountains of Arizona, where Abert's squirrels naturally occur (Hoffmeister 1956). This may indicate that Mt. Graham red squirrels exhibit reduced territorial defense.

Milder Winters

Extended or harsh winter conditions in northern areas of Abert's squirrels' range could prevent colonization of mixed-conifer and spruce-fir forests. The Pinaleño Mountains may have milder winters due to their southerly location, allowing Abert's squirrels to colonize and survive in mixed-conifer and spruce-fir forests. Prolonged periods of snow cover are correlated with increased mortality in Abert's squirrels (Stephenson and Brown 1980; Dodd et al. 2003). The absence of Abert's squirrels from ponderosa pine forests north of southern Wyoming and the Sierra

Nevada may be due to higher snowfall in these areas. However, lack of Abert's squirrels in these areas could be due to low dispersal rates or barriers (Davis and Brown 1989).

Low Density

High densities of red squirrels may prevent Abert's squirrels from establishing home ranges in mixed-conifer and spruce-fir elevation forests. The low density of red squirrels in the Pinaleño Mountains could allow Abert's squirrels to become established in mixed-conifer and spruce-fir forests. Densities of red squirrels and middens in the Pinaleño Mountains are approximately four times lower than the closest population in the White Mountains of Arizona (J. L. Koprowski, unpublished data).

Source/Sink

Viable populations of Abert's squirrels only exist in ponderosa pine forests. Observations of Abert's squirrels outside of ponderosa pine forests represent non-viable sink populations that consist of dispersing individuals from source populations. The strong elevational gradients in the Pinaleño Mountains decrease the distance between ponderosa pine forests and higher-elevation mixed-conifer and spruce-fir forests (Brown 1982). The close proximity of forest types allows Abert's squirrels from ponderosa pine forests to disperse farther into mixed-conifer and spruce-fir forests than normally occurs in their natural range. A source/sink relationship in the Pinaleño Mountains is unlikely because the ponderosa pine forests are poorly developed and not extensive or continuous in distribution, probably due to steep slopes at elevations (1,830–2,745 m) where ponderosa pines are found (Johnson 1988). In addition, Abert's squirrels in mixed-conifer and spruce-fir forests appear to be resident animals that reproduce (Hutton et al. 2003).

These hypotheses are not mutually exclusive, and several processes could be interacting to allow Abert's squirrels to invade mixed-conifer and spruce-fir forests in the Pinaleño Mountains. Understanding why Abert's squirrels occupy mixed-conifer and spruce-fir forests in the Pinaleño Mountains will assist in making appropriate management decisions about this species.

Potential Impacts of Abert's Squirrels on Mt. Graham Red Squirrels

Hoffmeister (1956) believed that Mt. Graham red squirrels were more common prior to introduction of Abert's squirrels, and several biologists have suggested that Abert's squirrels might competitively exclude Mt. Graham red squirrels from some habitats (Minckley 1968; Brown 1984; Hoffmeister 1986). A list of possible direct and indirect impacts of Abert's squirrels on red squirrels includes interference or exploitation competition (food, nest sites, space); raiding of middens; modification of seed crops and forest structure; misidentification by hunters; disease transmission; and demographic influences. Below, we review these potential impacts, including supporting evidence and recommended research directions.

Dietary overlap occurs between Abert's squirrels and Mt. Graham red squirrels; both species rely heavily on conifer cones and fungi as food sources (Nash and Seaman 1977; Steele 1998). In the Pinaleño Mountains, Abert's squirrels have been observed feeding on epigeous and hypogeous fungi and cones of Douglas-fir, corkbark fir, Engelmann spruce, southwestern white pine, and ponderosa pine (Hutton et al. 2003; Edelman and Koprowski 2005b), all of which are also eaten by Mt. Graham red squirrels (Froehlich 1990). Based on what little is known, it is possible that Abert's squirrels may limit the number of conifer cones or fungi available for red squirrels. In ponderosa pine forests, Abert's squirrels can reduce the cone crop an average of 20 percent (0.3–74.7 percent) a year through cone eating (Larson and Schubert 1970). On two sites in Arizona, Abert's squirrels ate 41 percent and 97 percent of available ponderosa pine cones (Lema 2001). Cone crop levels could also affect the impact of Abert's squirrel seed feeding. In years of cone crop failures, Abert's squirrels may remove a large proportion of available cones; however, in mast years, their effect on cone resources may be minimal. In addition, in years of poor cone crops, Abert's squirrels might have a survival advantage over red squirrels in being able to subsist on inner bark. The prolonged drought that has affected Arizona over the past decade may have further favored Abert's squirrels with reduced snowfall and decreased cone crops upon which red squirrels depend. An analysis of stomach contents during autumn found that fungi made up 60 percent of Abert's squirrels' diet and only 30 percent of red squirrels' diet (States and Wettstein 1998). The heavy use of fungi by Abert's squirrels could reduce the amount of fungi available for Mt. Graham red squirrel use. The impact of Abert's squirrels feeding on fungi resources may be most detrimental during drought years when fungi levels decrease due to lack of moisture (Kotter 1981). A multi-year study documenting Abert's squirrels' impact on food resources in mixed-conifer and spruce-fir forests would be valuable in determining their effect on Mt. Graham red squirrels.

Squirrels use nests for resting, avoiding predators, cover during inclement weather, and raising of young (Young et al. 2002). If suitable nest sites for Mt. Graham red squirrels are limited, then survival and reproduction could be affected. Interspecific overlap in nest-site selection between introduced Abert's and red squirrels could reduce the number of potential nest sites for either species. Nest surveys in mixed-conifer and spruce-fir forests of the Pinaleño Mountains found moderate overlap in nest-tree and nest-site characteristics between these species (Morrell et al. in this vol.). Differences between species were most prominent in average nest-tree size and tree density. Nest types found also differed between species, with red squirrels using drey, cavity, and ground nests and Abert's squirrels using only dreys (Morrell et al. in this vol.); however, recent radio tracking of Abert's squirrels in mixed-conifer forests has documented use of cavity nests (Edelman and Koprowski 2006a). Currently, we are comparing nest-site characteristics between Abert's and red squirrels to assess the potential for nest competition. Future research needs to address whether nest sites are a limiting resource for Mt. Graham red squirrels.

Space is a critical resource that squirrels require to find food, nest sites, cache sites, and mates (Gurnell 1987). Abert's and red squirrels could coexist through interspecific partitioning of space. In Britain and Italy, however, no spatial partitioning occurs between syntopic introduced eastern gray squirrels and native Eurasian red squirrels (Wauters et al. 2000; Wauters, Gurnell et al. 2002). North American red squirrels differ from these *Sciurus* species in their highly territorial behavior within western coniferous forests (Steele 1998). Abert's squirrels are non-territorial and have overlapping home ranges (Farentinos 1979; Halloran and Bekoff 2000). This difference in behavior may allow red squirrels to exclude Abert's squirrels from midden areas. Anecdotal information supports the idea of behavioral dominance of red squirrels over Abert's squirrels despite the disparity in body size. In the Pinaleño Mountains, red squirrels were observed chasing Abert's squirrels 74 times, while only once were Abert's squirrels seen chasing red squirrels (Hutton et al. 2003). Currently, we are radio tracking syntopic Abert's and Mt. Graham red squirrels to examine whether interspecific spatial partitioning occurs. Even if these two species are partitioning space, a removal study will be necessary to determine if red squirrels would use areas previously occupied by Abert's squirrels. Although adult red squirrels may be able to dominate Abert's squirrels, juveniles could be detrimentally affected by Abert's squirrels. In England, Eurasian red squirrels that have a high core-area overlap with eastern gray squirrels have lower juvenile recruitment rates (Wauters et al. 2000). Juvenile Mt. Graham red squirrels do not establish a territory until after dispersal, and they disperse greater distances than other populations (Kreighbaum and Van Pelt 1996). The lack of a residence, paucity of experience, increased dispersal distance, or other factors could make young red squirrels less dominant in interspecific agonistic encounters. Future studies could focus on the effect of Abert's squirrels on dispersal of juvenile Mt. Graham red squirrels.

The conspicuousness structure of red squirrel middens may facilitate cone stealing. Abert's squirrels steal cones from red squirrel middens, but the frequency of this behavior is unknown (Ferner 1974; Hall 1981; Hutton et al. 2003). In a small study using marked cones and remote cameras, we observed no Abert's squirrels raiding Mt. Graham red squirrel middens (Edelman et al. 2005). Red squirrels will chase Abert's squirrels away from middens (Hutton et al. 2003), indicating that territorial behavior may reduce cone stealing and that red squirrels must expend energy towards interspecific territorial defense. Introduced eastern gray squirrels will steal seed caches of Eurasian red squirrels, possibly resulting in reduced energy intake for the native squirrel (Wauters, Tosi, and Gurnell 2002). If Abert's squirrels are frequently raiding middens, then Mt. Graham red squirrels may also experience reduced energy intake. Even if red squirrels successfully defend middens against Abert's squirrel incursions, there may still be detrimental effects. Increased territorial defense could lead to greater energetic costs, less foraging time, and more exposure to predators. Future research should quantify how frequently Abert's squirrels raid middens and how much time Mt. Graham red squirrels spend defending territories from Abert's squirrels.

Abert's squirrels may also indirectly affect red squirrels by altering forest productivity through herbivory. From late fall to early spring, Abert's squirrels clip branches from trees to obtain inner bark. Ponderosa pine trees used by Abert's squirrels for inner bark have lower photosynthetic surface areas (15 percent) and growth rates (Soderquist 1987; Allred and Gaud 1994) than non-used trees. Abert's squirrels can also reduce the cone crop of a ponderosa pine forest by 55 percent over two years through the excision of shoots and cones (Allred et al. 1994). Abert's squirrels in mixed-conifer forests of the Pinaleño Mountains feed on inner bark of mainly Engelmann spruce and, to a lesser extent, southwestern white pine, ponderosa pine, and corkbark fir (Edelman and Koprowski 2005b). This herbivory may reduce future cone crop levels, which in turn would decrease the number of cones available for red squirrels. Researchers could assess the Abert's squirrel's impact on mixed-conifer and spruce-fir forests by comparing productivity of trees used by Abert's squirrels for inner bark with trees that are not eaten.

Hunting of Mt. Graham red squirrels has been prohibited since 1986 (U.S. Forest Service 1988); however, hunting of Abert's squirrels is still permitted in the Pinaleño Mountains. Hunting is legal from October–November (Arizona Game and Fish Department 2003) when tree squirrels have winter coats. Mt. Graham red squirrels in winter have a grayish cast to their sides (J. L. Koprowski, personal observation), which may resemble Abert's squirrels to novice hunters and even experienced hunters in low-light conditions. Currently, no published data for Abert's or red squirrel harvest are available for the Pinaleño Mountains. Check stations and hunting surveys would provide an estimate of Abert's and red squirrels killed each year in the Pinaleño Mountains. These data would allow wildlife managers to determine the potential impact of hunting on red squirrels. If deemed necessary, an elevation restriction for hunting (e.g., under 2,280 m) could be instituted to reduce the chances of the accidental killing of red squirrels.

Introduced Abert's squirrels could serve as vectors for transmission of diseases to Mt. Graham red squirrels. In the United Kingdom, introduced eastern gray squirrels are vectors for a parapoxvirus that is lethal to native Eurasian red squirrels. This disease may be responsible for the widespread replacement of Eurasian red squirrels by eastern gray squirrels (Tompkins et al. 2003). There is no evidence to suggest that diseases have contributed to the decline of Mt. Graham red squirrels. With the high level of human incursion into the Pinaleño Mountains, there is the possibility that novel diseases could be introduced in the future. Interspecific transmission of diseases from Abert's to red squirrels could occur directly during interactions between species. Abert's squirrels may also act as a vector for diseases to move from lower elevations to higher elevations where red squirrels could be infected. Parasite-borne illness could pass indirectly through parasites that are common to both species (Hill and Duszynski 1986; Patrick and Wilson 1995). Continual monitoring of Mt. Graham red squirrels would allow managers to quickly act in the event of a disease outbreak.

Lastly, Abert's squirrels could negatively affect red squirrel reproduction and survival through several mechanisms. Abert's squirrels could kill red squirrel young or interfere with mating. Cannibalism and infanticide has been occasionally observed in eastern gray and fox squirrels (D. Allen 1943; Thompson 1976). Abert's squirrels have never been observed to attack adult or juvenile red squirrels, although only intense observation would discount Abert's squirrels as a possible source of mortality. Fox squirrels have been documented in eastern gray squirrel mating chases (Koprowski 1991). No interspecific involvement in mating chases has been observed in the 20 Mt. Graham red squirrel (Munroe et al. in this vol.) and 10 Abert's squirrel mating chases recorded. Abert's squirrels could reduce red squirrel survival by increasing predator densities. The large-bodied Abert's squirrel is a main food item for northern goshawks during the winter months when many migratory avian prey species are absent (Drennan and Beier 2003). Northern goshawks are common in the Pinaleño Mountains, and attempted predation of Mt. Graham red squirrels by this species has been observed (Schauffert et al. 2002). Other species present in the Pinaleño Mountains that are known or potential predators of both Abert's and red squirrels include great-horned owl (*Bubo virginianus*), red-tailed hawk (*Buteo jamaicensis*), Cooper's hawk (*Accipiter cooperii*), gray fox (*Urocyon cinereoargenteus*), and bobcat (*Lynx rufus*) (Farentinos 1972b; Hall 1981; Schauffert et al. 2002). The additional prey source provided by Abert's squirrels might increase predator survival, reproduction, and density, leading to higher predation rates for red squirrels. Conversely, Abert's squirrels could decrease per capita predation on red squirrels by serving as an alternate food source. Further study of mortality patterns of Mt. Graham red squirrels and Abert's squirrels are needed.

Conclusions

The potential impact of Abert's squirrels on Mt. Graham red squirrels has largely been ignored. As we have discussed above, Abert's squirrels could directly and indirectly affect Mt. Graham red squirrels in numerous ways. We believe the most probable mechanisms for Abert's squirrels to negatively impact red squirrels are directly through resource competition for food and space or indirectly by increasing predator density. However, the effects of interspecific interactions are often difficult to determine (Schoener 1983). Further studies on Abert's and red squirrels in the Pinaleño Mountains and comparisons with other populations may assist in elucidating the relationship between these two species. Currently, we are examining nest- and space-use overlap between Abert's and Mt. Graham red squirrels. Small-scale removal of Abert's squirrels may also assist in determining whether red squirrels are negatively impacted by this introduced species. Given the current precarious status of the Mt. Graham red squirrel and continued loss of habitat from fire (King and Koprowski in this vol.) and insect outbreaks (Koprowski et al. 2005), it would be shortsighted to ignore the interactions between these two species.

Literature Cited

Allen, D. L. 1943. *Michigan fox squirrel management.* Game Division Publication 100. Department of Conservation, Lansing, Michigan.

Allen, J. A. 1895. On a collection of mammals from Arizona and Mexico, made by Mr. W. W. Price, with field notes by the collector. *Bulletin of the American Museum of Natural History* 7:193–258.

Allred, W. S., and W. S. Gaud. 1994. Effects of Abert squirrel herbivory on foliage and nitrogen losses in ponderosa pine. *Southwestern Naturalist* 39:350–353.

Allred, W. S., W. S. Gaud, and J. S. States. 1994. Effects of herbivory by Abert squirrels (*Sciurus aberti*) on cone crops of ponderosa pine. *Journal of Mammalogy* 75:700–703.

Arizona Game and Fish Department. 2003. *2003–2004 Arizona hunting regulations.* Arizona Game and Fish Department, Phoenix.

Baker, R. H., and J. K. Greer. 1962. Mammals of the Mexican state of Durango. *Publications of the Museum, Michigan State University, Biological Series* 2:25–154.

Brown, D. E., ed. 1982. Biotic communities of the American Southwest: United States and Mexico. *Desert Plants* 4:1–342.

——. 1984. *Arizona's tree squirrels*. Arizona Game and Fish Department, Phoenix.

Cooper, D. J. 1987. Abert's squirrel above treeline on the San Francisco Peaks, Arizona. *Southwestern Naturalist* 32:507.

Davis, R., and D. E. Brown. 1988. Documentation of the transplanting of Abert's squirrels. *Southwestern Naturalist* 33:490–492.

——. 1989. Role of post-Pleistocene dispersal in determining the modern distribution of Abert's squirrel. *Great Basin Naturalist* 49:425–434.

Dodd, N. L., J. S. States, and S. S. Rosenstock. 2003. Tassel-eared squirrel population, habitat condition, and dietary relationships in north-central Arizona. *Journal of Wildlife Management* 67:622–633.

Drennan, J. E., and P. Beier. 2003. Forest structure and prey abundance in winter habitat of northern goshawks. *Journal of Wildlife Management* 67:177–185.

Edelman, A. J. 2004. The ecology of an introduced population of Abert's squirrels in a mixed-conifer forest. Master's thesis, University of Arizona, Tucson.

Edelman, A. J., and J. L. Koprowski. 2005a. Selection of drey sites by Abert's squirrels in an introduced population. *Journal of Mammalogy* 86:1220–1226.

——. 2005b. Diet and tree use of Abert's squirrels (*Sciurus aberti*) in a mixed-conifer forest. *Southwestern Naturalist* 50:461–465.

——. 2006a. Characteristics of Abert's squirrel (*Sciurus aberti*) cavity nests. *Southwestern Naturalist* 51:64–70.

——. 2006b. Seasonal changes in home ranges of Abert's squirrels: Impact of mating season. *Canadian Journal of Zoology* 84:404–411.

Edelman, A. J., J. L. Koprowski, and J. L. Edelman. 2005. Kleptoparasitic behavior and species richness at Mt. Graham red squirrel middens. Pages 395–398 in G. J. Gottfried, B. S. Gebow, L. G. Eskew, and C. Edminster, compilers. *Connecting mountain islands and desert seas: Biodiversity and management of the Madrean Archipelago II*. Proceedings RMRS-P-36. U.S. Dept. of Agriculture, Forest Service, Rocky Mountain Research Station, Fort Collins, Colorado.

Farentinos, R. C. 1972a. Nests of the tassel-eared squirrel. *Journal of Mammalogy* 53:900–903.

——. 1972b. Observations on the ecology of the tassel-eared squirrel. *Journal of Wildlife Management* 36:1234–1239.

——. 1979. Seasonal changes in home range size of tassel-eared squirrels (*Sciurus aberti*). *Southwestern Naturalist* 24:49–62.

Farentinos, R. C., P. J. Capretta, R. E. Kepner, and V. M. Littlefield. 1981. Selective herbivory in tassel-eared squirrels: Role of monoterpenes in ponderosa pines chosen as feeding trees. *Science* 213:1273–1275.

Ferner, J. W. 1974. Habitat relationships of *Tamiasciurus hudsonicus* and *Sciurus aberti* in the Rocky Mountains. *Southwestern Naturalist* 18:470–473.

Findley, J. S., A. H. Harris, D. E. Wilson, and C. Jones. 1975. *Mammals of New Mexico*. University of New Mexico Press, Albuquerque.

Forbes, R. B. 1997. Subnivean foraging by Abert's squirrels. Pages 287–290 in T. L. Yates, W. L. Gannon, and D. E. Wilson, eds. *Life among the muses: Papers in honor of James*

S. *Findley*. University of New Mexico, Museum of Southwestern Biology Special Publication 3, Albuquerque.

Froehlich, G. F. 1990. Habitat use and life history of the Mount Graham red squirrel. Master's thesis, University of Arizona, Tucson.

Gurnell, J. 1987. *The natural history of squirrels*. Facts on File, New York.

Hall, J. G. 1981. A field study of the Kaibab squirrel in Grand Canyon National Park. *Wildlife Monographs* 75:1–54.

Halloran, M. E., and M. Bekoff. 1994. Nesting behavior of Abert squirrels (*Sciurus aberti*). *Ethology* 97:236–248.

——. 2000. Home range use by Abert squirrels: A comparative analysis. *Southwestern Naturalist* 45:253–257.

Hill, T. P., and D. W. Duszynski. 1986. Coccidia (Apicomplexa: Eimeriidae) from Sciurid rodents (*Eutamias, Sciurus, Tamiasciurus* spp.) from the western United States and northern Mexico with descriptions of two new species. *Journal of Protozoology* 33:282–288.

Hoffmeister, D. F. 1956. Mammals of the Graham (Pinaleno) Mountains, Arizona. *American Midland Naturalist* 55:257–288.

——. 1971. *Mammals of Grand Canyon*. University of Illinois Press, Chicago.

——. 1986. *Mammals of Arizona*. University of Arizona Press, Tucson.

Hutton, K. A., J. L. Koprowski, V. L. Greer, M. I. Alanen, C. A. Schauffert, and P. J. Young. 2003. Use of mixed-conifer and spruce-fir forests by an introduced population of Abert's squirrels (*Sciurus aberti*). *Southwestern Naturalist* 48:257–260.

Johnson, W. T. 1988. Flora of the Pinaleno Mountains, Graham County, Arizona. *Desert Plants* 8:147–191.

Keith, J. O. 1965. The Abert squirrel and its dependence on ponderosa pine. *Ecology* 46:150–163.

Koprowski, J. L. 1991. Mixed-species mating chases of fox squirrels, *Sciurus niger*, and eastern gray squirrels, *S. carolinensis*. *Canadian Field-Naturalist* 105:117–118.

Koprowski, J. L., M. I. Alanen, and A. M. Lynch. 2005. Nowhere to run and nowhere to hide: Response of endemic Mt. Graham red squirrels to catastrophic forest damage. *Biological Conservation* 126:491–498.

Kotter, M. M. 1981. Interrelationships of tassel-eared squirrels, ponderosa pine and hypogeous mycorrhizal fungi. Master's thesis, Ohio State University, Columbus.

Kreighbaum, M. E., and W. E. Van Pelt. 1996. *Mount Graham red squirrel juvenile dispersal telemetry study*. Nongame and Endangered Wildlife Program Technical Report 89. Arizona Game and Fish Department, Phoenix.

Lange, K. I. 1960. Mammals of the Santa Catalina Mountains, Arizona. *American Midland Naturalist* 64:436–458.

Larson, M. M., and G. H. Schubert. 1970. *Cone crops of ponderosa pine in central Arizona including the influence of Abert squirrels*. Research Paper RM-58. U.S. Dept. of Agriculture, Forest Service, Rocky Mountain Forest and Range Experiment Station, Fort Collins, Colorado.

Lema, M. F. 2001. Dynamics of Abert squirrel populations: Home range, seasonal movements, survivorship, habitat use, and sociality. Master's thesis, Northern Arizona University, Flagstaff.

Lomolino, M. V., J. H. Brown, and R. Davis. 1989. Island biogeography of montane forest mammals in the American Southwest. *Ecology* 70:180–194.

Manchester, S. J., and J. M. Bullock. 2000. The impacts of non-native species on UK biodiversity and the effectiveness of control. *Journal of Applied Ecology* 37:845–864.

Minckley, W. L. 1968. Possible extirpation of the spruce squirrel from the Pinaleno (Graham) Mountains, south-central Arizona. *Journal of the Arizona Academy of Science* 5:110.

Murphy, S. M., and Y. B. Linhart. 1999. Comparative morphology of the gastrointestinal tract in the feeding specialist *Sciurus aberti* and several generalist congeners. *Journal of Mammalogy* 80:1325–1330.

Nash, D. J., and R. N. Seaman. 1977. *Sciurus aberti. Mammalian Species* 80:1–5.

Patrick, M. J., and W. D. Wilson. 1995. Parasites of the Abert's squirrel (*Sciurus aberti*) and red squirrel (*Tamiasciurus hudsonicus*) of New Mexico. *Journal of Parasitology* 81:321–324.

Patton, D. R., and W. Green. 1970. Abert's squirrels prefer mature ponderosa pine. Research Note RM-169. U.S. Dept. of Agriculture, Forest Service, Rocky Mountain Forest and Range Experiment Station, Fort Collins, Colorado.

Pederson, J. C., R. C. Farentinos, and V. M. Littlefield. 1987. Effects of logging on habitat quality and feeding patterns of Abert squirrels. *Great Basin Naturalist* 47:252–258.

Pederson, J. C., R. N. Hasenyager, and A. W. Heggen. 1976. *Habitat requirements of the Abert squirrel* (Sciurus aberti navajo) *on the Monticello District, Manti-La Sal National Forest of Utah*. Utah State Division of Wildlife Resources Publication 76–9, Salt Lake City.

——. 1978. Use of artificial nest boxes by Abert's squirrels. *Southwestern Naturalist* 23:700–702.

Polechla, P. J., Jr. 2005. Mammals. Pages 169–195 in R. Julyan and M. Stuever, eds. *Field guide to the Sandia Mountains*. University of New Mexico Press, Albuquerque.

Ratcliff, T. D., D. R. Patton, and P. F. Ffolliott. 1975. Ponderosa pine basal area and the Kaibab squirrel. *Journal of Forestry* 73:284–286.

Reynolds, H. G. 1966. Abert's squirrels feeding on pinyon pine. *Journal of Mammalogy* 47:550–551.

Schauffert, C. A., J. L. Koprowski, V. L. Greer, M. I. Alanen, K. A. Hutton, and P. J. Young. 2002. Interactions between predators and Mt. Graham red squirrels (*Tamiasciurus hudsonicus grahamensis*). *Southwestern Naturalist* 47:498–501.

Schoener, T. W. 1983. Field experiments on interspecific competition. *American Naturalist* 122:240–285.

Snyder, M. A. 1992. Selective herbivory by Abert's squirrel mediated by chemical variability in ponderosa pine. *Ecology* 73:1730–1741.

——. 1993. Interactions between Abert's squirrel and ponderosa pine: The relationship between selective herbivory and host plant fitness. *American Naturalist* 141:866–879.

Soderquist, T. R. 1987. The impact of tassel-eared squirrel defoliation on ecotonal ponderosa pine. *Journal of Mammalogy* 68:398–401.

Spicer, R. B. 1985. *Status of the Mount Graham red squirrel,* Tamiasciurus hudsonicus grahamensis *(Allen), of southeastern Arizona.* U.S. Fish and Wildlife Service, Albuquerque, New Mexico.

States, J. S., and P. J. Wettstein. 1998. Food habits and evolutionary relationships of the tassel-eared squirrel (*Sciurus aberti*). Pages 185–194 in M. A. Steele, J. F. Merritt, and D. A. Zegers, eds. *Ecology and evolutionary biology of tree squirrels.* Virginia Museum of Natural History Special Publication 6, Martinsville.

Steele, M. A. 1998. *Tamiasciurus hudsonicus. Mammalian Species* 586:1–9.

Steele, M. A., and J. L. Koprowski. 2001. *North American tree squirrels.* Smithsonian Institution Press, Washington, D.C.

Stephenson, R. L. 1975. Reproductive biology and food habits of Abert's squirrels in central Arizona. Master's thesis, Arizona State University, Tempe.

Stephenson, R. L., and D. E. Brown. 1980. Snow cover as a factor influencing mortality of Abert's squirrels. *Journal of Wildlife Management* 44:951–955.

Stromberg, J. C., and D. T. Patten. 1991. Dynamics of the spruce-fir forests on the Pinaleño Mountains, Graham Co., Arizona. *Southwestern Naturalist* 36:37–48.

Thompson, D. C. 1976. Accidental mortality and cannibalization of a nestling gray squirrel. *Canadian Field-Naturalist* 90:52–53.

Tompkins, D. M., A. R. White, and M. Boots. 2003. Ecological replacement of native red squirrels by invasive greys driven by disease. *Ecology Letters* 6:189–196.

U.S. Fish and Wildlife Service. 1993. *Mount Graham Red Squirrel* (Tamiasciurus hudsonicus grahamensis) *Recovery Plan.* Albuquerque, New Mexico.

U.S. Forest Service. 1988. *Mount Graham red squirrel: An expanded biological assessment.* Safford Ranger District, Coronado National Forest, Safford, Arizona.

Wauters, L. A., J. Gurnell, A. Martinoli, and G. Tosi. 2002. Interspecific competition between native Eurasian red squirrels and alien grey squirrels: Does resource partitioning occur? *Behavioral Ecology and Sociobiology* 52:332–341.

Wauters, L. A., P. W. W. Lurz, and J. Gurnell. 2000. Interspecific effects of grey squirrels (*Sciurus carolinensis*) on the space use and population demography of red squirrels (*Sciurus vulgaris*) in conifer plantations. *Ecological Research* 15:271–284.

Wauters, L. A., G. Tosi, and J. Gurnell. 2002. Interspecific competition in tree squirrels: Do introduced grey squirrels (*Sciurus carolinensis*) deplete tree seeds hoarded by red squirrels (*S. vulgaris*)? *Behavioral Ecology and Sociobiology* 51:360–367.

Young, P. J., V. L. Greer, and S. K. Six. 2002. Characteristics of bolus nests of red squirrels in the Pinaleño and White Mountains of Arizona. *Southwestern Naturalist* 47: 267–275.

CHAPTER 25

A Risk Assessment of Multiple Impacts on the Endangered Mt. Graham Red Squirrel

ELAINE K. HARDING, DANIEL F. DOAK, CYNTHIA HARTWAY, TERRY MOORE FREDERICK, AND GENICE FROEHLICH

THE MT. GRAHAM RED SQUIRREL (MGRS) is a classic rare and endemic species with both a small range and a limited total population size (Rabinowitz 1981). As such, the continued existence of the MGRS is subject to a variety of potential threats, including habitat destruction and modification from road and astronomical observatory building, forest fires (or the lack thereof), erratic food supply, climatic warming, forest management practices, potential competition with introduced Abert's squirrels, and simple demographic stochasticity. Concern for the MGRS arose when extensive developments associated with astronomy facilities were proposed and then built within its limited habitat (Pennisi 1989). While the well-definable threats anticipated from this direct habitat destruction spurred interest in the species, the consequences of less-delineated and predictable impacts, as well as natural threats to viability, must also be considered in any analysis of the subspecies' chances of survival and in the development of a safe management strategy for the squirrel and its habitat.

In particular, temporal variation in habitat quality can have drastic effects on the demography and population sizes of red squirrels (Kemp and Keith 1970; Klenner and Krebs 1991; Young 1995). Two environmental forces are likely to be especially important for the MGRS: the relatively short-term influence of conifer masting on food availability and the longer-term effects of fires. From 1986 to 1998, the total MGRS population has varied between approximately 161 and 462 individuals (Snow in this vol.). Much of this variation in numbers has likely been driven by the masting of the primary cone-producing trees of the squirrel's habitats, at least at the higher elevations (Young 1995; Young et al. 2000).

Fires can result in rapid but long-lasting changes in habitat quality for red squirrels (Hatten in this vol.) that require relatively high densities of older trees that have greater cone production for maximum habitat quality (Stromberg and Patten 1995). The conifer forests of the Pinaleños are naturally fire-adapted communities, and thus have been dramatically altered by a century of fire suppression and grazing (U.S. Forest Service 1988). The results of these land management practices have been to limit the spread of low-intensity fires, leading to an average fire interval of approximately 100 years in two of the MGRS's major habitats, mixed-conifer and ecotone forest (Grissino-Mayer and Fritts 1995). In contrast, the mean fire interval for these forests during the period 1584 to 1880 was 4.2 years (Grissino-Mayer et al. 1994, 1995). These fires were of low intensity, often burning fairly large areas of forest (Grissino-Mayer et al. 1994). Thus, fire was an extremely common occurrence, which may have served to keep the mixed-conifer stands more open (Grissino-Mayer et al. 1995), while promoting higher numbers of fire-tolerant species such as ponderosa pine and Douglas-fir (Wright and Bailey 1982). In contrast, the third major MGRS habitat, spruce-fir forest, has a much longer natural fire interval of 300–400 years, with the last unsuppressed fire occurring about 300 years ago. This catastrophic fire burned 800 ha and killed many Engelmann spruce and a lower proportion of Douglas-fir. Since then, corkbark fir has become increasingly common in the spruce-fir habitat (Grissino-Mayer et al. 1994). More recently, the Clark Peak fire of 1996 burned approximately 2,500 ha of spruce-fir and mixed-conifer (Young et al. 2000). Continued fire suppression and fuel build-up have set the stage for even more intense and widespread fires in the future (U.S. Forest Service 1988).

The damage to conifer-bearing trees incurred by fire will depend on the intensity of the fire, with effects ranging from reduced cone production, to weakening tree defenses against insect attacks, to widespread tree mortality. Within the MGRS habitats, it has been shown that fire (e.g., the Clark Peak fire of 1996) can reduce cone production for several years (Young et al. 2000). However, a secondary, and indirect, influence of medium-intensity fires is to allow the infestation of insects that feed on weakened conifer trees. After the Clark Peak fire, in 1997 and 1998, large outbreaks of defoliating looper moths colonized the spruce-fir habitat. Subsequent to this insect outbreak, population explosions of two species of bark beetles occurred, which continued to spread in the spruce-fir habitats, killing large numbers of trees. The cumulative effect of these insect outbreaks was a likely reduction in cone-bearing trees (Hatten in this vol.; Lynch in this vol.), with documented declines in the MGRS populations (Young et al. 2000).

To determine how variation in the two critical environmental factors of cone crop abundance and fire risk may combine to threaten the viability of the MGRS, we asked the following questions in this study: (1) What is the "natural" pattern of environmental (spatial and temporal) variation? (2) How are the overall population dynamics of the MGRS governed by these environmental patterns? and (3) What are the short and long-term risks to squirrels from the environmental variation produced by these combined patterns, both now and in the future? As with most other studies of rare species, some of the data available to address these questions is excellent, while some is almost lacking. Due to this heterogeneous set of information, we have used three modeling approaches, operating at different temporal and spatial scales, to address the future viability of the MGRS population. First, we used estimates of relative population size over a 10-year period in relation to conifer-masting events to evaluate the variation due to natural, short-term temporal and spatial variation. Second, we conducted extinction risk analyses to assess the importance of medium-term (20–100 yr) fluctuations in habitat quality in driving MGRS dynamics (Lande and Orzack 1988; Dennis et al. 1991). Finally, we developed a more speculative, spatially explicit, habitat-based model to explore how MGRS dynamics might be altered in relation to changes in conifer masting and to alterations in fire spread and intensity over longer time frames (200 yrs). Together, these approaches allow us to use the diverse and often uncertain information that is available for the MGRS better than would be possible within a single modeling framework. Finally, we use our results to assess management priorities and opportunities for this species in light of the natural and the human-caused uncertainties in its future environment.

Methods

Short-Term Population Dynamics in Relation to Cone Masting

All of our analyses of MGRS population trends and cone production rely upon data provided by the Arizona Game and Fish Department (AGFD), U.S. Forest Service (USFS), and Paul Young of the University of Arizona. One important characteristic of these data is that all population estimates are ultimately based upon occupancy rates of known squirrel middens, rather than upon direct counts of individuals. We used occupancy rate as a measure of relative population size. Surveys of midden occupancy were conducted during spring and fall across the three forest types; our calculations are separately based on either fall or spring occupancy rates, which were available from 1987–1997. Occupancy was determined by visiting a random sample of known (including new) middens and checking for signs of recent caching or use. We used occupancy rates to estimate both habitat-specific (spruce-fir, ecotone, and mixed-conifer) and overall geometric mean population growth rates (λ_r values), calculated as midden occupancy in the current year divided by occupancy in the prior year.

To relate occupancy rates of the MGRS to cone crop abundance, we used USFS cone data from each of the three habitats during the period 1989 to 1997 (excluding 1990, for which no cone data were available). Each year, USFS personnel estimate cone abundance of six species of conifer at approximately 26 known squirrel middens found across the three habitat zones, with different middens chosen each year (G. Froehlich, Safford Ranger District, USFS, unpublished data). The trees closest to the midden are rated on a 1–4 scale (1 = no new cones and 4 = heavily loaded with cones). Additionally, each area is rated on a scale of 1–5 for general productivity (1 = no productivity and 5 = high productivity of cones). In our analyses, we used the mean ratings for a species, averaged across all individual trees within a habitat for each year.

Before relating cone production to MGRS densities, we used Pearson correlation coefficients to determine the degree of synchronicity in conifer cone production for the major tree species found in MGRS habitat: Douglas-fir (DF), southwestern white pine (WP), white fir (WF), Engelmann spruce (ES), and corkbark fir (CF). We eliminated ponderosa pine from our analyses because there were few trees present in any habitat, they often had no cones for many years, and red squirrels rarely use them. We performed these correlations for the average cone production of each tree species within and among habitats.

Next, we assessed the relationship between cone crop abundance and squirrel occupancy rates, analyzed across all habitats. It was not clear *a priori* whether the occupancy rate of squirrel middens in one year would be most influenced by masting in that year or in the previous year. For example, if squirrel populations increase due to enhanced over-winter survival from having a good crop of stored cones, then the prior year's crop is likely to be most influential. Conversely, if conifer buds are used as a significant food source during reproduction in spring and summer, then higher populations would be seen the same year that the cones mature. To consider both possibilities, we performed two separate multiple regressions (Sokal and Rohlf 1995) (stepwise backward, with a criterion for retention of $P = 0.15$) on occupancy rates of average cone ratings, using data for each of the three habitats over seven years. For the first analysis, we used data for both squirrels and cones from the same year, and for the second, cone data from the prior year was used in conjunction with squirrel data.

Analysis of Extinction Probabilities (Medium-Term Risk)

While the above analyses use the midden data to make inferences about population stability and growth, they do not directly address the issue of extinction risk. Although midden occupancy is not a direct measure of population size, it meets the most basic criteria needed to make inferences about population viability from count data using diffusion approximations for future population growth rates and extinction risks (Dennis et al. 1991; Morris and Doak 2002): midden occupancy estimates are likely to be an unbiased estimate of relative population size. Therefore, we used the methods of Lande and Orzack (1988) and Dennis et al. (1991) to estimate short- (20-yr) and medium-term (100-yr) extinction times from the midden data. While this method can be inaccurate when based on limited data (Fieberg and Ellner 2001; Morris and Doak 2002), it nonetheless provides valuable measures of relative population risk that are highly useful for conservation management (Elderd et al. 2003).

To make these extinction risk estimates, we used the mean (m) and variance (s^2) across time of the ln (l) values for occupancy in each habitat and for all habitats together. We defined two alternative quasi-extinction thresholds, 10 and 50, below which we consider the population to be in grave danger of extinction (Morris and Doak 2002). We used the conservative estimate of population size in 1998 (462 squirrels) as our initial population size. While some of our analyses only consider the squirrels in one habitat, using a single current population size allows easier comparison of the intrinsic stability, or risk, predicted by different data sets or within different habitats. The methods to use these parameters to estimate extinction risk in the absence of density dependence are described in detail in Dennis et al. (1991) and Morris and Doak (2002).

Significant negative correlations in the midden data between current occupancy and subsequent l for both the spring and fall census data (Doak et al. 2001) suggest the possibility of substantial density dependence in squirrel dynamics. However, the sequences of masting/non-masting years of major cone-producing tree species probably generate these correlations, rather than density dependence per se. Untangling these two effects would require considerably more years of data than are currently available. To assess the influence of density effects on extinction risk estimates, we estimated the parameters for stochastic Ricker models for spring and fall censuses (these models were supported better than either density-independent or theta-logistic models using AIC values). We then ran 25,000 simulations for each data set and a quasi-extinction threshold (10 or 50) to estimate extinction risks, starting with an initial population size of 462 squirrels.

Spatially Explicit Model (Long-Term Risk)

We also designed an individual-based, spatially explicit simulation model to project habitat and population dynamics of the MGRS (fig. 25.1). Several aspects of this model cannot be parameterized with accurate quantitative information, given the limited data available for the MGRS and its habitat. Nonetheless, the importance of long-term forest dynamics, especially fire cycles, in MGRS habitat makes some consideration of the processes we model in this simulation necessary for any careful viability assessment. Therefore, we targeted two key issues that our other analyses cannot adequately address:

1. What is the threat to MGRS population viability under different fire management strategies?
2. What is the relative importance of different habitats for the long-term stability of the MGRS population?

The model is designed to simulate squirrel populations within a landscape subdivided into the three habitat types utilized by the MGRS: mixed-conifer, spruce-fir, and ecotone. This approach allows us to explicitly state the percentage of the landscape made up of each of the three habitat types, the number of middens in each habitat type (and as such, the density of middens/habitat type), and the initial number of occupied middens in each habitat.

In addition to tracking information about squirrel territories (middens), the model also maintains information about each individual squirrel (age, sex, and litter size). Thus, we can use this model to infer how local (midden) and habitat-level (fire, masting, and insect outbreaks) processes combine with individual demographic rates to affect the dynamics of the squirrel population (fig. 25.1). In each year, the simulation uses a Monte Carlo process to determine if fire or insect outbreaks will occur, where they will be, and how large an area they will cover. Monte Carlo methods are also used to determine whether individual squirrels survive, whether a female occupant gives birth, and the number of offspring/litter. After demographic information has been calculated for each squirrel, surviving offspring disperse from their mother's territory to search for new territories, with the distances and directions moved governed by a realistic stochastic process.

We simulated a range of possible values governing estimates of future fire frequency and size, and time for recovery of forests to useful squirrel habitat following fires, based on past fire history and the opinions of experts about likely future fire dynamics (Grissino-Mayer et al. 1994, 1995). In the model, we simulate fire ignition, spread, and intensity as functions of a single parameter: the time elapsed since the last fire (TLF). To simulate stochastic fire ignition and spread, a single random cell is selected in each year of a simulation as the possible ignition site.

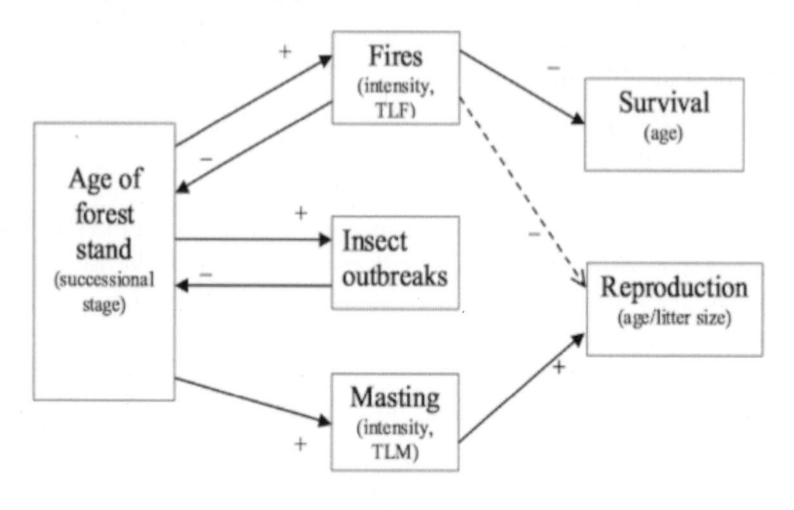

Figure 25.1. A graphic depicting the relationship between landscape and local-scale environmental conditions and squirrel demographics used in the spatial model. Local conditions may vary on a scale as fine as one hectare in the model landscape. Solid lines show a direct effect, whereas dashed lines show an indirect relationship. Increased forest age increases the probability of fires and insect outbreaks, while these factors, in turn, set back the successional stage of forests. Successional stage positively influences mast intensity. Masting has a positive effect on squirrel fecundity, while fires have a direct negative impact on survival.

From the habitat type and TLF of this cell, we calculate the mean (M) and variance (V) for potential fire sizes (expressed as a fraction of the total landscape size), using these relationships:

SF Habitat:

$$M_{sf} = B_{sf}(TLF)^{E_{sf}} \qquad V_{sf} = \sqrt{M_{sf}(1-M_{sf})}/4$$

MC and EC Habitats:

$$M_{mcec} = B_{mcec}(TLF)^{E_{mcec}} \qquad V_{mcec} = \sqrt{M_{mcec}(1-M_{mcec})}/3$$

where B_i and E_i are parameters governing fire size (with $i = sf$ for SF habitat, and *mcec* for both MC and EC habitats). Equations for the variance scale the variation in fire sizes as a fraction of the maximum possible variance (given the mean of the distribution). For each habitat, we also cap the possible values for mean fire size at a value, C. We used a range of baseline values for the four parameters, with B_{mcec} = 0.005–0.040 and E_{mcec} = 0.2125–0.85 for the MC and EC habitats; the values of E_{sf} equal to that of E_{mcec} in each simulation; and the value of B_{sf} equal to 75 percent of B_{mcec}. We chose these relationships and parameters to make mean fire size a gradually increasing function of TLF and to keep fire sizes considerably smaller in SF than in MC or EC habitats. In all simulations we report here, we chose C to equal 0.6, meaning that at most 60 percent of the landscape may burn in any single fire event. Finally, using the mean and variance governed by these parameters, a random beta variate is selected to determine the proportion of landscape burned. This area is a square centered on the initially selected fire ignition site (or offset if the fire abuts one of the landscape edges).

Overall, these relationships result in fire frequencies and size distributions that roughly correspond to the patterns of occurrences seen historically (Grissino-Mayer et al. 1994). However, any formal model validation procedure is essentially impossible, given that the existing analysis of fire history in the Pinaleño Mountains does not allow a full characterization of fire sizes and frequencies (Grissino-Mayer et al. 1994). Similarly, with no information on MGRS dynamics through even a small fraction of a single inter-fire interval, direct validation of the modeled population dynamics is not possible. Thus, we are left assessing the sensitivity of model outputs to different parameter estimates, rather than directly confirming their values. Initial model runs showed that the parameters controlling fire behavior have by far the largest impacts on viability estimates (Doak et al. 2001). Therefore, we ran a total of 50 simulations, each of 200 years, for 16 combinations of B_{mcec} and E_{mcec} values (and the corresponding values for B_{sf} and E_{sf}). We summarize the results of these simulations in two ways: the risk and speed of extinction, and the mean and variance in the sizes of populations that do not become extinct. We used a quasi-extinction threshold of 20 animals, corresponding to Goodman's (1987) threshold for effects on the viability of populations under the influence of demographic stochasticity.

A detailed description of the different components of the spatial model and the parameter values used is provided in Doak et al. (2001).

Results

Short-Term Population Dynamics in Relation to Habitat and Cone Production

We found moderate to strong temporal correlations in cone production among all tree species other than WP. This synchrony in masting was strongest between CF and ES ($r = 0.77$, $P = 0.10$, $n = 8$ years) and DF and WF ($r = 0.72$, $P = 0.19$, $n = 8$ years). Because WF does not occur in spruce-fir habitat and was in lower abundance in other habitats, we did not include this species in the following analyses. Given the high correlation in masting of WF and DF, we assumed that trends seen for DF are similar to WF.

Among habitats, conifer production showed a high degree of synchrony between years, with good mast years likely to occur in all three habitats within the same year. These relationships were approximately linear, although with only eight years of complete data, the shape of the relationships is difficult to test. This relationship is strongest for ES, for which the pair-wise Pearson correlation coefficients of cone production intensity among habitats ranged from 0.93–0.87 ($n = 8$ years). For most other species and habitat pairs, correlations in masting intensities were 0.68 or greater ($n = 8$ years); the exceptions were WP and CF in the ecotone versus mixed-conifer habitats. Overall, the correlated cone production across habitats and among tree species tends to reinforce high temporal variation in food availability; that is, if cone production was asynchronous or negatively correlated, variability in food supply would be lower.

Interestingly, cones of different conifer species appear to have affected squirrel occupancy differently. ES and CF were retained in the regression model of effects on the next year's squirrel occupancy ($P = 0.11$ and 0.05 respectively). Yet, together these species had only a moderate effect on occupancy rate (multiple regression $r^2 = 0.32$). In a separate regression for same-year occupancies, WP was strongly influential ($P = 0.04$) and was the only species retained (multiple regression $r^2 = 0.24$). Overall,

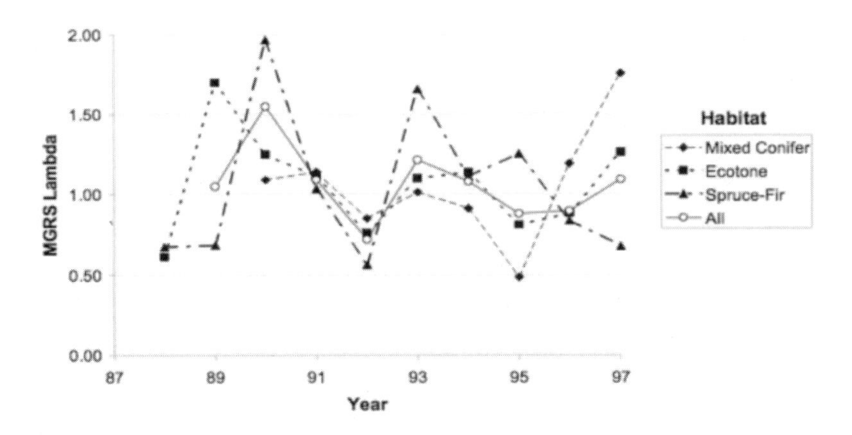

Figure 25.2. Annual growth rates (lambda) of MGRS for each habitat and over all habitats. Lambda is equal to fall midden occupancy in one year divided by fall occupancy the prior year.

the species of conifer having the strongest links to squirrel population dynamics (as indicated by midden occupancy rates) were white pine (WP), Engelmann spruce (ES), and corkbark fir (CF).

By examining population change (l) from 1988 to 1997, it is clear that the overall MGRS population is on average stable, geometric mean lambda (λ_G = 1.04), but highly fluctuating (fig. 25.2). Further, mixed-conifer (λ_G = 1.00, sd = 0.49) and spruce-fir (λ_G = 0.96, sd = 0.47) exhibit the most temporal variation and the ecotone the least (λ_G = 1.02, sd = 0.31) of all subpopulations. Combining these results with cone production analyses, it is clear that the masting cycles of WP, ES and CF, which drive most of the variation in occupancy rates, will greatly influence MGRS dynamics, both spatially and temporally, in the short-term.

Medium-Term Extinction Analysis

We separately calculated the mean (μ) and variance (σ^2) of the natural log of annual population growth rates ($\ln(\lambda)$) based upon both spring and fall surveys for each habitat as well as for the entire population (table 25.1). From these analyses, we estimated extinction risks over 20- and 100-year

Table 25.1. Mean (μ) and variance (σ^2) in stochastic population growth ($\ln(\lambda)$) for extinction risk calculations based upon midden occupancy rates in the spring and fall for the three habitats

MC = Mixed-conifer, EC = Ecotone, SF = Spruce-fir, All = all three habitats. The remaining columns show predicted probability of extinction by 20 and 100 years, based upon two pseudo-extinction thresholds (N_{ex} = 10 and 50 animals). Growth rates near zero are stable; values below zero indicate a decreasing population, and values above zero indicate an increasing population.

Data set	Stochastic population growth rate			P_{ex} 20 yrs		P_{ex} 100 yrs	
	Mean (μ)	95% Conf. Limits (μ)	Var. (σ^2)	N_{ex} = 10	50	N_{ex} = 10	50
Spring MC	−0.029	−0.352, 0.294	0.233	0.100	0.368	0.622	0.801
Spring EC	0.005	−0.216, 0.226	0.109	0.005	0.099	0.180	0.425
Spring SF	−0.016	−0.457, 0.425	0.430	0.196	0.463	0.621	0.783
Spring All	0.011	−0.234, 0.256	0.133	0.009	0.123	0.186	0.419
Fall MC	0.008	−0.236, 0.249	0.117	0.006	0.103	0.172	0.408
Fall EC	0.022	−0.191, 0.231	0.089	0.001	0.040	0.051	0.204
Fall SF	−0.037	−0.338, 0.259	0.176	0.070	0.336	0.663	0.839
Fall All	0.041	−0.113, 0.193	0.047	<0.001	0.001	<0.001	0.011

time periods using two different quasi-extinction thresholds. Growth rates based on spring data are more variable than those for fall, perhaps due to changing over-winter survival or greater observation errors because midden activity is more obvious in fall. Nonetheless, this difference suggests that extinction estimates will be more accurate when based on fall data.

For the fall data, only the spruce-fir habitat analysis predicts a substantial extinction risk during a 20-year horizon when using a quasi-extinction threshold of 10 animals. However, all three single-habitat extinction-risk estimates are non-trivial at 20 years with quasi-extinction set at 50. For 100-year time frames, the single habitat risk estimates range from substantial (0.051) to extreme (0.839). However, the analyses based on middens across all habitats are likely to be most accurate, and these indicate only small (0.011 spring) to moderately high (0.419 fall) risks of extinction at a quasi-extinction threshold of 10 animals. It is important to emphasize the imprecision of these results, as indicated by the large confidence limits around the best estimate of the μ values (table 25.1).

Extinction predictions from the density-independent models are confirmed by our density-dependent simulations. Models parameterized with fall data showed less than a 1 percent chance of extinction after 100 years for either quasi-extinction threshold. In contrast, simulations using spring census data showed substantial extinction risks that broadly match those of the density-independent models, but are more extreme. For example, the 100-year risk of extinction predicted by these models is 0.50 and 0.93 for thresholds of 10 and 50, respectively, compared with risk of 0.19 and 0.42 predicted from the simpler analyses (table 25.1).

Our results also highlight consistent differences in relative extinction risk of portions of the MGRS population occupying different habitat types. In particular, the dynamics of MGRS in the spruce-fir habitat are less stable and hence more extinction prone than populations in either the mixed-conifer or ecotone portions of the population. This result is clearest from analysis of fall census data. However, we caution that this conclusion is based only on the variability in population growth seen during the study period, which included only a single significant fire event.

Long-Term Dynamics Predicted by Spatial Modeling

As the spatial simulation is essentially an analysis of the sensitivity of MGRS viability to fire regime, we present most of the results from our spatial model in terms of the two parameters that govern fire size. B_{mcec} governs minimum fire size, while E_{mcec} adjusts the sensitivity of fire size to the time since last fire (TLF). Larger values of both parameters indicate greater risks of larger fires. Two-hundred-year projections indicated that the risk of extinction for simulated MGRS populations was highly sensitive to fire regime (fig. 25.3). For median values of B_{mcec} and E_{mcec} (0.01 and 0.425, respectively), 27 percent of simulated populations became extinct (fig. 25.3A), with a mean time to extinction of 80 years (fig. 25.3B). Increases in either parameter value lead to increasing extinction risk and decreasing mean time to extinction (fig. 25.3B). Extinction risk asymptotes at about a 70–80 percent probability of extinction over a 200-year time horizon, and a mean time to extinction of approximately 20–30 years. While higher probabilities of larger fires greatly increase the chances of extinction, reductions in the risk of large fires (small values of the two parameters) lead to very low extinction risks. This confirms the average results of our last analyses, which indicated that the combined effects of randomly varying survival rates and cone production on reproduction would not alone seriously threaten population viability of the MGRS.

There are also large effects of fire regime on population numbers in the absence of extinctions. For simulations that did not result in extinction, we calculated the mean and coefficient of variation (CV) of population sizes for each combination of fire regime parameters. Population size was largest and least variable with the smallest fire sizes and the smallest increase in fire size with time since last fire (TLF). However, using the largest parameter values for fire size, mean population size (of surviving populations) was greater than 200 animals. Whereas this suggests high enough numbers to avoid Allee effects and inbreeding problems, the high CV for these population sizes show that populations repeatedly fall below the mean, and thus the problems associated with very small population numbers are likely to be a compounding threat if a large fire were to occur. It is important to note that we did not account for Allee effects or inbreeding depression in our model; the high variation in numbers

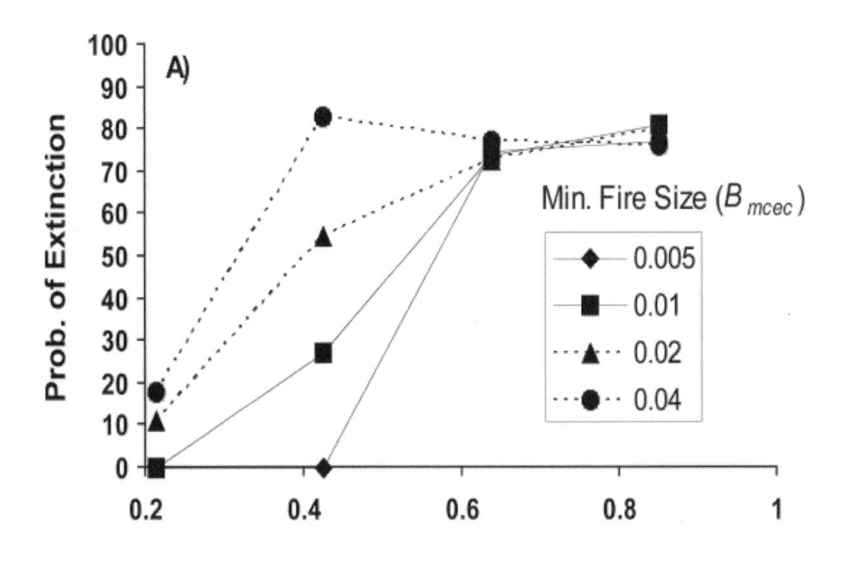

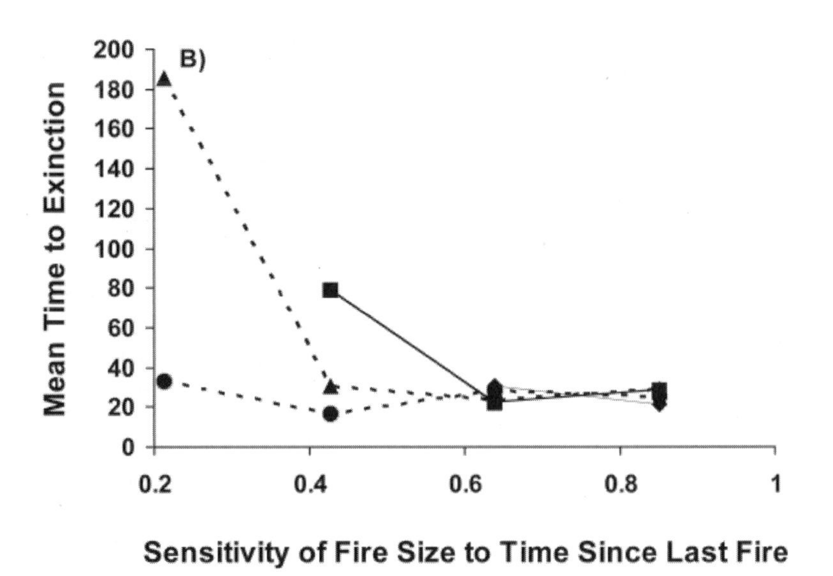

Figure 25.3. (A) Probabilities of extinction and (B) mean time to extinction in spatial simulations, as functions of the two parameters controlling fire size (E_{mcec} = sensitivity of fire size to time since last fire (TLF), and B_{mcec} = minimum fire size). For both parameters, higher values indicate greater fire size.

of surviving populations suggests that these factors could significantly worsen the predicted viability of the MGRS in the face of large and infrequent fires, over and above the threats indicated by our simulations.

To assess the importance of different habitats in maintaining MGRS viability, we also tracked the mean and variance in population numbers within each habitat over the course of our simulations. For mean population size, the striking result is the high sensitivity of the populations in mixed-conifer and ecotone habitats to fire regime, in comparison to the apparent insensitivity of numbers in SF habitat. Whereas total numbers consistently remained lowest in SF habitat, this was the most stable segment of the population when compared across different fire regime scenarios. This result is in strong contrast to our short-term extinction analyses (table 25.1), which showed the opposite pattern over times scales that did not include fires.

Understanding these contrasting results is clarified by examining the CV in population sizes across years within each habitat. If fire sizes are small, we see a result that parallels the pattern seen in the midden census data, with the population occupying spruce-fir forest being somewhat less stable through time ($B_{mcec} = 0.005$ and $E_{mcec} = 0.2125$). However, when fires of larger size occur, there is a rapid increase in the variation of the mixed-conifer and ecotone populations, but little change in the variability of the spruce-fir component of the MGRS population. This numerical result bolsters the intuitive idea that spruce-fir habitat, which is much less fire-prone than other MGRS habitats, may be of key importance in maintaining a viable population over longer time spans when fire becomes a key variable driving population dynamics. However, even this stability is not enough to offset the threats posed by very large fires (fig. 25.3).

Discussion

Our population analysis of the MGRS has confirmed that this subspecies is clearly threatened by both natural and human-caused processes within the next 100–200 years. The results indicate that although changes in food availability due to temporal variation in cone production are highly correlated with squirrel dynamics, alone they are not likely to seriously threaten MGRS within the current configuration of habitat. However, there are consistent differences in the relative extinction risk of the portions of the MGRS population occupying different habitat types due to the cyclic nature of their primary food source. In particular, the dynamics of spruce-fir habitat are more variable and therefore less stable in the short-term and more extinction prone than are either the mixed-conifer or ecotone portions of the population. However, when extinction risk is viewed over the longer term with a consideration of wildfires, the spruce-fir habitat appears less vulnerable, due to a reduced likelihood of highly devastating fire events.

The exact fire regime that will occur in the future will have large effects on the viability of MGRS, with even our baseline estimates of fire-regime parameters yielding a 20 percent probability of extinction. As we noted when introducing the spatial model, it requires many parameters—for fires and many other factors—for which there are no precise empirical estimates. Because of this, we do not place great confidence in the exact extinction probabilities generated by the model. However, the prediction that viability is highly sensitive to the mean and variance in fire sizes is likely to be highly robust, as indicated by the very consistent patterns in fire-size effects on extinction and numbers. This sensitivity to fire regime suggests that more research into past fire history and also modeling by fire behavior experts on the probable future fire regimes in MGRS habitat could be highly useful in anticipating how and to what extent future management should attempt to alter fire effects so as to maximize the chances of MGRS survival.

While more research on fire history is needed, our results clearly indicate that careful management to prevent large fires, at least over 100–200-year time horizons, can greatly increase the viability of the MGRS population. However, it is important to note that in our model this result arises in part because small fires do occur, limiting the likelihood and severity of larger blazes. Thus, the likely result of continued fire suppression is an increase in the likelihood that these very large fires might occur, which is what drive squirrels to extinction in the simulation model. Although prior work, such as the MGRS Biological Assessment (U.S. Forest Service 1988), identified fire, and particularly fire suppression, as a substantial risk to the population, it has proven difficult to determine the parameters governing this risk and develop solutions for fire management that can be implemented.

Due to high fuel loads and the highly flammable nature of Engelmann spruce and corkbark fir (Wright and Bailey 1982), without active management, the spruce-fir habitat is likely to experience an intense and widespread fire in the near future (Grissino-Mayer et al. 1995). Such intense fires are likely to destroy many mature, cone-bearing trees and thus severely impact MGRS. Because of past fire-suppression policies, the mixed-conifer and ecotone habitats are now also susceptible to intense fires that would affect large areas (Grissino-Mayer et al. 1995). Studies in other coniferous habitats in the northern United States have shown that midden densities declined 48–66 percent after an intense and catastrophic fire (Mattson and Reinhart 1996).

Thus, large stand-replacement fires are likely to have a key role in influencing MGRS dynamics in the future. However, detailed field studies of the MGRS have only been conducted since 1987. While this time period encompasses the Clark Peak fire in 1996 in which the direct loss of at least 30 middens led to similar declines in squirrel abundance within spruce-fir and ecotone forest types (Young et al. 2000), the long-term effects of this fire on squirrel population dynamics are still largely unknown. Consequently, efforts to understand fire effects on MGRS population viability must proceed using incomplete information.

Because of the increased likelihood of catastrophic fires, it is likely that there will be much higher variability in population performance of MGRS in the future than during the period for which we currently have census data. This makes the future possibility of reaching very low population numbers considerably higher than estimated from the direct occupancy data that we employ in our medium-term extinction analyses. Thus, we would emphasize that our results suggest that high variance in population sizes each year, combined with negative density dependence and the certainty of fire effects in the future, mean that there is a significant risk of extinction to the MGRS. While our simulation model shows this risk, the data on which it is based are currently insufficient to exactly quantify this threat, making continued data collection essential for more robust analyses of extinction risks in the future.

Repeated annual surveys of population size, such as the annual midden censuses, are the correct type of data needed for estimating extinction risk. However, fairly long time series are needed for even the best estimates of population size to confidently assess risk—exactly how long depends upon the inherent variability of the population being studied

(Fieberg and Ellner 2001; Elderd et al 2003). In addition, observation error in population size estimates (i.e., added variation in estimates due to inaccuracy in estimation) will both decrease the reliability of risk estimates and tend to inflate the estimated risk of extinction (Morris and Doak 2002). With these considerations as background, below we detail the most important management implications from our analysis.

Management Implications

The amount of fuel load across the study area is very high (Grissino-Mayer et al. 1995), and our results indicate that there is an extremely high risk to the squirrel population from intense, widespread fires. It is particularly worrisome that a large fire could spread into spruce-fir habitat, as the stability of this segment of the forest population appears to be key for long-term MGRS population viability, due to its usual invulnerability to fire disturbances. Fire poses the greatest risk to the MGRS population, so it is critical that a carefully designed plan to reduce fuel load is implemented within the next several years. This procedure could occur in steps, with small areas in each habitat type being tested for removal techniques and monitoring of squirrel populations to ascertain the potential impacts of this form of disturbance.

While fires will drive larger and more infrequent disturbances, tree masting is the most important short-term determinant of MGRS dynamics. Abundant conifer cone crops may improve annual reproduction, as shown in studies where supplemental food was used (Klenner and Krebs 1991), and our analyses clearly link increased midden occupancy to masting events. Furthermore, our extinction calculations show that with its current small numbers, the MGRS population is at a moderate risk of extinction due to the fluctuations caused by masting alone, with no consideration of fires. This risk is especially worrisome in light of climate change effects that will certainly alter the relative abundances and productivities of tree species relied upon by the MGRS. Understanding the real importance of different food trees for squirrel demography and dynamics is currently limited by the data available. In particular, the current census method (random sampling for middens) makes it difficult to separate the effects of temporal and spatial variation in population trends, thereby confounding a clear picture of cause and effect resulting from varying environmental conditions. We therefore recommend sampling the same middens and adjacent trees each year as a subset of the total sampling effort.

Our analysis of quasi-extinction over 20- and 100-year time periods suggests that the risk of extinction is moderate. However, examining the data by habitat, we found that the populations in spruce-fir were highly vulnerable, while ecotone populations have a reduced risk. These results are partially a function of the differences in year-to-year variability in numbers shown in these habitats. However, over longer time spans, variation in fire effects is far more important than masting alone in driving variance in squirrel numbers, and the spruce-fir part of the population is most stable. Furthermore, the importance of this stability is likely to be underestimated by our spatial model, as we assumed a constant adult survivorship across habitats. Adult survival has the highest effects on population growth rates (Doak et al. 2001) and is very likely to vary across habitats, but current data do not allow us to estimate these differences. Together, these results and data limitations mean that it is crucial to better understand the true importance of different habitats for MGRS demographics and hence population viability in order to correctly prioritize management efforts for the species.

Our analyses emphasize what can and cannot be concluded based upon the limited data in hand for the MGRS—data that are typical of many endangered species. While we must be cautious in making any exact predictions of viability from our results, this uncertainty should not mask the most important conclusion: a rapid and active new fire management strategy is essential for minimizing the risk faced by this species. Indeed, given the small habitat area for the MGRS relative to likely fire sizes, better data will likely do no more than re-emphasize the need to consider carefully the ways in which fire risks can be reduced and controlled if the viability of the population is to be maintained. Our contrasting results from the use of different data sets and different modeling methods indicates a finer point: every method of predicting the future comes with limitations and assumptions, and these features can result in sometimes striking differences in predictions (Mills et al. 1996). We feel that the diverse approaches taken in this viability assessment complement one another and serve to emphasize that we do not need certainty in predictions in order to make a persuasive case for certain management actions. While we recognize that effecting management change can be considerably more difficult than suggesting its need, we hope that this work can serve to spur efforts on behalf of this unique species.

Acknowledgments

Funding for this study was provided by the University of Arizona through the Mt. Graham Red Squirrel Study Committee, which is comprised of representatives from the Arizona Game and Fish Department, the University of Arizona, the U.S. Forest Service, and the U.S. Fish and Wildlife Service. Paul Young at the University of Arizona generously provided data also used in this project. We would also like to thank Reed Sanderson, the chairman of the MGRS Study Committee, for his generous support throughout all phases of this work and Dr. Michael Wisdom for his input during early discussions of this project.

Literature Cited

Dennis, B., P. L. Munholland, and J. M. Scott. 1991. Estimation of growth and extinction parameters for endangered species. *Ecological Monographs* 61:115–143.

Doak, D. F., E. K. Harding, and C. Hartway. 2001. *Demographic analysis for the Mt. Graham red squirrel,* Tamiasciurus hudsonicus grahamensis. A final report submitted to the Mt. Graham Red Squirrel Study Committee, University of Arizona, Tucson. Report is on file at the University of Arizona Library, Special Collections.

Elderd, B., P. Shahani, and D. F. Doak. 2003. The problems and potential of count-based population viability analyses. Pages 173–202 in C. Brigham and M. Schwartz, eds. *Population viability analysis in plants: Conservation, management, and modeling of rare plants.* Spring-Verlag, Berlin.

Fieberg, J., and S. P. Ellner. 2001. Stochastic matrix models for conservation and management: A comparative review of methods. *Ecology Letters* 4:244–266.

Goodman, D. 1987. The demography of chance extinction. Pages 11–34 in M. E. Soulé, ed. *Viable populations for conservation.* Cambridge University Press, Cambridge.

Grissino-Mayer, H. D., C. H. Baisan, and T. W. Swetnam. 1994. *Fire history and age structure analyses in the mixed-conifer and spruce-fir forests of Mt. Graham.* A final report submitted to the Mt. Graham Red Squirrel Study Committee, University of Arizona, Tucson.

——. 1995. Fire history in the Pinaleño Mountains of southeastern Arizona: Effects of human-related disturbances. Pages 399–407 in L. F. DeBano, P. F. Ffolliott, A. Ortega-Rubio, G. J. Gottfried, R. H. Hamre, and C. B. Edminster, technical coordinators. *Biodiversity and management of the Madrean Archipelago: The sky islands of southwestern United States and northwestern New Mexico: September 19-23, 1994,*

Tucson, Arizona. General Technical Report RM-GTR-264. U.S. Dept. of Agriculture, Forest Service, Rocky Mountain Forest and Range Experiment Station, Fort Collins, Colorado.

Grissino-Mayer, H. D., and H. C. Fritts. 1995. Dendrochronology and dendroecology in the Pinaleno Mountains. Pages 100–122 in C. A. Istock and R. S. Hoffman, eds. *Storm over a desert mountain: Conservation biology and the Mt. Graham affair.* University of Arizona Press, Tucson.

Kemp, G. A., and L. B. Keith. 1970. Dynamics and regulation of red squirrel (*Tamiasciurus hudsonicus*) populations. *Ecology* 51(5):763–779.

Klenner, W., and C. J. Krebs. 1991. Red squirrel population dynamics. I. The effect of supplemental food on demography. *Journal of Animal Ecology* 60:961–978.

Lande, R., and S. H. Orzack. 1988. Extinction dynamics of age-structured populations in a fluctuating environment. *Proceedings of the National Academy of Sciences, USA* 85:7418–7421.

Mattson, D. J., and D. P. Reinhart. 1996. Indicators of red squirrel (*Tamiasciurus hudsonicus*) abundance in the whitebark pine zone. *Great Basin Naturalist* 56(3):272–275.

Mills, L. S., S. G. Hayes, C. Baldwin, M. J. Wisdom, J. Citta, D. J. Mattson, and K. Murphy. 1996. Factors leading to different viability predictions for a grizzly bear data set. *Conservation Biology* 10:863–873.

Morris, W. F., and D. F. Doak. 2002. *Quantitative conservation biology: The theory and practice of population viability analysis.* Sinauer Associates, Sunderland, Massachusetts.

Pennisi, E. 1989. Biology versus astronomy: The battle for Mount Graham. *BioScience* 39(1):10–13.

Rabinowitz, D. 1981. Seven forms of rarity. Pages 205–217 in H. Synge, ed. *The biological aspects of rare plant conservation.* John Wiley and Sons, Chichester, England.

Sokal, R., and F. J. Rohlf. 1995. *Biometry: The principles and practice of statistics in biological research.* W. H. Freeman and Co., New York.

Stromberg, J. C., and D. T. Patten. 1995. Vegetation dynamics of the spruce- fir forests of the Pinaleno Mountains. Pages 89–99 in C. A. Istock and R. S. Hoffman, eds. *Storm over a desert mountain: Conservation biology and the Mt. Graham affair.* University of Arizona Press, Tucson.

U.S. Forest Service. 1988. *Mount Graham red squirrel: An expanded biological assessment.* Safford Ranger District, Coronado National Forest, Safford, Arizona.

Wright, H. A., and A. W. Bailey. 1982. *Fire ecology.* John Wiley and Sons, New York.

Young, P. S. 1995. Monitoring the Mt. Graham red squirrel. Pages 226–246 in C. A. Istock and R. S. Hoffman, eds. *Storm over a desert mountain: Conservation biology and the Mt. Graham affair.* University of Arizona Press, Tucson.

Young, P. S., V. L. Greer, J. E. Lowry, E. Bibles, N. Ferguson, and E. Point. 2000. *The Mount Graham red squirrel monitoring program: 1989–1998.* Report submitted to Steward Observatory and the Office of the Vice President for Research, University of Arizona, Tucson.

Appendix A

Further Reading of Recent Research on Mt. Graham Red Squirrels and Their Sky Island Allies

The recipe for perpetual ignorance is:
Be satisfied with your opinions
and content with your knowledge.
—Elbert Hubbard

Edelman, A. J., and J. L. Koprowski. 2005. Diet and tree use of Abert's squirrels (*Sciurus aberti*) in a mixed-conifer forest. *Southwestern Naturalist* 50:461–465.

———. 2006a. Characteristics of Abert's squirrel (*Sciurus aberti*) cavity nests. *Southwestern Naturalist* 51:64–70.

———. 2006b. Seasonal changes in home ranges of Abert's squirrels: Impact of mating season. *Canadian Journal of Zoology* 84:404–411.

———. 2006c. Selection of drey sites by Abert's squirrels in an introduced population. *Journal of Mammalogy* 86:1220–1226.

———. 2007. Communal nesting in asocial Abert's squirrels: The role of social thermoregulation and breeding strategy. *Ethology* 113:147–154.

———. In press. Does niche overlap for nest sites occur between coexisting native and exotic tree squirrels? *Journal of Mammalogy* 90.

Edelman, A. J., J. L. Koprowski, and J. L. Edelman. 2005. Kleptoparasitic behavior and species richness at Mt. Graham red squirrel middens. Pages 395–398 in G. J. Gottfried, B. S. Gebow, L. G. Eskew, and C. Edminster, compilers. *Connecting mountain islands and desert seas: Biodiversity and Management of the Madrean Archipelago II.* Proceedings RMRS-P-36. U.S. Dept. of Agriculture, Forest Service, Rocky Mountain Research Station, Fort Collins, Colorado.

Hutton, K. A., J. L. Koprowski, V. L. Greer, M. I. Alanen, C. A. Schauffert, and P. J. Young. 2003. Use of mixed-conifer and spruce-fir forests by an introduced population of Abert's squirrels (*Sciurus aberti*). *Southwestern Naturalist* 48:257–260.

Koprowski, J. L. 2002. Handling tree squirrels with an efficient and safe restraint. *Wildlife Society Bulletin* 30:101–103.

——. 2005a. Annual cycles in body mass and reproduction in the endangered Mt. Graham red squirrel, a montane island endemic. *Journal of Mammalogy* 86:309–313.

——. 2005b. Impacts of fragmentation on tree squirrels: A review and synthesis. *Animal Conservation* 8:369–376.

——. 2005c. Management and conservation of tree squirrels: The importance of endemism, species richness, and forest condition. Pages 245–250 in G. J. Gottfried, B. S. Gebow, L. G. Eskew, and C. Edminster, compilers. *Connecting mountain islands and desert seas: Biodiversity and Management of the Madrean Archipelago II*. Proceedings RMRS-P-36. U.S. Dept. of Agriculture, Forest Service, Rocky Mountain Research Station, Fort Collins, Colorado.

——. 2005d. Pine squirrel (*Tamiasciurus hudsonicus*): A technical conservation assessment. U.S. Forest Service, Rocky Mountain Region <http://www.fs.fed.us/r2/projects/scp/assessments/ pinesquirrel.pdf> (accessed 11 September 2008).

Koprowski, J. L., M. I. Alanen, and A. M. Lynch. 2005. Nowhere to run and nowhere to hide: Response of endemic Mt. Graham red squirrels to catastrophic forest damage. *Biological Conservation* 126:491–498.

Koprowski, J. L., and M. C. Corse. 2005. Activity, time budgets, and behavior of Mexican fox squirrels. *Journal of Mammalogy* 86:947–952.

Koprowski, J. L., A. J. Edelman, B. S. Pasch, and D. J. Buecher. 2005. A dearth of data on mammals in the Madrean Archipelago: What we think we know and what we do know. Pages 412–415 in G. J. Gottfried, B. S. Gebow, L. G. Eskew, and C. Edminster, compilers. *Connecting mountain islands and desert seas: Biodiversity and Management of the Madrean Archipelago II*. Proceedings RMRS-P-36. U.S. Dept. of Agriculture, Forest Service, Rocky Mountain Research Station, Fort Collins, Colorado.

Koprowski, J. L., S. R. B. King, and M. J. Merrick. 2008. Expanded home ranges in a peripheral population: Space use by endangered Mt. Graham red squirrels. *Endangered Species Research* 4:227–232.

Koprowski, J. L., K. M. Leonard, C. A. Zugmeyer, and J. L. Jolley. 2006. Direct effects of fire on endangered Mt. Graham red squirrels. *Southwestern Naturalist* 51:59–63.

Koprowski, J. L., N. Ramos, B. S. Pasch, and C. A. Zugmeyer. 2006. Observations on the ecology of the endemic Mearns's squirrel (*Tamiasciurus mearnsi*). *Southwestern Naturalist* 51:426–430.

Leonard, K. M., and J. L. Koprowski. In press. A comparison of habitat use and demography of red squirrels at the southern edge of their range. *American Midland Naturalist* 161.

——. In press. Effects of fire on endangered Mt. Graham red squirrels (*Tamiasciurus hudsonicus grahamensis*): Responses of individuals with known fates. *Southwestern Naturalist* 53.

Lurz, P. W. W., J. L. Koprowski, and D. J. A. Wood. 2008. The use of GIS and modelling approaches in squirrel population management and conservation: A review. *Current Science* 95(7):918–922.

Merrick, M. J., S. R. Bertelsen, and J. L. Koprowski. 2007. Nest site selection by endangered Mt. Graham red squirrels. *Journal of Wildlife Management* 71:1958–1963.

Palmer, G. H., J. L. Koprowski, and T. Pernas. 2008. Tree squirrels as invasive species: Conservation and management implications. Pages 273–282 in G. L. Witmer, W. C. Pitt, and K. A. Fagerstone, eds. *Managing vertebrate invasive species: Proceedings of an international symposium*. USDA/APHIS Wildlife Services, National Wildlife Research Center, Fort Collins, Colorado.

Pasch, B. S., and J. L. Koprowski. 2005. Ecology and conservation of Chiricahua fox squirrels (*Sciurus nayaritensis chiricahuae*), a sky island endemic. Pages 426–428 in G. J. Gottfried, B. S. Gebow, L. G. Eskew, and C. Edminster, compilers. *Connecting mountain islands and desert seas: Biodiversity and Management of the Madrean Archipelago II*. Proceedings RMRS-P-36. U.S. Dept. of Agriculture, Forest Service, Rocky Mountain Research Station, Fort Collins, Colorado.

——. 2006. Sex differences in space use of Chiricahua fox squirrels. *Journal of Mammalogy* 87:380–386.

Rushton, S. P., D. J. A. Wood, P. W. W. Lurz, and J. L. Koprowski. 2006. Modelling the population dynamics of the Mt. Graham red squirrel: Can we predict its future in a changing environment with multiple threats? *Biological Conservation* 131:121–131.

Schauffert, C., J. Koprowski, V. Greer, M. Alanen, and K. Hutton. 2002. Potential predators of Mt. Graham red squirrels. *Southwestern Naturalist* 47:498–501.

Steele, M., and J. L. Koprowski. 2001. *North American tree squirrels*. Smithsonian Institution Press, Washington, D.C.

Wood, D. J., S. Drake, S. P. Rushton, D. Rautenkranz, P. W. W. Lurz, and J. L. Koprowski. 2007. Fine-scale analysis of Mount Graham red squirrel habitat following disturbance. *Journal of Wildlife Management* 71:2357–2364.

Wood, D. J., J. L. Koprowski, and P. Lurz. 2007. Tree squirrel introduction: A theoretical approach with population viability analysis. *Journal of Mammalogy* 88:1271–1279.

Zugmeyer, C. A., and J. L. Koprowski. 2007. Avian nest predation by endangered Mt. Graham red squirrels. *Southwestern Naturalist* 52:155–157.

——. In press. Severely insect-damaged forest may function as a temporary trap for red squirrels. *Forest Ecology and Management* 256.

——. In press. Unaltered habitat selection after a severe insect infestation: Concerns for forest-dependent species. *Journal of Mammalogy* 90.

Index

I

J

K

L

M

N

R

S

T

About the Editors

H. REED SANDERSON (retired) was the resident biologist on the Mt. Graham Biological Programs at the University of Arizona (UA) from 1992 to 2007. Previously, he was a research scientist for the U.S. Forest Service (USFS) where he did range and wildlife habitat research for 30-plus years in California, West Virginia, and Oregon. His research projects ranged from bitterbrush propagation, mountain meadow ecology, and fox and gray squirrel habitat to the impact of livestock grazing management strategies on herbage and browse, water quality and quantity, and economic, social, and other natural resources. He retired from the USFS in December 1989 and moved to southeast Arizona. Reed and his wife, Georgie, initially lived at the Florida Canyon Headquarters of the Santa Rita Experimental Range where they volunteered for the USFS until the range was turned over to the UA in 1991. Reed then volunteered as the acting manager of the experimental range for the UA until 1996 when they moved to Green Valley, Arizona. Reed has a B.S. degree from Humboldt State University (Wildlife Management, 1957) and a M.S degree from Colorado State University (Range Management, 1959). He did additional studies in animal behavior and reproductive physiology at West Virginia University. Reed's parting words upon leaving the UA were: "It has been my pleasure to work with everyone on the Mt. Graham project. We began as colleagues working for the benefit of the Mt. Graham red squirrel, and we are now friends. I hope you, the readers, benefit from the results of this publication as much as I have enjoyed bringing it to fruition."

JOHN L. KOPROWSKI is a professor of wildlife conservation and management in the School of Natural Resources at the University of Arizona where he has taught courses in natural resources ecology, sky island biogeography, and mammalian conservation since 2000. Previously, John served on the faculty of Willamette University in Salem, Oregon. His efforts have garnered awards for outstanding teaching, publications, and research. He holds degrees in zoology from Ohio State University (B.S.) and Southern Illinois University (M.A.), and in biology from the University of Kansas (Ph.D.). John is the director of the Mt. Graham Red Squirrel Monitoring Program and previously was director of the Desert Southwest Cooperative Ecosystems Studies Unit. His research program has been funded by, among others, the National Geographic Society, the Arizona Game and Fish Department, the U.S. Forest Service, and the National Park Service and focuses on the ecology and conservation of mammals with a particular emphasis on tree and ground squirrels. John believes that tree squirrels can often serve much like the miners' canary as harbingers of changes occurring within our forested ecosystems. He has worked on each of the 10 species of tree squirrel found in North America over his 20 years of research. He has authored more than 70 articles on the ecology of tree squirrels and a book, *North American Tree Squirrels* (co-authored with Michael Steele), published in 2001 by the Smithsonian Institution Press. John lives in Tucson, Arizona, with his wife, Nancy, and two children.

About the Contributors

MARIT I. ALANEN received her B.A. in environmental science from Wesleyan University in 1992. After volunteering for the National Park Service (Apostle Islands National Lakeshore) and working for the Wisconsin Department of Natural Resources, she went on to receive her M.S. in wildlife science from the University of Arizona in 1998. Since then, she has worked for the Florida Fish and Wildlife Conservation Commission, the University of Arizona's Mt. Graham Red Squirrel Monitoring Program, and the Arizona Game and Fish Department. She currently works for the U.S. Fish and Wildlife Service as a fish and wildlife biologist in Tucson, Arizona.

R. SCOTT ANDERSON is presently a professor of environmental and quaternary sciences and director of the Laboratory of Paleoecology at Northern Arizona University, Flagstaff, where he has also served as director of environmental sciences and education. A native of Colorado, he obtained his B.A. in environmental ecology from the University of Colorado, his M.S. in paleoecology from the University of Maine, and his Ph.D. in paleoecology and geology from the University of Arizona. Throughout his career, he has maintained an interest in the interaction of climate, vegetation, and ecosystem disturbance over long timescales and has worked on projects located in the western North America and Mexico, Alaska, New England, Argentina, and Spain.

DIANE ANGELL is currently an assistant professor in the Biology Department at St. Olaf College in Northfield, Minnesota. She received her B.A. from Macalester College in 1985 and her Ph.D. in ecology and evolutionary biology from Brown University in 1993. She has broad interests in evolution and small mammal conservation. Her current research interests involve the isolation of small mammals in prairie remnants of southern

Minnesota. In this area, islands of prairie are isolated amongst vast fields of corn and soy. Although apparently common in other states, one species in particular, the prairie vole (*Microtus ochrogaster*), has experienced dramatic declines here. She regularly teaches courses in evolution and diversity, ecology, and environmental health along with an upper-level research course at the Minnesota Zoo.

CHRISTOPHER BAISAN is a senior research specialist at the Laboratory of Tree-Ring Research, the University of Arizona. He received a B.S. degree in renewable natural resources from the University of Arizona in 1991. Christopher has participated in numerous projects to develop dendrochronology-based fire histories, which have resulted in several published papers. His field collections have resulted in the development of a new network of millennial-length tree-ring chronologies that are being used in climate reconstructions and as archaeological dating controls, including a 2,300-year-long tree-ring chronology in central Utah. He has also developed techniques to extract more refined seasonal reconstructions of environmental variables.

PAUL J. BARRETT received his B.S. in biology from Bates College, in Lewiston, Maine, and his M.S. in zoology from Arizona State University, Tempe. He then began working for the Arizona Ecological Services Office of the U.S. Fish and Wildlife Service (USFWS) in Phoenix where he specialized in instream flow studies. After four years, he left USFWS to pursue a Ph.D. at the University of Arizona, where his research focused on habitat use by native fishes in the presence and absence of non-native species. Upon completion of his Ph.D. studies, he worked in several USFWS offices throughout the western United States. Eventually, he returned to work for USFWS in Tucson where he serves as the endangered species listing and recovery coordinator. He has taught classes throughout the United States on endangered species recovery planning for USFWS's National Conservation Training Center.

SADIE R. BERTELSEN worked on the University of Arizona's Mt. Graham Red Squirrel Monitoring Program for four years, while finishing her undergraduate degree in wildlife biology at the University of Arizona.

During that time, she focused her work on habitat preference and nesting habits of the red squirrel. She left the profession to become a mother in 2004 and now works as a massage therapist in Tucson.

CAROL A. COATES received her B.S. in environmental and forest biology from the State University of New York, College of Environmental Science and Forestry in Syracuse and her M.S. in wildlife ecology from the University of Arizona. Prior to getting her master's degree, Carol worked as a wildlife biologist for the University of Arizona's Mt. Graham Red Squirrel Monitoring Program. Carol has worked in the field of natural resources for 15 years and has a broad work experience. She worked as a chemical analyst for an environmental laboratory and as an environmental planner for a private consulting firm and has conducted fieldwork for several research projects that dealt with issues related to threatened and endangered species.

STEPHANIE ROTAN COX attended Clemson University from 1998 to 2002 and received a B.S. She received her medical degree in May 2007 from the Wake Forest University School of Medicine and entered into an emergency medicine residency at Wake Forest Baptist Medical Center.

JAMES C. DEVOS JR. recently retired as the chief of research for the Arizona Game and Fish Department after serving in that capacity for 19 years. In addition, Jim has been active in management and research on southwestern wildlife for more than 30 years. His primary research activities have focused on ungulates; however, he has also worked on projects as diverse as rosy boas and rainbow trout. Much of the recent research he was involved in is related to wildlife responses to habitat restoration, impacts of highways and methods to maintain connectivity, and the impacts from developing artificial waters to benefit wildlife. Jim has published more than 300 articles dealing with wildlife in the Southwest, with the most recent being an article on Sonoran pronghorn survival in an era of above-average rainfall, published in *The Wildlife Society Bulletin*, and an article on mountain lion populations and factors that relate to their increase in the United States. Jim continues to be involved with wildlife as he teaches classes at Arizona State University.

DANIEL F. DOAK received his B.A. from Swarthmore College in 1983 and his Ph.D. from the Department of Zoology, University of Washington, in 1990. He is currently a professor in the Department of Zoology and Physiology, University of Wyoming. His research interests include stochastic demography, life history theory, population viability analysis, community interactions, and the population ecology of arctic and alpine plants. With William F. Morris, he is author of a guide to the use and construction of population viability analyses, *Quantitative Conservation Biology: The Theory and Practice of Population Viability Analysis* (Sinauer Associates, Sunderland, MA, 2002). He has been involved in viability assessments for a wide variety of animal and plant species.

ANDREW J. EDELMAN has broad research interests in the behavior, ecology, and conservation of small mammals. He received his B.S. in biology from Willamette University in 1999 and his M.S. in wildlife ecology from the University of Arizona in 2004. His master's thesis examined the diet, nesting habits, and space use of introduced Abert's squirrels (*Sciurus aberti*) in the Pinaleño Mountains. He also served as a research assistant with the University of Arizona's Mt. Graham Red Squirrel Monitoring Program for two and a half years. Currently, he is pursuing a Ph.D. in biology at the University of New Mexico. His dissertation research focuses on parental investment in offspring by banner-tailed kangaroo rat (*Dipodomys spectabilis*) mothers.

CRAIG FRANK is currently an associate professor in the Department of Biological Sciences at Fordham University. He received his B.S. from the State University of New York at Albany in 1984, a M.S. from Kansas State University in 1987, and a Ph.D. from the University of California at Irvine in 1992. His research interests are in the behavioral, physiological, and biochemical adaptations of mammals to extreme environments. He is particularly interested in how these adaptations integrate to prepare mammals for thermal stress. His laboratory is presently focusing on three areas of research: (1) hibernation, (2) food storage, and (3) the combined use of hibernation and food storage as an over-winter strategy. His most recent publication is: Frank, C. L., W. R. Hood, and M. C. Donnelly. 2004. The role of α-linolenic acid (18:3) in mammalian torpor. Pages

71– 80 in B. M. Barnes and H. V. Carey, eds. *Life in the Cold*. Institute of Arctic Biology Press.

TERRY MOORE FREDERICK graduated with a B.S. in wildlife management from Humboldt State University in 1986. She received her M.S. from the University of Arizona in 1995. She coordinated the interagency Mt. Graham red squirrel surveys for the Arizona Game and Fish Department from 1995 to 1998. Her interests include wildlife habitat relationships, conservation of sensitive and declining wildlife species, and conservation planning. She runs a consulting firm in Sierra Vista, Arizona.

GENICE F. FROEHLICH received her B.S. in wildlife and fisheries biology from the University of California, Davis, in 1987 and a M.S. in wildlife management from the University of Arizona in 1990. Her thesis topic was "Habitat use and life history of the Mt. Graham red squirrel." She began her career with the U.S. Forest Service in 1990 on the Coronado National Forest working on the Mount Graham Red Squirrel Recovery Plan. In 1992, she moved to Safford, Arizona, as the district biologist and continued monitoring the red squirrel, its habitat, and other threatened species, such as the Mexican spotted owl. Currently, Genice is the planning team leader for the Lakeside Ranger District of the Apache-Sitgreaves National Forest. She continues to work with the Mt. Graham Red Squirrel Recovery Team as well as on other environmental issues, such as conservation biology as it relates to forest and range management.

KATHLEEN A. GRANILLO received her B.S. degree in forestry with an emphasis in wildlife management from the University of California, Berkeley, in 1979 and her M.S. degree in wildlife resources from the University of Idaho in 1985. Her thesis was on habitat use of songbirds and small mammals in the Centennial Mountains (located on the Idaho/Montana border just west of Yellowstone National Park). Kathy worked for the Research Branch of the Forest Service for five years and then moved into management. From 1988 through 1990, she was the district biologist on the Safford Ranger District, Coronado National Forest and, from 1990 through 1992, the forest biologist on the Klamath National

Forest. She joined the U.S. Fish and Wildlife Service in Albuquerque, New Mexico, in November 1992 where she is currently the regional refuge biologist. Her interests center on excelling at leadership and communication, which are key to successful conservation efforts.

VICKI L. GREER has been with the University of Arizona's Mt. Graham Red Squirrel Monitoring Program since its inception in 1989. She received a B.S. in fisheries and wildlife science and a B.A. in ecology from the University of Missouri. She supervises a team of biologists and technicians, oversees the population censuses of squirrels on the monitored areas, manages many long-term databases for the project, and prepares reports for cooperating agencies as well as manuscripts for peer-reviewed journals and presentations at professional conferences. Before coming to Arizona to chase squirrels professionally, she studied pileated woodpecker behavior and home ranges in Missouri, limnology in Alaska, and breeding biology of mallards in North Dakota.

MARTHA GRINDER (deceased) was posthumously awarded her Ph.D. degree from the University of Arizona, School of Natural Resources. Martha died July 16, 1999, from injuries sustained from a hit-and-run accident. Martha studied the distribution and movement of urban coyotes around the Tucson area, and she analyzed the DNA of coyotes to determine their diet. She had completed her dissertation "Coyotes in Tucson, Arizona" and was scheduled to defend it when the accident occurred. Martha was an inspiration to all who knew her. An endowment and one-year graduate college fellowship in her memory was established. Previously, Martha was a wildlife biologist on the University of Arizona's Mt. Graham Red Squirrel Monitoring Program. Martha received a B.A. from Hope College, Michigan, in 1989 and a M.S. in animal ecology from Iowa State in 1992.

HENRI D. GRISSINO-MAYER is an associate professor of geography at the University of Tennessee, Knoxville. His research interests focus on analyzing past climate using tree rings, reconstructing pre-settlement fire regimes in both the eastern and western United States, and searching for long-term patterns of climate forcing on wildfire activity. He has over 50 publications in diverse journals (ecology, forestry, dendrochronology, and archaeology), given over 170 presentations at national and international meetings, and been featured in television documentaries numerous times (History Channel, Discovery Channel, and CNN). Dr. Grissino-Mayer developed a 2,129-year-long reconstruction of precipitation for western New Mexico, which shed new light on important responses of native peoples to decadal-scale climate change. Along with his colleague Chris Baisan and mentor Tom Swetnam, Dr. Grissino-Mayer conducted initial studies in 1989 and 1996 on the potential of mixed-conifer trees on Mt. Graham to provide information on both past climate and past wildfire activity.

ELAINE K. HARDING received a B.A. in 1989 and a M.S. in 1991, both in biological science, from San Jose State University, California, and a Ph.D. in environmental studies from the University of California, Santa Cruz, in 2000. She is currently a consultant and adjunct lecturer in the School of Tropical Environment Studies and Geography, James Cook University in Cairns, Queensland, Australia. Her research interests include population and community ecology of birds and mammals, wetland diversity and functioning, and approaches for linking human well-being to ecosystem health.

CYNTHIA HARTWAY is a post-doctoral researcher with Dr. Carla Cáceres at the University of Illinois. She received her Ph.D. from the University of California, Santa Cruz. Her research interests span behavioral, evolutionary, and population ecology, including the evolution of life history traits in variable environments and the interplay of individual behaviors and population dynamics.

JAMES R. HATTEN is currently a research geographer at the U.S. Geological Survey's Columbia River Research Lab, located in the Columbia River Gorge, Cook, Washington. Jim obtained a bachelor's degree in environmental studies from Western Washington University in 1984 and a master's degree in geography at Northern Arizona University in 1996. Some of Jim's previous positions include senior GIS analyst with the Arizona Game and Fish Department's Nongame and Endangered Wildlife Program, habitat biologist for the Hoh Indian Tribe, and foreign fishery observer with National Marine Fisheries Service. Jim has spent the last 10 years analyzing the distribution patterns and habitat requirements of sensitive and endangered species. Jim has created habitat suitability models for the Mt. Graham red squirrel, jaguar, southwestern willow flycatcher, Little Colorado spinedace, white sturgeon, and Chinook salmon. Jim uses cell-based modeling and remote sensing techniques to create habitat suitability models and to quantify changes in critical habitats.

LAWRENCE L. C. JONES received a B.S. in zoology and a M.S. in biology from California State University, Long Beach. During the first part of his 20-year tenure with the U.S. Forest Service, he was involved in research on small mammals, birds, martens, and amphibians at the Pacific Northwest Research Station. Later, he moved to Arizona to work for the National Forest System. In Arizona, he worked as a district biologist in Safford, and then went to the supervisor's office in Tucson, where he is currently the assistant wildlife program manager. He is the liaison for the U.S. Forest Service on the Mt. Graham Red Squirrel Recovery Team. Although he is interested in all living things, he leans toward herpetology and has written about the herpetofauna of the Coronado National Forest, which boasts about 110 species of reptiles and amphibians.

KELLY HUTTON KIMPLE received B.S. degrees in wildlife biology in 1998 and in botany in 2000 from the University of Montana and a M.S. degree in wildlife biology from the University of Arizona in 2005. She studied resource competition between native and non-native cavity-nesting birds along an urban gradient in Tucson, Arizona. Prior to receiving her M.S. degree, she was a wildlife biologist for the University of Arizona's Mt. Graham Red Squirrel Monitoring Program. She also worked on many seasonal projects throughout college studying birds across the United States. Kelly worked for the Rocky Mountain Bird Observatory in May 2004 researching playas in eastern Colorado. She was the assistant director of their monitoring division from 2005 to June 2007. Kelly is presently the development manager for Partners Mentoring Youth.

SARAH KING studied for her B.Sc. and Ph.D. at Queen Mary, University of London. Her Ph.D. researched the behavioral ecology of the Przewalski's horse (*Equus ferus przewalskii*) reintroduced to Mongolia—the first scientific study of this animal in the wild. Before conducting her

Ph.D., Sarah worked as an intern for the U.S. Fish and Wildlife Service on the red wolf project in North Carolina. Her interest in endangered species and in how the behavior of animals relates to their habitat led her to Arizona where she worked for the Mt. Graham Red Squirrel Monitoring Program between 2003 and 2005. Sarah is currently working in Mongolia again, where she is running a project for the Zoological Society of London training Mongolians in ecological techniques.

ANN M. LYNCH has been a research entomologist with the Rocky Mountain Research Station of the U.S. Forest Service for 20 years, currently in Tucson, Arizona. Her research interests include insect disturbance ecology in high-elevation forested ecosystems, the effects of climate change on forest insect ecology, and the development of quantitative tools for resource managers. She has conducted research on forest insects in the southwestern United States since 1987 and began working on spruce aphid and other insects in the Pinaleño Mountains high-elevation ecosystems in 1999. She earned her Ph.D. at the University of Michigan and serves as adjunct associate professor in the Laboratory of Tree-Ring Research at the University of Arizona, and as a faculty affiliate in the Department of Bioagricultural Sciences and Pest Management at Colorado State University.

WILLIAM H. MILLER is an associate professor of applied biological sciences at Arizona State University Polytechnic Campus, Mesa. He has a B.S. and M. S. in forest and range sciences with emphasis in wildlife and a Ph.D. in animal science (nutrition) from Washington State University. Dr. Miller has been involved in wildlife nutrition and wildlife habitat research in Arizona for over 23 years in a wide range of species including reptiles, birds, and small and large mammals. Currently, Dr. Miller is involved in studies of the year-round nutritional status of pronghorn antelope and the nutritional status of mule deer on the Kaibab winter range.

THOMAS E. MORRELL received his B.S. in wildlife biology from the University of Montana, an M.S. in wildlife management from the University of Nevada, and a Ph.D. in wildlife science from Penn State University. Some of his recent publications include: “Breeding bird communities in burned and unburned sites in a mature Indiana oak forest”

(*Proceedings Indiana Academy of Science*); "Predictive occurrence models for bat species in California" (*Journal of Wildlife Management*); "Diurnal bed site selection of urban-dwelling javelina in Prescott, Arizona" (*Journal of Wildlife Management*); and "Nest site habitat characteristics of red-tailed tropic birds on Rose Atoll, American Samoa" (*Journal of Field Ornithology*). Currently, he is a biology instructor at Imperial Valley College. His research interests include investigating the effects of management activities on wildlife, and wildlife-habitat relationships. Presently, he is involved in developing predictive occurrence models for bats in northern California using landscape-scale characteristics.

KAREN E. MUNROE is a Ph.D. student at the University of Arizona studying the social and behavioral ecology of round-tailed ground squirrels (*Spermophilus tereticaudus*) at the Casa Grande Ruins National Monument. She received her master's degree in ecology from Purdue University and her bachelor's degree in biology from Arizona State University. She has been studying small mammals for over 13 years.

ERIC A. POINT received a B.S. from Murray State University and was a wildlife biologist on the University of Arizona's Mt. Graham Red Squirrel Monitoring Program from 1997 to 1999.

JOHN R. RATJE received his B.S. in electrical engineering from New Mexico State University, Las Cruces, New Mexico, in 1974. He is also licensed as a professional mechanical engineer. Amateur astronomy has been a hobby since junior high school. Since 1984, he has been employed by Steward Observatory, the research arm for the Department of Astronomy at the University of Arizona, initially as operations manager and currently as assistant director for operations. He resides in Safford, Arizona, and is the site manager for the Mt. Graham International Observatory. Since 1985 he has been involved with all aspects (environmental analysis, design, construction and operation) of the observatory.

SUSAN SMITH is presently a senior research scientist and laboratory manager for the Laboratory of Paleoecology, Bilby Research Center, Northern Arizona University, Flagstaff. Susan is a native of California and obtained her B.S. in geology from Humboldt State University and her M.S. in paleoecology from Northern Arizona University. Her interests include determining vegetation changes from analysis of pollen assemblages in sediment cores and determining the human use of plants and human impact on the environment from pollen recovered at archaeological sites. She has worked extensively on sites throughout western North America and has served to further the field of paleoethnobotany.

TIM K. SNOW has a B.S. degree in renewable natural resources from the University of Arizona, Tucson, 1987, and is a nongame specialist II with the Arizona Game and Fish Department in Tucson. As a regional nongame specialist (1999–present), Tim is responsible for implementing the department's conservation and management strategies for Nongame Threatened and Endangered Wildlife in southeastern Arizona. He has been with the department for the past 14 years. Some of his responsibilities include coordinating and participating in the bi-annual Mt. Graham red squirrel survey, participating in the capture and collaring of juvenile red squirrels, and serving as a liaison to the technical team of the current Mt. Graham Red Squirrel Recovery Team. When not working with squirrels and other wildlife, his other passion is coaching youth sports, and you can usually find him at some baseball field coaching one of his three sons.

PATRICIA M. SPOERL (retired) served with the U.S. Forest Service for over 25 years, spending most of her career with the Coronado National Forest. She served first as forest archaeologist, then as Recreation and Lands Staff Officer, and finally as the Forest Tribal Liaison and Heritage Resources Program Leader. She holds a Ph.D. in anthropology from Southern Illinois University with an emphasis in southwestern archaeology. She has been involved in archaeological and tribal issues concerning Mt. Graham since the observatory was first proposed to the Coronado National Forest in the mid-1980s. Her most recent research on Mt. Graham resulted in a determination that the entire mountain is eligible for listing in the National Register of Historic Places as a traditional cultural property to Western Apache communities. Pat, now retired from the U.S. Forest Service, continues to be active in archaeological and historic preservation matters.

ROBERT J. STEIDL is an associate professor of wildlife and fisheries science at the University of Arizona in Tucson. Dr. Steidl is a vertebrate population ecologist whose research interests focus on conservation biology and applied quantitative ecology. Specifically, he and his students study a range of issues that center on quantifying the effects of human activities on wildlife populations, especially rare and endangered vertebrates. He teaches courses in applied statistics, research design, and population analysis, and his service activities include membership on the Mt. Graham Red Squirrel Recovery Team and the Desert Tortoise Scientific Advisory Team.

THOMAS W. SWETNAM is a professor of dendrochronology and director of the Laboratory of Tree-Ring Research at the University of Arizona. Tom received his B.S. in biology from the University of New Mexico in 1977. He worked as a wildland fire fighter in the Gila Wilderness of southern New Mexico and then attended graduate school at the University of Arizona from 1980 to 1987. He received his M.S. and Ph.D. in watershed management in 1983 and 1987, respectively. Tom uses tree rings to study the long-term history of wildfires and insect outbreaks, and the effects of climatic change on forest ecosystems. His interests include the application of historical and ecological knowledge to land management. As director of the Tree-Ring Lab, Tom works with faculty, staff, and students to maintain the excellence of this premier and largest laboratory in the world dedicated to tree-ring studies of environmental and cultural change.

JACK WARD THOMAS holds the following degrees: B.S., Texas A&M University (Wildlife Biology); M.S., West Virginia University (Wildlife Ecology); University of Massachusetts, M.F. and Ph.D. (Forestry). He has been named a Distinguished Alumnus by those institutions. He has received Honorary Doctorates from Lakehead University, Lewis and Clark College, and North Carolina State University. He was employed by the Texas Parks and Wildlife Department (1957–1966), U.S. Forest Service (1966–1996), and the University of Montana (1997–2008). He is Chief Emeritus of the U.S. Forest Service and Professor Emeritus of the University of Montana. He has received numerous awards and honors, including the Aldo Leopold Medal from The Wildlife Society, the Distinguished Service Award from the Department of Agriculture, and the Distinguished Achievement Award from the Society for Conservation Biology. He has authored or edited over 500 articles and books, including the award-winning books *The Elk of North America* and *Wildlife Habitats in Managed Forests.*

WILLIAM E. YODER received his B.S. and M.S. degrees from Arizona State University with emphasis in wildlife habitat management and wildlife nutrition. He is currently working towards a Ph.D. at Arizona State University developing a new method of wildlife habitat assessment using GIS and remote sensing. His research projects have involved wildlife inventories, habitat analysis, and some restoration-related work. Specific species focused on in these studies have varied from birds and reptiles to small and large mammals. He is employed by USDA Animal and Plant Health Inspection Service, Plant Protection and Quarantine as a GIS analyst.

PAUL J. YOUNG obtained his B.S. degree in biology from the University of Wisconsin, Stevens Point, in 1976, a M.S. degree in biological sciences from Texas Tech University in 1979, and a Ph.D. in zoology from the University of Alberta, Edmonton, Canada, in 1988. He served as the supervisor for the University of Arizona's Mt. Graham Red Squirrel Monitoring Program from its beginning in 1989 until 2000. He currently resides in northeast Iowa, where he divides his time between vegetable farming and serving as the president and chief biologist of a non-profit research organization (Prairie Ecosystems Research Group) conducting research and management consulting.